BASIC TRANSPORT PHENOMENA IN BIOMEDICAL ENGINEERING

Ronald L. Fournier, Ph.D., PE
Professor and Chair
Department of Bioengineering
The University of Toledo

TAYLOR & FRANCIS
Founded 1798

USA	Publishing Office:	Taylor & Francis 325 Chestnut Street Philadelphia, PA 19106 Tel: (215) 625-8900 Fax: (215) 625-2940
	Distribution Center:	Taylor & Francis 1900 Frost Road, Suite 101 Bristol, PA 19007-1598 Tel: (215) 785-5800 Fax: (215) 785-5515
UK		Taylor & Francis 1 Gunpowder Square London EC4A 3DE Tel: +44 171 583 0490 Fax: +44 171 583 0581

BASIC TRANSPORT PHENOMENA IN BIOMEDICAL ENGINEERING

1 2 3 4 5 6 7 8 9 0

Printed by Edwards Brothers, Lillington, NC, 1998.
A CIP catalog record for this book is available from the British Library.
⊗ The paper in this publication meets the requirements of the ANSI Standard Z39.48-1984 (Permanence of Paper).

Library of Congress Cataloging-in-Publication Data

Fournier, Ronald L.
Basic transport phenomena in biomedical engineering / Ronald L. Fournier.
p. cm.
Includes bibliographical references and index.
ISBN 1-56032-708-1 (alk. paper)
1. Biological transport. 2. Biomedical engineering. 3. Biotechnology. I. Title.
R857.B52F68 1998
571.6′4 – dc21 98-18733
CIP

ISBN 1-56032-708-1

BASIC TRANSPORT PHENOMENA IN BIOMEDICAL ENGINEERING

To my wife Lynn, and our sons Joshua Adam and Ryan Michael

CONTENTS

PREFACE

The challenge in presenting a text on biomedical engineering is the fact that it is a very broad field that encompasses a variety of different engineering disciplines and life sciences. There are many recent edited works that provide specialized treatments and summaries of current research in various areas of biomedical engineering. These works tend to be written more for the sophisticated biomedical researcher than the student entering the field perhaps for the first time. There is therefore a need for an entry level book that introduces some of the basic concepts. This book is designed to meet this goal. This work brings together fundamental engineering and life science principles to provide a focused coverage of key transport concepts in biomedical engineering.

In accomplishing this task within a finite volume, the coverage was limited to those areas that emphasize chemical and physical transport processes with applications towards the development of artificial organs, bioartificial organs, controlled drug delivery systems, and tissue engineering. With this focus, the book first covers such topics as the body fluids, capillary solute transport, the physical and flow properties of blood, tissue oxygen transport, and pharmacokinetic models. This is followed by application of these principles to extracorporeal devices and the relatively new areas of tissue enginering and bioartificial organs. Throughout the book considerable emphasis is placed on developing a quantitative understanding of the underlying physical, chemical, and biological phenomena. Therefore, mathematical models are developed using the conceptually simple "shell balance" or compartmental approaches. Numerous examples are presented based on these mathematical models and they are compared in many cases with experimental data. In this way the student should be able to gain confidence in the development of mathematical models for relatively simple systems, and then be able to apply these concepts to a wide variety of problems of even greater complexity.

A notable feature is that nearly all of the examples are worked out using the mathematics processing software Mathcad®. This allows for the development and solution of sophisticated mathematical models while at the same time making the mathematics less formidable. The Mathcad solutions that are given provide the reader with the opportunity to explore various aspects of the solution on their own, or apply these techniques as starting points for the solution to their own problems. I have found that students easily become quite adept at using Mathcad® to solve a variety of complex scientific and engineering problems. Mathcad® is a product of MathSoft, Inc. and is available in both student and professional versions. More information on Mathcad® is available at their Web site, www.mathsoft.com/. It is hoped that this book is timely and useful to engineers and researchers in the biomedical community, as well as to students in chemical engineering, mechanical engineering, biotechnology, bioengineering, and the life sciences.

Ronald L. Fournier

ACKNOWLEDGMENTS

I have had the opportunity over the years to work with many fine people at both the University of Toledo and the Medical College of Ohio. Their friendship, guidance, and support played a significant role in the writing of this book. Specifically, I would like to express my gratitude to Dr. Stephen L. Rosen, who got me off on the right foot at the start of my academic career, and Drs. Sasidhar Varanasi and Patricia Relue who have always been there to listen, provide critical comments, and lend their support. My colleagues at the Medical College of Ohio; Drs. Peter J. Goldblatt, James M. Horner, James A. Hampton, Steven H. Selman, and Steven L. Britton deserve special thanks for opening many doors and providing a stimulating intellectual environment where engineers, physicians, and life scientists could work together to solve health related problems. I also want to express my thanks to the many dedicated students I have worked with over the past years and whose efforts and desire to learn contributed immensely to the development of this book. Special thanks go out to my former graduate students Dr. Jeffrey G. Sarver, Dr. James P. Byers, Dr. Guodong Chen, and Dr. Robert J. Mellon, as well as Zhan Ding, Jeffery Henning, Andrew Ryan Baker, and Tamara Phares. Several undergraduate students also deserve special mention, including Jennifer Long, Nick Abu-Absi, Karen Green, Elizabeth Dubich, Michael Lehman, Alexander Tarjanyi, David Whalen, and Julie Krohn. Support for an earlier version of this work from the University of Toledo *Center for Teaching Excellence* is also greatly appreciated.

CHAPTER

ONE

PHYSICAL PROPERTIES OF THE BODY FLUIDS AND THE CELL MEMBRANE

1.1 BODY FLUIDS

To begin our study of transport phenomena in biomedical engineering, we first must examine the physical properties of the fluids that are within the human body. Many of our engineering calculations or the development and design of new procedures, devices, or treatments will either involve or affect the fluids that are within the human body. Therefore, we will focus our initial attention on the types and characteristics of the fluids that reside within the body.

The body has three types of fluids. These are the *extracellular*, *intracellular*, and *transcellular* fluids. As shown in Table 1.1, nearly 60% of the body weight for an average 70-kg male is comprised of these body fluids, resulting in a total fluid volume of about 40 liters.

The largest fraction of the fluid volume, or about 36% of the body weight, consists of *intracellular fluid*, that is the fluid contained within the body's cells—for example the fluid found within red blood cells, muscle cells, liver cells, etc. *Extracellular fluid* consists of the *interstitial fluid* that comprises about 17% of the body weight and the *blood plasma* that comprises around 4% of the body weight. Interstitial fluid circulates within the spaces (*interstitium*) between cells. The interstitial fluid space represents about one-sixth of the body volume. The interstitial fluid is formed as a filtrate from the plasma within the *capillaries*. The capillaries are the smallest element of the cardiovascular system and represent the site where the exchange of vital substances occurs between the blood and the tissues surrounding the capillary. We shall see that the interstitial fluid composition is very similar to that of the plasma.

Table 1.1 Body fluids

Fluid	Fluid volume	% of Body weight
Intracellular	25 liters	36 wt %
Extracellular	15 liters	21 wt %
Interstitial	12 liters	17 wt %
Plasma	3 liters	4 wt %
Transcellular	—	
Total	40 liters	57 wt %

Data from Guyton, A. C. 1991. *Textbook of medical physiology*, 8th ed., Philadelphia: W. B. Saunders.

The blood volume is 5 liters, with 3 liters of the blood volume consisting of plasma and the remaining 2 liters representing the volume of the cells that are in the blood, primarily the red blood cells. The 2 liters of cells found in the blood are filled with intracellular fluid. The fraction of the blood volume due to the red blood cells is called the *hematocrit*. The red blood cell volume fraction is measured by centrifuging a given volume of blood in order to find the "packed" red cell volume. Since a small amount of plasma is trapped between the packed red blood cells, the true hematocrit (H) is about 96% of the measured hematocrit (*Hct*). The true hematocrit is about 40% for a male and about 36% for a female. The *transcellular fluids* are those fluids found only within specialized compartments and include the cerebrospinal, intraocular, pleural, pericardial, synovial, sweat, and digestive fluids. Some of these compartments have membrane surfaces that are in proximity to one another and have a thin layer of lubricating fluid between them.

The measurement of these fluid volumes can be done by using "tracer materials" that have the unique property of remaining in specific fluid compartments. The fluid volume of a specific compartment can then be found by adding a known mass of a tracer to a specific compartment and after an appropriate period of time for dispersal, measuring the concentration (mass/volume) of the tracer in the fluid compartment. The compartment volume is then given by the ratio of the tracer mass that was added and the measured tracer concentration, i.e., mass/mass/volume = volume.

Examples of tracers used to measure fluid volumes include radioactive water for measuring *total body water* and radioactive sodium, radioactive chloride, or inulin for measuring *extracellular fluid volume*. Tracers that bind strongly with plasma proteins or radioactive proteins may be used for measuring the *plasma volume*. *Interstitial fluid volume* may then be found by subtracting the plasma volume from the extracellular fluid volume. Subtraction of the extracellular fluid volume from the total body water provides the *intracellular fluid volume*.

1.2 FLUID COMPOSITIONS

The compositions of the body fluids are presented in Table 1.2 in concentration units of *milliosmolar* (mOsmole/liter of solution). We will provide a definition of milliosmolar

Table 1.2 Osmolar solutes found in the extracellular and intracellular fluids

Solute	Plasma	Interstitial	Intracellular
	(mOsm/L)	(mOsm/L)	(mOsm/L)
Na^+	143	140	14
K^+	4.2	4.0	140
Ca^{++}	1.3	1.2	0
Mg^{++}	0.8	0.7	20
Cl^-	108	108	10
HCO_3^-	24	28.3	10
HPO_4^{--}, $H_2PO_4^{--}$	2	2	11
SO_4^{--}	0.5	0.5	1
Phosphocreatine			45
Carnosine			14
Amino acids	2	2	8
Creatine	0.2	0.2	9
Lactate	1.2	1.2	1.5
Adenosine triphosphate			5
Hexose monophosphate			3.7
Glucose	5.6	5.6	
Protein	1.2	0.2	4
Urea	4	4	4
Others	4.8	3.9	11
Total (mOsmoles/liter)	302.8	301.8	302.2
Corrected osmolar activity (mOsmoles/liter)	282.5	281.3	281.3
Total osmotic pressure at 37°C (mm Hg)	5450	5430	5430

Data from Guyton, A. C. 1991. *Textbook of medical physiology*, 8th ed., 277. Philadelphia: W. B. Saunders.

in Section 1.6. For now the definition is not important for the following discussion. Of particular interest is the fact that nearly 80% of the total osmolarity of the interstitial fluid and plasma is produced by sodium and chloride ions. As we discussed earlier, the interstitial fluid arises from the filtration of plasma through the capillaries. We would therefore expect the composition of these two fluids, shown in Table 1.2, to be very similar. We find this to be true with the exception that the protein concentration in the interstitial fluid has been significantly reduced in comparison to its value in the plasma.

1.3 CAPILLARY PLASMA PROTEIN RETENTION

The retention of proteins by the walls of the capillaries during filtration of the plasma, is readily explained by comparing the molecular sizes of typical plasma protein molecules to the size of the pores within the capillary walls. Figure 1.1 illustrates the relative size of various solutes as a function of their molecular weight. The wall of a capillary, illustrated in Figure 1.2, consists of a single layer of *endothelial cells* that are surrounded on their outside by a *basement membrane*. The basement membrane is a mat-like cellular support structure, or extracellular matrix, that consists primarily of a protein called *type IV collagen*, and is joined to the cells by the glycoprotein called *laminin*. The basement membrane is about 50–100 nm thick. The total thickness of the capillary wall is about 0.5 microns.

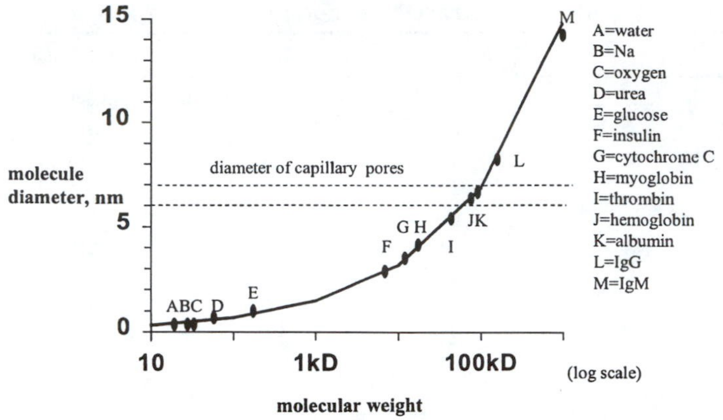

Figure 1.1 Approximate diameter of molecules as a function of their molecular weight (kD = 1000 daltons).

As shown in Figure 1.2, there are several mechanisms that allow for the transport of solutes across the capillary wall. These include the *intercellular cleft* and *pinocytotic vesicles and channels*. The intercellular cleft is a thin slit or slit-pore that is formed at the interface between adjacent endothelial cells. The size of this slit is about 6–7 nm, just sufficient to retain albumin and other larger proteins. The collective surface area of these openings represents less than 1/1000 of the total capillary surface area.

The plasma proteins are generally larger than the capillary slit-pores. Although ellipsoid–shaped proteins, for example the clotting protein fibrinogen, may have a minor axis that is smaller than that of the capillary slit-pore, the streaming effect caused by the fluid motion within the capillary orients the major axis of the proteins parallel to the flow axis and prevents their entry into the slit-pore. This streaming effect is shown in Figure 1.3. Therefore, the smaller substances found in the plasma, such as ions, glucose, and metabolic waste products, will readily pass through the slit-pores of the capillary wall, whereas the plasma proteins will be retained in the lumen of the capillary. The composition of the plasma and interstitial fluid should therefore only differ in their protein content. This is shown in Table 1.2.

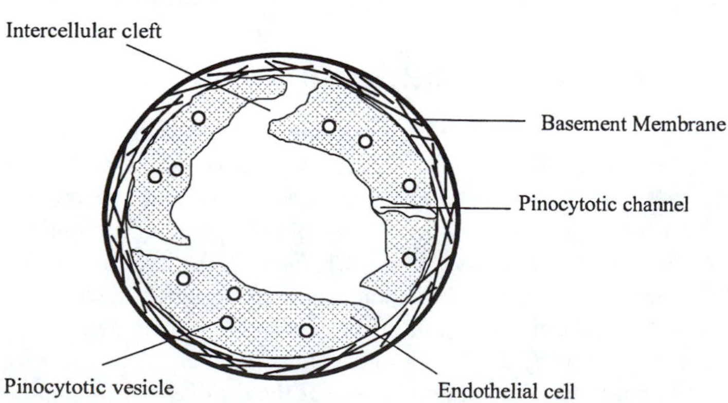

Figure 1.2 Cross-section of a capillary.

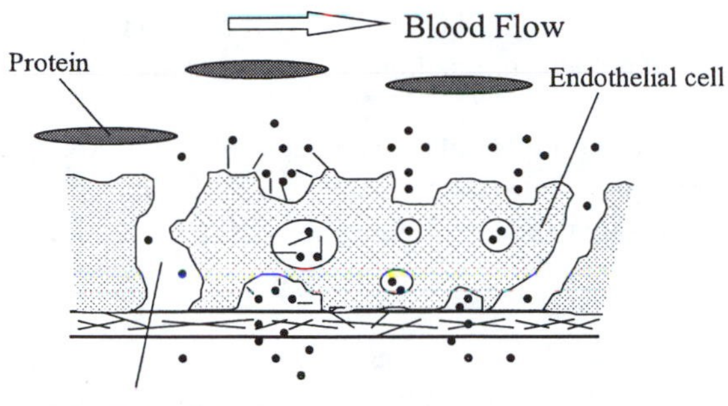

Figure 1.3 Orientation of ellipsoid-shaped proteins by streaming prevents their entry into the capillary pores.

1.4 OSMOTIC PRESSURE

Retention of proteins in the plasma in comparison to the interstitial fluid creates what is known as *osmotic pressure* between the plasma and the interstitial fluid. This concept is illustrated in Figure 1.4 where a semipermeable membrane separates region A that contains a nondiffusing solute (i.e., plasma proteins) and region B that contains only water (i.e., the interstitial fluid). Water will diffuse from region B into region A in an attempt to satisfy the requirement that at equilibrium the chemical potential or fugacity of water on each side of the membrane must be equal. This movement of water from region B to region A is called *osmosis* and will tend to increase the level of region A until its increase in hydrostatic pressure stops the flow of water from region B and equilibrium is attained. The pressure difference between regions A and B at this point is the osmotic pressure of region A.

Since the plasma proteins are the only constituent of the plasma that do not readily pass through the capillary wall, it is the plasma proteins that are responsible for the formation of the osmotic pressure between the interstitial and plasma fluids. The osmotic pressure created by the proteins is given the special name of *colloid osmotic pressure* or *oncotic pressure*. For human plasma, the colloid osmotic pressure is about 28 mm Hg;

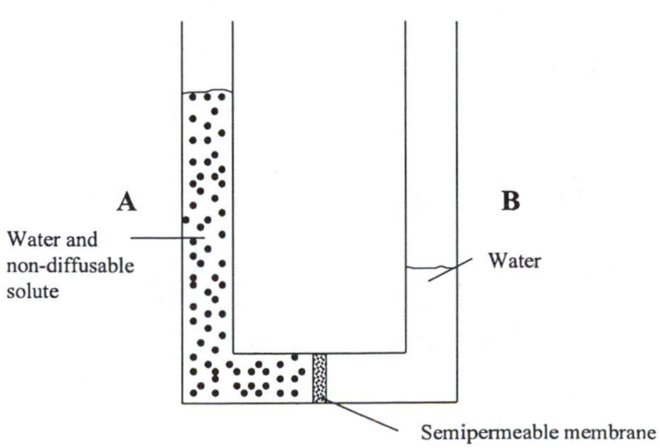

Figure 1.4 Concept of osmotic pressure.

19 mm Hg caused by the plasma proteins and 9 mm Hg caused by the cations within the plasma that are retained through electrostatic interaction with the negative surface charges of the proteins.

The colloid osmotic pressure is actually quite small in comparison to the osmotic pressure that is developed when a cell is placed in pure water. In this case, it is assumed that all of the species present within the intracellular fluid are retained by the cell membrane. As shown in Table 1.2, the total osmotic pressure of the intracellular fluid in this case would be 5450 mm Hg at 37°C.

If a cell is placed within a solution that has a lower concentration of solutes or osmolarity, then the cell is in a *hypotonic solution*, and establishment of osmotic equilibrium requires the osmosis of water into the cell. This influx of water into the cell results in swelling of the cell and a subsequent decrease in its osmolarity. On the other hand, if the cell is placed in a solution with a higher concentration of solutes or osmolarity, i.e., *hypertonic solution*, then osmotic equilibrium requires osmosis of water out of the cell, concentrating the intracellular solution and resulting in shrinkage of the cell. An *isotonic* solution is a fluid that has the same osmolarity of the cell. When cells are placed in an isotonic solution, there is neither swelling or shrinkage of the cell. A 0.9 weight percent solution of sodium chloride or a 5 weight percent solution of glucose is just about isotonic with respect to a cell.

1.5 THE THERMODYNAMICS OF OSMOSIS

A quantitative understanding of osmosis is needed for a variety of calculations and concepts involving solutions and semipermeable membranes. We can easily derive an expression for the osmotic pressure of a solution by starting with the fundamental requirement that at equilibrium the temperature and fugacity of water in regions A and B in Figure 1.4 must be equal (Sandler 1989):

$$\bar{f}_W^A(T, P^A) = f_W^B(T, P^B) \tag{1.1}$$

In this equation, $f_W^B(T, P^B)$ denotes the fugacity of water as a pure component and $\bar{f}_W^A(T, P^A)$ is the fugacity of water as it exists in solution with the other solutes. Equation $\bar{f}_W^A(T, P^A)$ can be written in terms of the water activity coefficient (γ_W^A), the water mole fraction (x_W^A), and the water pure component fugacity ($f_W^A(T, P^B)$) using the *Poynting factor* (the exponential term in Equation 1.2) to correct for the effect of pressure on the pure component fugacity. $\underline{V}_W^L$ is the water molar volume.

$$\gamma_W^A x_W^A f_W^A(T, P^B) \exp\left[\underline{V}_W^L (P^A - P^B)/RT\right] = f_W^B(T, P^B) \tag{1.2}$$

The pure component water fugacities in regions A and B, $f_W^A(T, P^B)$ and $f_W^B(T, P^B)$, are the same, and the osmotic pressure ($P^A - P^B$) can be readily solved for to give the following result. This pressure difference is given the symbol π and represents the osmotic pressure of region A.

$$\pi = (P^A - P^B) = \frac{-RT}{\underline{V}_W^L} \ln\left(\gamma_W^A x_W^A\right) \tag{1.3}$$

This equation may be simplified for the special case where the water and solute form an ideal solution. In this case $\gamma_W^A = 1$ and since x_W^A is close to unity, we can rewrite Equation 1.3 as follows:

$$\pi = -\frac{RT}{\underline{V}_W^L} x_S^A = RTC_S \tag{1.4}$$

The mole fraction of the solute is represented by x_S^A. Since the solute mole fraction is generally quite small, we may approximate x_S^A as $\underline{V}_W^L C_S$, where C_S is the concentration of the solute in gmoles/liter of solution. This ideal dilute solution osmotic pressure, described by Equation 1.4, is known as *van't Hoff's law*. If the solution contains N ideal solutes, then the total osmotic pressure of the solution would be the summation of the osmotic pressure generated by each solute according to Equation 1.4:

$$\pi = RT \sum_{i=1}^{N} C_{Si} \tag{1.5}$$

Osmotic pressure is not determined on the basis of the mass of the solute in the solution, but rather on the number of particles that are formed by a given solute. Each nondiffusing particle in the solution contributes the same amount to the osmotic pressure regardless of the size of the particle. Thus if we take $1/1000$th of an Avogadro's number of glucose molecules or $1/1000$th of an Avogadro's number of albumin molecules, and form a one liter solution of each with water, we have a 1 mM (1 millimole/liter) solution of each solute. According to Equation 1.4, the osmotic pressure of these two solutions would be the same. However, the mass of glucose added to the solution would be 180 mg, whereas the mass of albumin added to the solution would be nearly 70 grams, demonstrating that it is not the mass of solute in the solution that is important in determining the osmotic pressure.

1.6 OSMOLARITY

Recognizing that it is the number of nondiffusing solute molecules that contributes to the osmotic pressure, we now must make the distinction between a nondiffusing substance that dissociates, and one that does not dissociate. The term nondiffusing is used to describe solutes that would be retained by a membrane that separates the solutions of interest. For example, compounds such as NaCl are strong electrolytes. In water they completely dissociate to form two ions (i.e. Na^+ and Cl^-). Each ion or particle formed will exert its own osmotic pressure. It is important to note that the charge of the ion has no effect on the osmotic pressure, so a sodium ion with a charge of $+1$ is equivalent to a calcium ion of charge $+2$. Substances such as glucose do not dissociate, and their osmotic pressure is based on their nondissociated concentration only.

The term *osmole* has been introduced to account for the effect of a dissociating solute. One osmole is therefore defined as one mole of a nondiffusing and nondissociating substance. Therefore, one mole of a dissociating substance such as NaCl is equivalent to two osmoles. However, one mole of glucose is the same as one osmole since glucose does not dissociate in solution. *Osmolarity* simply defines the number of osmoles per liter of solution.

For physiological solutions, it is convenient to work in terms of milliosmoles (mOsm) or milliosmolar (mOsM). At a physiological temperature of 37°C, equation 1.5 may be written as follows to give the osmotic pressure in mm Hg when the solute concentration of each nondiffusing species is expressed in mOsM:

$$\pi = 19.33 \sum_{i=1}^{N} C_{Si} \tag{1.6}$$

1.7 OTHER FACTORS THAT MAY AFFECT THE OSMOTIC PRESSURE

The previous discussion assumed the mixture of solutes formed an ideal solution. However, the osmotic pressure should take into account the various secondary solute interactions that occur within the solution due to their charge, size, shape, and other effects. These effects will either increase or decrease the osmotic activity of a particular solute and result in the corrected osmolar activity shown in Table 1.2. As shown in Table 1.2, we see that the ratio of the corrected osmolar concentration to the total osmolar concentration is about 0.93 for each of the body fluids. This value of 0.93 represents the overall average activity coefficient. In most cases, it is conventional practice to disregard the calculation of solute activity coefficients and calculate the osmotic pressure on the basis of solute concentration only, as discussed above. Since we will be primarily concerned with differences in the osmotic pressure and the generation of fluid flow across a membrane due to a difference in osmotic pressure, this constant of 0.93 will become absorbed in other constants that generally need to be determined experimentally. Thus, for our purpose, a small error is introduced simply by using Equation 1.6 to calculate the osmotic pressure on the basis of solute concentration only.

1.8 FORMATION OF THE INTERSTITIAL FLUID

The flow of fluid across the capillary wall or a porous membrane, is driven by a difference in pressure across the capillary wall or membrane. This pressure difference arises not only from hydrodynamic effects, but also from the difference in osmotic pressure between the fluids separated by the membrane.

Figure 1.5 shows the hydrodynamic and osmotic pressures in the capillary and the surrounding interstitial fluid. The arrows indicate for each pressure the corresponding direction of fluid flow induced by that pressure. We may write that the volumetric fluid transfer rate (J) across the capillary membrane is given by the following equation:

$$\frac{J}{L_P S} = [(P_C - P_{IF}) - (\pi_C - \pi_{IF})] = \overline{\Delta P} \tag{1.7}$$

where $\overline{\Delta P}$ is the effective pressure.

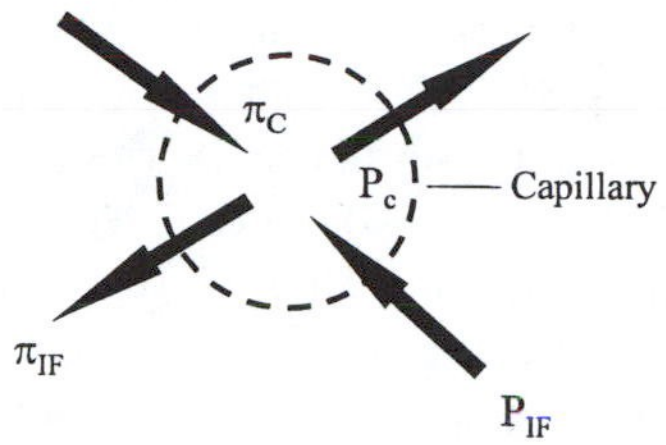

Figure 1.5 Forces acting to cause a flow of fluid across the capillary wall. Arrows indicate the direction of flow induced by each pressure.

In this equation, L_P represents the flow resistance of the capillary membrane and is called the *hydraulic conductance*. The hydraulic conductance is usually best determined by experiment. For example, Renkin (1977) reports values for L_P as low as 3×10^{-14} m^2 sec/kg for the tight junctions between the endothelial cells found in the capillaries of the rabbit brain, 5×10^{-12} m^2 sec/kg for nonfenestrated or continuous capillaries, and as high as 1.5×10^{-9} m^2 sec/kg for the capillaries in the kidney's glomeruli. S is simply the total circumferential surface area of the capillary wall. If the term in brackets is positive, then there is a net flow of fluid from the capillary into the interstitium. If the bracketed term is negative, then there will be a net flow of fluid from the interstitium into the capillary. The first difference term in the brackets represents the hydrodynamic pressure difference between the capillary and interstitial fluid. When this term is positive, fluid will leave the capillary. When this term is negative, fluid will flow from the interstitial fluid into the capillary. The second term within the brackets represents the difference in the osmotic pressure between the capillary and the interstitial fluid. When this term is positive, there will be an osmotic flow of water from the interstitial fluid into the capillary. If the first term in the bracket is positive, this net osmotic flow is in opposition to the flow induced by the hydrodynamic pressure difference.

Equation 1.7 represents the fluid flow across the capillary membrane. It can also be used to describe the flow across any semipermeable membrane regardless of whether or not it is of biological or synthetic origin. We will find, for example, that this equation applies to the dialysis membranes used in the artificial kidney, or hemodialyzer.

1.9 NET CAPILLARY FILTRATION RATE

We can use Equation 1.7 to estimate the *net capillary filtration rate* for the human body. To perform this calculation, we will first need to define nominal values for the pressures (note these are gauge pressures or relative to atmospheric pressure of 760 mm Hg) in Equation 1.7. At the arterial end of the capillary, the capillary pressure is about 30 mm Hg, and at the venous end of the capillary, the pressure is about 10 mm Hg. The mean capillary pressure is considered to be about 17.3 mm Hg and its bias to the lower end is based on the larger volume of the venous capillaries in comparison to the arterial capillaries. The interstitial fluid pressure has surprisingly been found to be subatmospheric and the accepted value is −3 mm Hg. As mentioned before, the colloid osmotic pressure for human plasma is 28 mm Hg and the value for the interstitial fluid is about 8 mm Hg.

At the arterial end of the capillary, we can calculate the flow rate of fluid across the capillary wall by Equation 1.7. Therefore, $J/L_P S = [(30+3) - (28-8)] = 13$ mm Hg, which is positive, indicating that there is a net flow of fluid from the capillary into the interstitium. At the venous end of the capillary, $J/L_P S = [(10+3) - (28-8)] = -7$ mm Hg, which is less than zero, indicating a net reabsorption of fluid from the interstitium back into the capillary. This flow of fluid out of the capillary at the arterial end, and its reabsorption at the venous end, is called *Starling flow* after E. H. Starling who first described it over a hundred years ago. Equation 1.7 is therefore also known as the *Starling equation*. If we base the filtration rate on the mean capillary pressure, then $J/L_P S = [(17.3+3) - (28-8)] = 0.3$ mm Hg. Hence on average, there is a slight imbalance in pressure resulting in more filtration of fluid out of the capillary than is reabsorbed.

Approximately 90% of the fluid that leaves at the arterial end of the capillary is reabsorbed at the venous end. However, the 10% that is not reabsorbed by the capillary, collects within the interstitium and enters the lymphatic system. The amount of this net filtration of fluid from the circulation for the human body can be estimated as illustrated in the following example.

Example 1.1 Calculate the normal rate of net filtration for the human body. Assume the capillaries have a total surface area of 500 m^2 and that the slit–pore surface area is 1/1000 of the total capillary surface area. Model the porous structure of the capillary wall as a series of parallel cylindrical pores with a diameter of 7 nm. Plasma filtrate may be considered to be a Newtonian fluid and has a viscosity of 1.2 centipoise. The mean net filtration pressure for the capillary was just calculated to be 0.3 mm Hg.

SOLUTION We will show in Chapter 3 that Poiseulle's equation provides a relationship between flow and pressure in a cylindrical duct. Therefore Equation 1.7 may be written:

$$J = L_P S\overline{\Delta P} = N\left(\frac{\pi r^4}{8\mu t_m}\right)\overline{\Delta P} \tag{1.8}$$

In this equation, N represents the total number of pores in the capillary system, t_m is the length of the pore of radius equal to r, and μ is the plasma filtrate viscosity. We assume that the tortuous pore length (t_m) is about twice the thickness of the capillary wall. The total cross-sectional area of the pores, A_P, is equal to $(N\pi r^2)$. We can then show by rearranging Equation 1.8 that $L_P = (A_P/S)r^2/8\mu t_m$. This allows us to now calculate the hydraulic conductance and the net filtration rate.

$$L_P = \left(\frac{0.5\,\text{m}^2}{500\,\text{m}^2}\right)\frac{(0.5\times 7\times 10^{-9}\,\text{m})^2}{8\times 1.2\,\text{cP}\times\frac{0.01\,\text{P}}{\text{cP}}\times\frac{1\,\text{g}}{\text{cm sec P}}\times\frac{100\,\text{cm}}{\text{m}}\times\frac{1\,\text{kg}}{1000\,\text{g}}\times 2\times 0.5\times 10^{-6}\,\text{m}}$$

$$= 1.28\times 10^{-12}\,\frac{\text{m}^2\text{sec}}{\text{kg}}$$

Note that this prediction of L_P is consistent with the values reported by Renkin (1977) for the hydraulic conductance of continuous capillaries.

$$J = 1.28 \times 10^{-12} \frac{\text{m}^2\text{sec}}{\text{kg}} \times 500\,\text{m}^2 \times 0.3\,\text{mm Hg} \times \frac{1\,\text{atm}}{760\,\text{mm Hg}} \times 101.33 \frac{\text{kPa}}{\text{atm}}$$

$$\times \frac{1\,\text{N/m}^2}{\text{Pa}} \times \frac{1000\,\text{Pa}}{\text{kPa}} \times \frac{\text{kgm/sec}^2}{\text{N}} \times \frac{60\,\text{sec}}{\text{min}} \times \frac{(100\,\text{cm})^3}{\text{m}^3}$$

$$J = 1.53\,\text{cm}^3/\text{min}$$

We find that the *total net filtration rate* due to the pressure imbalance at the capillaries for the human body is on the order of 2ml/min.

Going back to Equation 1.8, we can define the *filtration coefficient* as the ratio of J and $\overline{\Delta P}$ which is equal to $L_P S$.

$$\text{filtration coefficient} = \frac{J}{\overline{\Delta P}} = \frac{5.1\,\text{ml}}{\text{min mm Hg}} \tag{1.9}$$

The filtration coefficient is a constant and represents the properties of the pore, such as its radius, length, and the total area of the pores. The net filtration rate for any average pressure imbalance at the capillaries is given by simply multiplying the filtration coefficient by the pressure imbalance. For example, if the imbalance increased from 0.3 to 1.5 mm Hg, then the net filtration rate would increase by a factor of 5 to about 10 ml/min.

1.10 LYMPHATIC SYSTEM

We must now address where the net filtration of fluid from the capillaries, about 120 ml/hr for a human, ultimately goes. It clearly cannot continue to collect within the interstitium since this would lead to *edema*, or excess fluid within the tissues of the body.

Two types of edema can occur. The first type involves excess extracellular fluid and is caused either by too much filtration of fluid from the capillaries as just discussed, a failure to drain this excess fluid from the interstitium, or retention of salt and water as a result of impaired kidney function. Edema can also be caused by intracellular accumulation of fluid as a result of cellular metabolic problems or inflammation. In these cases either the sodium-ion pumps are impaired (see Section 1.13) or the cell membrane permeability to sodium is increased. In either case, the excess sodium in the cell causes osmosis of water into the cell.

The *lymphatic system* is an accessory flow system in the body that drains excess fluid from the interstitial spaces and returns it to the blood. The lymphatics consist of a system of lymphatic capillaries and ducts that empties into the venous system at the junctures of the left and right internal jugular and subclavian veins. This system is also responsible for the removal from the interstitial space of large proteins and other matter that cannot be reabsorbed into the capillary. Like the vascular system capillaries, the lymphatic capillaries are also formed by endothelial cells. However, the lymphatic endothelial cells have much larger intercellular junctions and the cells overlap in such a manner to form a valve-like structure. Interstitial fluid can force the valve open to flow into the lymphatic capillaries; however, backflow of lymphatic fluid out from the lymphatic capillaries is prevented.

1.11 SOLUTE TRANSPORT ACROSS THE CAPILLARY ENDOTHELIUM

Lipid-soluble substances such as oxygen and carbon dioxide can diffuse directly through the endothelial cells that line the capillary wall without the use of the slit-pores. Accordingly, their rate of transfer across the capillary wall is significantly higher than water-soluble, but lipid-insoluble substances, such as sodium ions, chloride ions, and glucose, for which the cell membrane of the endothelial cell is essentially impermeable. The transport of these latter substances across the capillary wall is through the use of the capillary slit-pores.

In addition to the slit pores, there are two other pathways that can provide additional routes for the transport of large lipid-insoluble solutes such as proteins across the endothelium of the capillary wall. These pathways are called *pinocytosis* and *receptor-mediated transcytosis* (Lauffenberger and Linderman 1993).

Pinocytosis is not solute specific and as shown in Figure 1.6, involves the ingestion by the cell of the surrounding extracellular fluid and its associated solutes. A small portion of the cell's plasma membrane forms a pocket containing the extracellular fluid. This pocket grows in size and finally pinches off to form the intracellular pinocytotic vesicles. This ingestion of extracellular material by the cell is also known as *endocytosis*. The pinocytotic vesicles are either processed and their contents used internally, or they migrate through the cell and reattach to the cell membrane on the opposite side where they release their contents to the surrounding milieu. This release of material from the pinocytotic vesicle is known as *exocytosis*. Because of its nonspecific nature, pinocytosis is usually not a significant solute transport mechanism.

However, receptor-mediated transcytosis can provide for significant transport of specific solutes across the capillary endothelium. This process is also shown in Figure 1.6. The solute, referred to as a *ligand*, first binds with complementary receptors that are located on the surface of the cell membrane. Unlike the nonspecific process of pinocytosis, receptor-mediated transcytosis can concentrate a particular ligand by many orders of magnitude. This solute-concentrating mechanism by the cell-surface receptors is responsible for the significant transport rates that can be achieved by this process. The ligand-receptor complexes are then endocytosed forming transcytotic vesicles. These

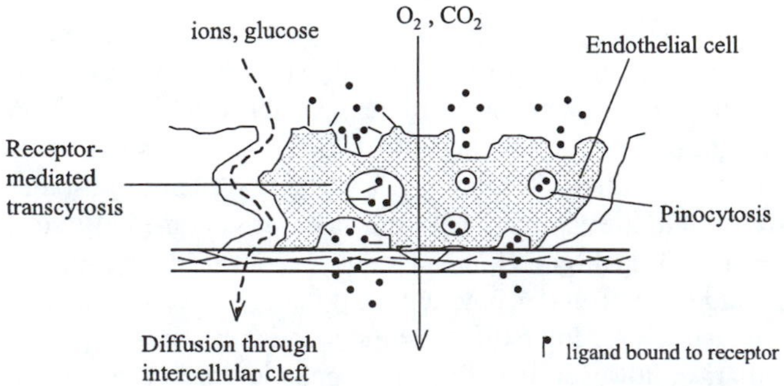

Figure 1.6 Mechanisms for solute transport across the capillary endothelium.

vesicles are either processed internally or they can move through the cell and reattach themselves to the opposite side of the cell. The transcytotic vesicle then releases its contents by exocytosis.

The importance of receptor-mediated transcytosis of large lipid insoluble solutes is provided by the following example. In vitro studies of insulin transport across vascular endothelial cells have shown that 80% of the insulin was transported by receptor-mediated processes. The remaining 20% was transported either through the slit pores between the endothelial cells or by nonspecific pinocytosis (Hachiya et al. 1988).

1.12 THE CELL MEMBRANE

A question that needs to be addressed is why smaller molecules simply do not diffuse across the endothelial cells that line the capillaries. Water can diffuse through the endothelial cell; however, much more can simply pass through the slit-pores. For other substances, we must take a closer look at the chemical nature of the *cell membrane*.

The cell membrane, illustrated in Figure 1.7, is composed mainly of a *lipid bilayer*. Lipid molecules are insoluble in water but readily dissolve in organic solvents such as benzene. Three classes of lipids are found in cellular membranes. These are *phospholipids*, *cholesterol*, and *glycolipids*. The lipid bilayer results because the lipid molecule has a head and tail configuration as shown in Figure 1.7. The head of the lipid molecule is polar and thus hydrophilic, whereas the tail of the lipid molecule is nonpolar and hydrophobic (typically derived from a fatty acid). Such molecules are also called *amphipathic* because the molecule has both hydrophilic and hydrophobic properties. The term lipid bilayer indicates their tendency to form bimolecular sheets when surrounded on all sides by an aqueous environment as is the case for a cell membrane. Thus, the hydrophilic heads of the lipid molecules face into the aqueous environment *inside and outside of the cell*, and the hydrophobic tails are sandwiched between the heads of the lipid molecules.

The lipid bilayer forms the basic structure of the cell membrane. However, other molecules such as proteins are scattered throughout the lipid bilayer of the cell membrane. These protein molecules serve many important functions. For example, special proteins allow for the transport of specific molecules across the cell membrane. Other proteins have catalytic activity and mediate chemical reactions that occur within the cell

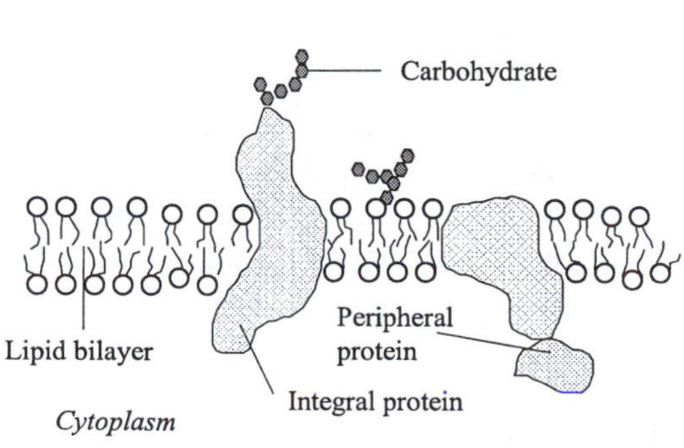

Figure 1.7 Cell membrane structure.

membrane. These proteins are known as enzymes. Still other proteins provide structural support to the cell or provide connections to surrounding cells or other extracellular materials. Some proteins found in the cell membrane act as receptors to extracellular substances or chemical signals, and through transduction of these signals, control intracellular events. Other proteins present foreign materials to the immune system or identify the cell as self.

Protein molecules associated with the cell membrane can be classified into two broad categories. The *transmembrane proteins* are also amphipathic and extend through the lipid bilayer. They typically have hydrophobic regions that may travel across the membrane several times, and hydrophilic ends that are exposed to water on either side of the membrane. Integral membrane proteins are transmembrane proteins that are held tightly within the cell membrane through chemical linkages with other components of the cell membrane. Integral proteins have major functions related to the transport of water-soluble but lipid-insoluble substances across the cell membrane. The *peripheral membrane proteins* are not located within the plasma membrane but associate on either side of the membrane with transmembrane or integral proteins. Peripheral proteins mostly function as enzymes.

For the most part, the cell membrane is impermeable to polar or other water-soluble molecules. Hence, large neutral polar molecules like glucose have very low cell membrane permeabilities. Charged molecules and ions such as H^+, Na^+, K^+, and Cl^- also have very low permeabilities. Hydrophobic molecules, such as oxygen and nitrogen, readily dissolve in the lipid bilayer and show very high permeabilities. Smaller neutral polar molecules such as CO_2, urea, and water, are able to permeate the lipid bilayer because of their much smaller size and neutral charge.

The transport of essential water soluble molecules across the cell membrane is achieved through the use of special transmembrane proteins that have a high specificity for a certain type or class of molecules. There are two basic types of membrane transport proteins, *carrier proteins* and *channel proteins*. The carrier proteins bind to the solute and then undergo a change in shape or conformation ("ding" to "dong") that allows the solute to traverse the cell membrane. The carrier protein therefore changes between two shapes, alternately presenting the solute-binding site to either side of the membrane. Channel proteins actually form water-filled pores that penetrate across the cell membrane. Solutes that cross the cell membrane by either carrier or channel proteins are said to be passively transported.

Figure 1.8 illustrates the passive transport of solutes through the cell membrane by either carrier proteins or channel proteins. Carrier proteins can transport only a single solute across the membrane, a process called uniport, or there can be transport of two different solutes, called coupled transport. The passive transport of glucose into a cell by glucose transporters is an example of a uniport. In coupled transport, the transfer of one solute occurs in combination with the transport of another solute. This coupled transport of the two solutes can occur with both solutes transported in the same direction, symport, or in opposite directions, antiport. An example of coupled transport is the sodium ion gradient–driven symport of glucose and sodium ions.

In general, the driving force for the passive transport of these solutes is due to the combined effect of their concentration gradient and the electrical potential difference

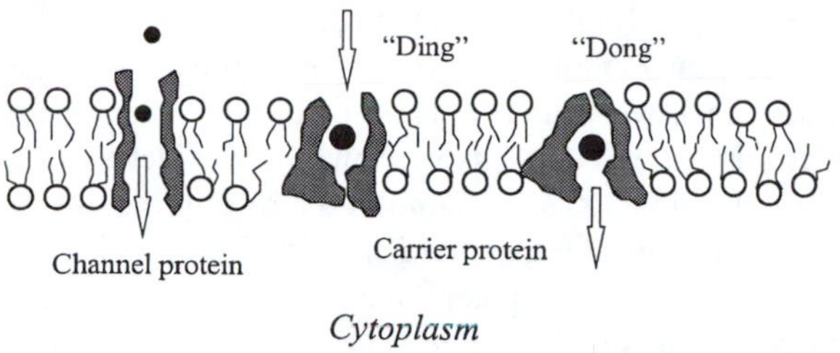

Figure 1.8 Membrane transport proteins. Carrier protein exists in two conformational states, "Ding" and "Dong."

that exists across the membrane. Neutral molecules diffuse from regions of high concentration to regions of low concentration. However, if the molecule carries an electrical charge, then both the concentration gradient and the electrical potential difference, or voltage gradient, across the cell membrane will affect the transport of the molecule. The *electrochemical gradient* is the term used to describe the combined effect of charge and solute concentration on the transport of a molecule. The voltage gradient for a cell membrane is such that the inside of the cell membrane is negative in comparison to the outside. This membrane potential (V_M) for cells at rest is about −90 mV.

The flow of charged molecules through channels in the cell membrane is responsible for the creation of the membrane potential. For example, the higher concentration of potassium ions within the cell relative to the surroundings will tend to cause a leakage of these ions out of the cell through the potassium ion–leak channels. The loss of these positive ions will make the interior of the cell negative in charge. This creates an electric field that is called the membrane potential. The growth of this membrane potential with continued loss of potassium ions will reach a point where the negative charge created within the cell begins to retard the loss of the positively charged potassium ions due to the difference in potassium concentration. When these two forces balance each other, there is no net flow of potassium ions. This balance or equilibrium of the concentration and voltage gradients for an ion is known as the *equilibrium membrane potential* for the cell. The following equation, known as the *Nernst equation*, can be used to calculate the equilibrium membrane potential for a particular ion.

$$V = \frac{RT}{zF} \ln \frac{C_0}{C_i} \tag{1.10}$$

In this equation, R represents the gas constant (1.987 cal/gmole °K), T is temperature in °K, z is the charge on the ion, and F is Faraday's constant (2.3×10^4 cal/V/gmol). At 37°C for a univalent ion, RT/zF is equal to 26.71 mV. Considering potassium ions, the intracellular concentration from Table 1.2 is 140 mOsM and the interstitial concentration is 4 mOsM. Therefore, the equilibrium membrane potential for this ion would be equal to about −95 mV.

The actual resting membrane potential is less than this value because of two opposing effects. First, there is a very slight leakage of sodium ions into the cell. Leakage of sodium ions into the cell will add a positive charge of about +9 mV. Secondly, the sodium-potassium pump (Section 1.13) has the effect of removing positive charge equivalent to about −4 mV. The net resting membrane potential is therefore about −90 mV.

We shall soon see that the cell must expend cellular energy in order to maintain these potassium and sodium ion gradients and thus maintain its resting membrane potential. Although the chloride ion concentration is very high outside of the cell, these ions are repelled by the negative charges within the cell. This is shown by the fact that the equilibrium membrane potential for chloride ions is the same as the resting membrane potential, hence there is no net driving force for the transport of chloride ions into the cell.

In nerve and muscle cells, the resting membrane potential can change very rapidly. This rapid change in the membrane potential is called an *action potential*, and provides for the conduction of a nerve signal from neuron to neuron or the contraction of a muscle fiber. Any stimulus to the cell that raises the membrane potential above a threshold value will lead to the generation of a self-propagating action potential. It is also important to note that the action potential is an all-or-nothing response. The development of an action potential is dependent on the presence of voltage-gated sodium and potassium channels. The voltage-gated sodium channels only open or become active when the membrane potential is less negative than during the resting state. They typically begin to open when the membrane potential is about −65 mV. These sodium gates remain open for only a few tenths of a millisecond after which time they close or become inactive. The voltage-gated sodium channels remain in this inactive or closed state until the membrane potential has returned to near its resting value. The voltage-gated potassium channels also open when the membrane potential becomes less negative than during the resting state; however, unlike the sodium channels, they open more slowly and become fully opened only after the sodium channels have closed. The potassium channels then remain open until the membrane potential has returned to near its resting potential. Figure 1.9 illustrates the events during the generation of an action potential.

The first stage of an action potential is a rapid *depolarization* of the cell membrane. The cell membrane becomes very permeable to sodium ions because of the opening of the voltage-gated sodium channels. This rapid influx of sodium ions—positive charge—increases the membrane potential in the positive direction, and in some cases, can result

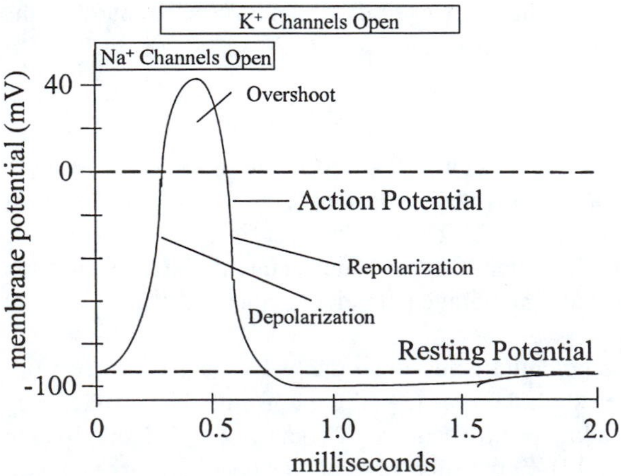

Figure 1.9 Action potential.

Table 1.3 Physical constants used to describe the electrical properties of the cell membrane

Basic electrical properties	Units
Charge	Coulomb (C), charge carried by 6.2×10^{18} univalent ions
Electric Potential	Volt (V), potential caused by separation of charges
Current	Ampere (A), flow of charge, C/sec
Capacitance	Farad (F), amount of charge needed on either side of a membrane to produce a given potential, C/V
Conductance	Siemens (S), ability of a membrane to conduct a flow of charge, A/V
Electrical properties of cells	Typical Values
Membrane potential	−20 to −200 mV
Membrane capacitance	~0.01 pF/μm^2 of cell membrane surface area
Conductance of a single ion channel	1 to 150 pS
Number of specific ion channels	~75/μm^2 of cell membrane surface area
Other relationships	
milli (m) = 10^{-3}, micro (μ) = 10^{-6}, nano (n) = 10^{-9}, pico (p) = 10^{-12}	

Data from Alberts et al. 1989. *The Cell*, 2nd ed, 1067. New York: Garland.

in a positive membrane potential (overshoot) for a brief period of time. This depolarization phase may last only a few tenths of a millisecond. Following the depolarization of the membrane, the sodium channels close and the potassium channels, which are now fully opened, allow for the rapid loss of positively charged potassium ions from the cell, thus re-establishing within milliseconds, the normal negative resting potential of the cell membrane (repolarization). However, for a brief period of time following an action potential, the sodium channels remain inactive and cannot open again for several milliseconds, regardless of external stimulation. This is known as the refractory period.

The net driving force for the transport of an ion due to the combined effect of the concentration gradient and the membrane potential, i.e., the electrochemical gradient, is proportional to the difference between the membrane potential, V_M, and the ion's equilibrium potential, V. If the quantity $(V_M - V)$ is greater than zero, then the direction of transport of the ion will be out of the cell, through either the channel protein pores or by membrane carrier proteins. However, if this quantity is less than zero, then the ion will be transported into the cell. The proportionality constant for the transport of an ion due to its electrochemical gradient is referred to as the *membrane conductance* (g), which is the inverse of the membrane resistance.

The transport of ions generates a current (i), that is given by the product, $g \times (V_M - V)$. This current, or flow of charge can be related to the flow of the ions themselves using the definitions summarized in Table 1.3. The following example illustrates the calculation of the flow of ions through a membrane channel.

Example 1.2 Calculate the flow of sodium ions through the voltage-gated sodium channels in a cell membrane during depolarization. Assuming the cell membrane has a surface area of 1 μm^2, how long would it take to change the membrane potential by 100 mV? Assume the equilibrium membrane potential for sodium ions is 62 mV, that the threshold membrane potential for the sodium channels is −65 mV, and that at the peak of the action potential the membrane potential is 35 mV.

SOLUTION Since the membrane potential rapidly changes during depolarization from the threshold value of −65 mV to the peak value of 35 mV, assume for the calculation of the sodium ion flow that the "average" membrane potential during this phase is −15 mV. The flow or current of sodium ions may then be calculated from the following equation.

$$i_{\mathrm{Na}} = g_{\mathrm{Na}} \times (V_M - V_{\mathrm{Na}}) \tag{1.11}$$

$$\begin{aligned} i_{\mathrm{Na}} &= \frac{4 \times 10^{-12}\,\mathrm{S}}{\mathrm{channel}} \times \frac{1\,\mathrm{A/V}}{\mathrm{S}} \times \frac{1\,\mathrm{C/sec/V}}{\mathrm{A}} \times \frac{6.2 \times 10^{18} \oplus}{\mathrm{C}} \times \frac{1\,\mathrm{V}}{1000\,\mathrm{mV}} \times \frac{1\,\mathrm{Na}}{\oplus} \\ &\quad \times \frac{75\,\mathrm{channels}}{\mu\mathrm{m}^2} \times (-15 - 62)\,\mathrm{mV} \\ &= -1.43 \times 10^8\,\mathrm{Na/sec}/\mu\mathrm{m}^2 \end{aligned}$$

This flow of sodium ions into the cell is also equivalent to a current of $-23\ \mathrm{pA}/\mu\mathrm{m}^2$. The flow of sodium ions transfers charge across the membrane and will therefore change the membrane potential. We can relate this change in charge and membrane potential to the membrane capacitance by the following relationship, $C_{\mathrm{membrane}} = i_{\mathrm{Na}} t / \Delta V_M$. Recall from Table 1.3 that the membrane capacitance is about .01 $\mathrm{pF}/\mu\mathrm{m}^2$. For the calculated sodium ion current and the 100 mV change in the membrane potential, we can then solve for the time required to achieve this change in membrane potential.

$$t = \frac{.01\,\mathrm{pF}/\mu\mathrm{m}^2 \times 1\,\mathrm{C/V/F} \times 1\,\mathrm{V}/1000\,\mathrm{mV} \times 100\,\mathrm{mV}}{23\,\mathrm{pA}/\mu\mathrm{m}^2 \times 1\,\mathrm{C/sec/A} \times 1\,\mathrm{sec}/1000\,\mathrm{msec}} = .043\ \text{milliseconds}$$

This example shows that the membrane potential is rapidly depolarized during the initial phase of the action potential.

1.13 ION PUMPS

Cells also have the ability to "pump" certain solutes against their electrochemical gradient. This process is known as active transport and involves the use of special carrier proteins. Since this is an "uphill" process, active transport requires the expenditure of cellular energy.

The large differences in the concentrations of sodium and potassium ions across the cell membrane will result in some leakage of potassium ions out of the cell and leakage of sodium ions into the cell. However, as we have discussed, proper functioning of the cell requires that these concentration differences be maintained in order to preserve the cell's resting membrane potential. Substances cannot diffuse against their own electrochemical gradient without the expenditure of energy. Therefore to compensate for the loss of potassium ions by leakage through the cell membrane, the cell must have a "pump" mechanism to shuttle potassium ions from the external environment into the cell. On the other hand, sodium ions leak into the cell and similarly the cell needs a "pump" to remove these ions. This process of shuttling substances across the cell membrane against their electrochemical gradient and at the expense of cellular energy is called *active transport.*

The energy for active transport is provided by the cellular energy storage molecule called *adenosine triphosphate* (ATP). ATP is a nucleotide and consists of the following three components: a base called adenine, a ribose sugar, and a triphosphate group. Through the action of the enzyme *ATPase*, a molecule of ATP can be converted into a molecule of ADP (adenosine diphosphate) and a free, high energy phosphate bond that can cause conformational changes in special cell membrane carrier proteins.

The best example of active transport is the *sodium-potassium (Na-K) pump* present in all cells. The Na-K pump transports sodium ions out of the cell and at the same time transfers potassium ions into the cell. Figure 1.10 illustrates the essential features of the Na-K pump. The carrier protein protrudes through both sides of the cell membrane. Within the cell, the carrier protein has three receptor sites for binding sodium ions and also has ATPase activity. On the outside of the cell membrane, the carrier protein has two receptor sites for binding potassium ions. When these ions are bound to their respective receptor sites, the ATPase then becomes activated, liberating the high energy phosphate bond from ATP. The energy in the phosphate bond causes a shape or conformational change in the carrier protein that allows for passage of the sodium and potassium ions. The Na-K ATPase pump is electrogenic; it removes a net positive charge from the cell equivalent to about −4 mV. Similar active transport carrier proteins are also available to transfer other ions such as Cl^-, Ca^{2+}, Mg^{2+}, and HCO_3^-. The only difference in the carrier proteins would be their preference for a specific ion.

Active transport of a solute can also be driven by ion gradients, a process referred to as secondary active transport. For example, the higher concentration of sodium ions outside of the cell can lead to a conformational change in a carrier protein that favors the symport of another solute. The other solute is therefore pumped into the cell against its own electrochemical gradient. The sodium ion gradient that drives this pump is maintained by the Na-K ATPase pump discussed earlier. Sodium ion antiports are also used to control intracellular pH. In this case, the removal of excess hydrogen ions generated by acid-forming reactions within the cell is coupled with the influx of sodium ions.

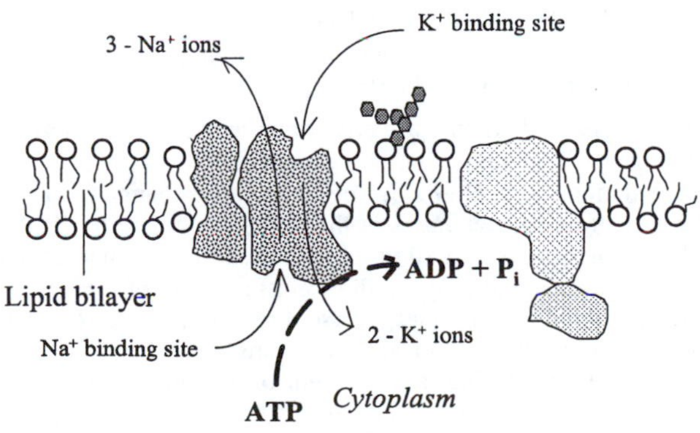

Figure 1.10 The sodium-potassium pump.

PROBLEMS

1. Derive Equation 1.6.

2. A membrane has pure water on one side of it, and a protein solution on the other side. The membrane has pores equivalent in size to a spherical molecule with a molecular weight (MW) of 100,000 daltons. The protein solution consists of albumin (40 g/L at MW = 69,000), globulins (70 g/L at MW = 150,000), and fibrinogen (60 g/L at MW = 340,000). Assuming the proteins form an ideal solution, what is the osmotic pressure of the protein solution? What would be the osmotic pressure of the protein solution if the membrane was completely impermeable to the proteins?

3. Explain why osmotic pressure is based on the molar concentration of the solute and not the solute mass concentration?

4. Explain the difference between a mole and an osmole. Cite some examples.

5. You are concentrating a solution containing a polypeptide of modest molecular weight by pressure filtration through a membrane. The solute concentration on the feed side of the membrane is 0.10 M and the temperature is 25°C. The applied pressure on the solute side of the membrane is 6 atm and the pressure on the opposite side of the membrane is atmospheric. If the polypeptide solute is completely rejected by the membrane, what is the "effective" pressure drop across the membrane? If the hydraulic conductance of the membrane is 3 ml/hr/m^2/mm Hg, what is the filtration rate of the solvent? Assume the membrane has a total surface area of 1 m^2. R = 0.082 liters atm/mol °K.

6. The following clean water flow rates were reported for a particular series of ultrafiltration membranes.

Nominal molecular weight cutoff (NMWCO), daltons	Clean water flow rate (ml/min/cm^2 at 50 psi)
10,000	0.90
30,000	3.00
100,000	8.00

From this data, calculate the hydaulic conductance for each case (L_P). Assuming the membranes have similar porosity and thickness, is there a relationship between the value of L_P and the NMWCO ? (Hint: base your analysis on Equation 1.8.)

7. Perform a literature search and write a short paper on the biomedical applications of liposomes.

8. In a biochemistry text (Stryer, 1988), look up the chemical structures of collagen, laminin, cholesterol, and the various lipids found in the cell membrane.

9. Search the literature and write a short paper on the biomedical applications of osmotic pumps.

10. How many potassium ions must a cell lose in order to produce a membrane potential of −95 mV for potassium? How does this compare to the number of potassium ions within the cell?

11. The Gibbs free energy (G) is a useful thermodynamic property for understanding biological reactions and processes. The Gibbs free energy is defined in terms of the enthalpy (H), temperature (T), and entropy (S) by the following equation, $G \equiv H - TS$. The free energy change (ΔG) of a reaction may be found by the difference of the free energies of the products and the reactants. At constant temperature, a biological reaction or process will proceed in the direction that results in a negative change in the free energy, i.e. $\Delta G < 0$. At equilibrium, the value of $\Delta G = 0$, and the reaction proceeds at the same rate in the forward and reverse directions. Processes such as diffusion and the movement of ions in an electric field also have changes in Gibbs free energy. For example, the change in Gibbs free energy due to movement of a solute by diffusion from a region of high concentration (C_0) to a region of low concentration (C_i) is given by the following equation:

$$\Delta G_{\text{conc}} = -RT \ln \frac{C_0}{C_i}$$

Similarly, the Gibbs free energy change for the movement of an ion across the cell membrane and into a cell at a voltage V relative to the outside is given by:

$$\Delta G_{\text{elec}} = zFV$$

where z is the charge on the ion, F is Faraday's constant (see Equation 1.10), and V the voltage representing the difference in potential between the inside and the outside of the cell. The Nernst equation (Equation 1.10) allows calculation of the equilibrium membrane potential for a particular ion. This equation is based on the requirement that at equilibrium, the concentration and voltage gradients are in balance. Using the above Gibbs free energy relationships, derive the Nernst equation.

12. The cell membrane is permeable to many different ions. Therefore, the equilibrium membrane potential for the case of multiple ions will depend not only on the concentrations of the ions within and outside the cell, but also on the permeability of the cell membrane to each ion P_i. The Goldman equation may be used to calculate the equilibrium membrane potential for the case of multiple ions. For example, considering Na^+ and K^+ ions, the Goldman equation may be written:

$$V = 26.71 \times \ln\left(\frac{P_{\text{Na}}C_{\text{Na},0} + P_K C_{K,0}}{P_{\text{Na}}C_{\text{Na},i} + P_K C_{K,i}}\right)$$

For a cell at rest, P_{Na} is much smaller than P_K. Typically P_{Na} is about $0.01 \times P_K$. Calculate the equilibrium membrane potential for a cell under these conditions. During the depolarization phase of an action potential, the sodium ion permeability increases dramatically due to the opening of the voltage-gated sodium ion channels. Under these conditions, P_K is about $0.5 \times P_{\text{Na}}$. Under these conditions, recalculate the value of the equilibrium membrane potential.

CHAPTER

TWO

SOLUTE TRANSPORT IN BIOLOGICAL SYSTEMS

2.1 DESCRIPTION OF SOLUTE TRANSPORT IN BIOLOGICAL SYSTEMS

In this chapter we will focus our discussion on the transport rate of solutes in biological systems. Solute transport occurs both through bulk fluid motion (convection) and by solute diffusion due to the presence of solute concentration gradients. In biological systems, the diffusion of a solute will also be affected by the presence of a variety of heterogeneous structures. For example, solutes will need to diffuse through porous structures such as the capillary wall, around or through cells within the extravascular space, and through the interstitial fluid containing a variety of macromolecules.

2.2 CAPILLARY PROPERTIES

We will first discuss solute transport through the capillary wall as a representative porous semipermeable membrane. Therefore, the following development is also generally applicable to synthetic membranes that are used in a variety of biomedical device applications. To begin our discussion of capillary wall solute transport, we first need to define the physical properties of a typical capillary. These properties are summarized in Table 2.1. Note that capillaries are very small, having diameters of about 10 microns and a length that is less than 1 millimeter. The residence time of blood in a capillary is also only on the order of 1–2 seconds. Therefore each capillary can only support a very small volume of tissue.

There are three types of capillaries. They are referred to as continuous, fenestrated, and discontinuous. The *continuous capillaries* are found in the muscle, skin, lungs, fat, the nervous system, and in connective tissue. The capillary lumen lies within a

Table 2.1 Capillary characteristics

Property	Value
Inside diameter (D_c)	10 microns
Length (L)	0.1 cm
Wall thickness (t_m)	0.5 microns
Average blood velocity (V)	0.05 cm/sec
Pore fraction	0.001
Wall pore diameter (d_p)	6–7 nm
Inlet pressure	30 mm Hg
Outlet pressure	10 mm Hg
Mean pressure (P_c)	17.3 mm Hg
Colloid osmotic pressure (π_P)	28 mm Hg
Interstitial fluid pressure (P_{if})	−3 mm Hg
Interstitial fluid colloid osmotic pressure (π_{if})	8 mm Hg

circumferential ring of several endothelial cells, as shown earlier in Figure 1.2. *Fenestrated capillaries* are much more permeable to water and small solutes in comparison to continuous capillaries. These capillaries are found in tissues that are involved in the exchange of fluid or solutes such as hormones. For example, within the kidney they are found in the glomerulus and the tubules. The endothelium is perforated by numerous small holes called *fenestrae*. The fenestrae are sometimes covered by a thin membrane that provides selectivity with regard to the size of solutes that are allowed to pass through. *Discontinuous capillaries* have large endothelial cell gaps that readily allow the passage of proteins, and even red blood cells.

2.3 CAPILLARY FLOW RATES

Some of the blood plasma that enters the capillary will be carried or filtered across the capillary wall by the combined effect of the hydrodynamic and oncotic pressure differences that exist between the capillary and the surrounding interstitial fluid. This perfusion of plasma across the capillary wall is also known as *plasmapheresis*. We can determine the total flow rate of this fluid across the capillary wall using the relationships developed in Chapter 1. Recall from Example 1.1 that the value of the hydraulic conductance, L_P, is given by the following equation:

$$L_P = \left(\frac{A_P}{S}\right)\frac{r^2}{8\mu t_m} \tag{2.1}$$

In this equation, S represents the circumferential surface area of a given capillary and A_P the total cross sectional area of the pores of radius r in the capillary wall. The ratio, A_P/S, is also referred to as the porosity of the capillary wall (or membrane) and is often given the symbol ε.

Example 2.1 Calculate the convective flowrate of plasma across the capillary wall.

SOLUTION Using the capillary properties provided in Table 2.1, and a plasma viscosity of 1.2 cP, the value of the hydraulic conductance, L_P, can be shown as in

Example 1.1 to be equal to 0.61 cm^3/hrm^2 mm Hg (or 1.28×10^{-12} m^{2}sec/kg). We can then calculate the filtration rate as follows using Equation 1.7.

$$J = \frac{0.61\,\text{cm}^3}{\text{hrm}^2\ \text{mm Hg}} \times \pi \times 10 \times 10^{-6}\,\text{m} \times 0.001\,\text{m}$$
$$\times [(17.3 - -3) - (28 - 8)]\,\text{mm Hg}$$
$$= 5.75 \times 10^{-9}\,\text{cm}^3/\text{hr} = 5.75 \times 10^{-6}\,\mu\text{L/hr}$$

We can compare this value of the plasma filtration flow rate across the capillary wall, to the total flow rate of blood entering the capillary; $Q_{\text{capillary}} = \frac{\pi}{4} D_c^2 V$; where V is the capillary blood velocity and D_c is the capillary diameter.

$$Q_{\text{capillary}} = \frac{\pi}{4}(10 \times 10^{-6}\,\text{m})^2 \times \frac{0.05\,\text{cm}}{\text{sec}} \times \left(\frac{100\,\text{cm}}{\text{m}}\right)^2 \times \frac{3600\,\text{sec}}{\text{hr}}$$
$$= 1.41 \times 10^{-4}\,\text{cm}^3/\text{hr} = 0.14\,\mu\text{L/hr}$$

We find that in comparison to the volumetric flow rate of blood entering the capillary, the total convective flow rate of plasma filtrate across the capillary wall is negligible. We may therefore assume that $Q_{\text{capillary}}$ is constant along the length of the capillary.

2.4 SOLUTE TRANSPORT BY CAPILLARY FILTRATION

The perfusion of the plasma across the capillary wall will carry with it a variety of solutes that are present in the blood. Because of the size of the pores in the capillary wall, this filtration of the plasma by the capillary wall will also tend to separate the solutes on the basis of their size. For example, smaller solutes like ions, glucose, and amino acids will readily pass through the capillary pores. However, larger proteins will be inhibited to varying degrees in their passage across the capillary wall by the size of the pores. The solute selectivity during this plasma filtration is described by the sieving coefficient (S_a), which is defined as the ratio of the solute concentration in the filtrate (C_f) to the solute concentration at the surface of the capillary on the blood side (C_{bs}). Theoretical expressions based on the motion of a spherical solute moving through a cylindrical pore have been developed to estimate the value of the sieving coefficient (Anderson and Quinn 1974; Deen 1987). The development of these expressions neglects secondary effects between the solute and the membrane pore, for example, electrostatic, hydrophobic, and van der Waals interactions. The following expression can be used to calculate the sieving coefficient.

$$S_a = \frac{C_f}{C_{bs}} = (1-\lambda)^2[2-(1-\lambda)^2]\left[1 - \frac{2}{3}\lambda^2 - 0.163\lambda^3\right] \tag{2.2}$$

In this equation, λ is defined as the ratio of the solute radius (a) to the capillary pore radius (r).

Under some conditions, the high filtration rate of the plasma leads to the formation of a layer of retained proteins that coat the nonfiltrate side of the membrane. This effect is also called *concentration polarization* and the layer of retained proteins will affect transport of the solutes. Concentration polarization is particularly important at the higher filtration rates used in commercial membrane-based plasmapheresis systems used for separating plasma from the cellular components of blood. The sieving coefficient for the case of concentration polarization (S_0) is defined as the ratio of the solute concentration in the filtrate (C_f) to that of the solute concentration in the bulk blood (C_b). The sieving coefficient under these conditions is given by the following equation (Mochizuki and Zydney 1992; Zydney 1995).

$$S_0 = \frac{C_f}{C_b} = \frac{S_a}{(1 - S_a)\exp(-j/k_m) + S_a} \tag{2.3}$$

In this equation, j, represents the filtration flux and is given by Equation 1.7, i.e., J/S. S_a is given by Equation 2.2. The film mass-transfer coefficient, k_m, is dependent on the fluid properties and the nature of the flow field. The mass-transfer coefficient is the proportionality constant that relates the mass or molar flux of a solute (N) to its concentration driving force, i.e., $N = k_m(C_{\text{high}} - C_{\text{low}})$. Since the flow of blood in the capillary is laminar, or not turbulent, k_m may be estimated from the following equation (Thomas 1992).

$$Sh = 3.66 + \frac{0.104\frac{ReSc}{L/D_c}}{1 + 0.016\left(\frac{ReSc}{L/D_c}\right)^{0.8}} \tag{2.4}$$

The Sh number in Equation 2.4 is the value averaged over the length L.

There are several dimensionless groups in this expression. The *Sherwood number* is defined as $Sh = k_m D_C/D$, $Re = \rho V D_C/\mu$ is the *Reynolds number* ($Re < 2000$ is laminar for tube flow), and $Sc = \mu/\rho D$ is the *Schmidt number*. D represents the solute diffusivity and is discussed in greater detail in the next section. The Sherwood number is a dimensionless group that compares the rate of solute transport by convection, i.e., fluid motion, to that by diffusion alone. The Reynolds number is simply the ratio of inertial forces to viscous forces and the Schmidt number compares the rates of momentum and diffusional transport.

Equation 2.4 is for the case where both the velocity and the concentration profiles are not fully developed. This means that the velocity and concentration profiles are changing with axial position within the capillary (or flow channel). When the flow is fully developed, the Sh number attains its asymptotic value of 3.66. For the hydrodynamic flow to be fully developed at any axial position z in the capillary, we must have that $z/D_C > 0.05Re$, and for the concentration field to be fully developed, we must have that $z/D_C > 0.05ReSc$. The Re for flow in capillaries is on the order of 0.001 to 0.005 and the Sc is on the order of 1000. Since L/D_C for a capillary is on the order of 100, we see that the criteria for fully developed hydrodynamic flow and concentration fields in a capillary are satisfied soon after the blood enters the capillary. Therefore for blood flow in a capillary we may assume that $Sh = 3.66$.

The length averaged mass-transfer coefficient for undeveloped laminar flow in cylindrical tubes and channels can also be estimated from the following equation (Thomas 1992).

$$Sh = 1.86\left(\frac{ReSc}{L/D_C}\right)^{1/3} \quad \text{for} \quad \frac{L/D_C}{ReSc} \leq 0.01 \tag{2.5}$$

For noncylindrical flow channels, the diameter (D_C) in the above equations is equal to the *hydraulic diameter*, D_H, which is defined by the following equation.

$$D_H = 4 \times \frac{\text{channel cross-sectional area}}{\text{wetted perimeter}} \tag{2.6}$$

2.5 SOLUTE DIFFUSION

In addition to the convective transport of the solute, for example as carried with the plasma filtrate across the capillary wall, the solute can also diffuse down its own concentration gradient. The transport of a solute by diffusion (J_S) is proportional to its concentration gradient. Diffusion of a solute in a bulk solution is described by *Fick's law* given by the following equation.

$$J_S = -DS\frac{\mathrm{d}C}{\mathrm{d}x} \tag{2.7}$$

In this equation, S is the surface area normal to the direction of solute diffusion and D is the *solute diffusivity*. Since the solute is assumed to be diffusing through a homogeneous medium, this solute diffusivity is sometimes referred to as the bulk diffusivity.

Figure 2.1 presents diffusivity data (Renkin and Curry 1979) for a variety of solutes in dilute aqueous solutions at 37°C as a function of solute molecular weight. The solid line through the data is the result of a linear least-squares regression. The following empirical equation based on the data in Figure 2.1 provides a relationship between the solute diffusivity and its molecular weight (MW).

$$D = 1.013 \times 10^{-4}\,(\text{MW})^{-.46}\ (\text{cm}^2/\text{sec}) \tag{2.8}$$

Cussler (1984) as well as Tyn and Gusek (1990) provide additional data on the diffusivity for a variety of solutes.

The diffusivity of solutes in dilute solutions can also be estimated from the Stokes-Einstein equation (Bird, Stewart, Lightfoot 1960; Cussler 1984):

$$D = \frac{RT}{6\pi \mu a N_A} \tag{2.9}$$

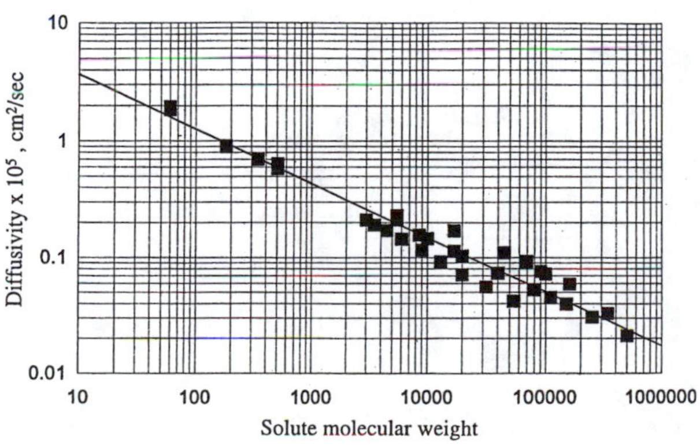

Figure 2.1 Solute diffusivity in water at 37°C (based on data from Renkin and Curry 1979).

In this equation, R is the ideal gas constant (8.314 J/gmole/K), T is the temperature in K, a is the solute radius, N_A is Avogadro's number (6.023×10^{23} molecules/gmole), and μ is the solution viscosity. The viscosity of water at 37°C is 0.76 cP and that of plasma is 1.2 cP (1 cP = 1 centipoise = 0.01 g/cm sec).

If the diffusivity for a solute is known, then Equation 2.9 may be used to obtain an estimate of the molecular radius (a) of the solute. If the diffusivity and size of the solute are not known, the solute size can first be estimated from the following equation. This equation assumes that the solute of molecular weight (MW) is a sphere with a density ($\rho \approx 1$ g/cm^3) equal to that of the solute in the solid phase.

$$a = \left(\frac{3\text{MW}}{4\pi \rho N_A} \right)^{1/3} \tag{2.10}$$

2.6 SOLUTE DIFFUSION WITHIN HETEROGENEOUS MEDIA

A unique aspect of biological systems is their heterogeneous nature. We therefore need to expand our understanding of solute diffusion to see how heterogeneous materials affect solute diffusion. Figure 2.2 illustrates the situation for diffusion of a solute from blood, through the cell-free plasma layer (see Chapter 3), across the capillary wall, and into the surrounding tissue space. Shown in the figure are the types of solute diffusivity we need to consider. These are the solute diffusivity in blood or tissue (D_T), the diffusivity in plasma (D_{plasma}), the diffusivity within the pores of the capillary wall (D_m), diffusivity in the interstitial fluid (D_0), and the diffusivity through cells (D_{cell}). The following discussion will relate these various solute diffusivities to the diffusivity of the solute in water (D).

We can use the Stokes-Einstein equation to estimate the solute diffusivity in plasma by adjusting the solution viscosity. The diffusivity of a solute in plasma at 37°C would therefore be given by the following relation: $D_{\text{plasma}} = D_{\text{water}} \times (0.76\,\text{cP}/1.2\,\text{cP}) = 0.63 D_{\text{water}}$.

For solute transport across the capillary wall, we must consider that the available surface area is not the total surface area S, but the pore area A_P. This assumes that the

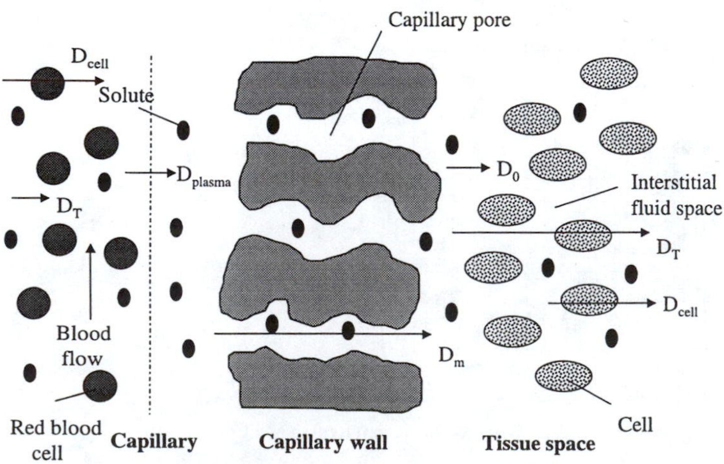

Figure 2.2 Solute transport from blood.

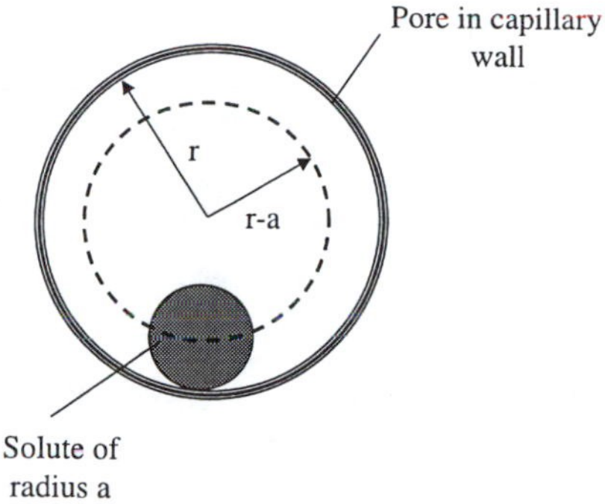

Figure 2.3 Steric exclusion.

solute itself is not soluble in the continuous nonporous phase. Furthermore, the solute must follow a path through the pores that is more tortuous, making the diffusion distance greater than the thickness of the membrane (t_m).

In many cases in pore diffusion, the solute radius is also comparable to that of the pore radius. This leads to two additional effects, both reducing the diffusion rate of the solute. The first effect, called *steric exclusion*, is illustrated in Figure 2.3. Steric exclusion restricts the ability of the solute from entering the pore from the bulk solution. In this figure, we see that a solute can get no closer to the pore wall than its radius (a). Therefore, only a fraction of the pore volume is available to the molecule. The fraction of pore volume available to the solute is given by the following ratio: $K = \pi(r-a)^2/\pi r^2 = (1-a/r)^2$, which is also known as the *partition coefficient*. Because of steric exclusion, the equilibrium concentration of solute is less within the pore mouth than in the bulk solution. This simple expression for K ignores any secondary attractive or repulsive interactions between the solute and the pore.

As the solute molecule diffuses through the pore, it experiences hydrodynamic drag caused by the flow of solvent over the surface of the solute. This is shown in Figure 2.4. As the solute radius increases relative to that of the pore radius, this hydrodynamic drag increases, reducing or restricting the diffusion of the solute through the pore compared to its motion in the bulk fluid. This drag effect on the solute in the pore results in a decrease in the solute diffusivity relative to its value in the bulk solution and is called *restricted* or *hindered diffusion*.

Beck and Schultz (1970) studied the diffusion of a variety of solutes across track-etched mica membranes. The mica membranes were bombarded with U^{235} fission fragments. The etched particle tracks that were formed resulted in relatively straight pores

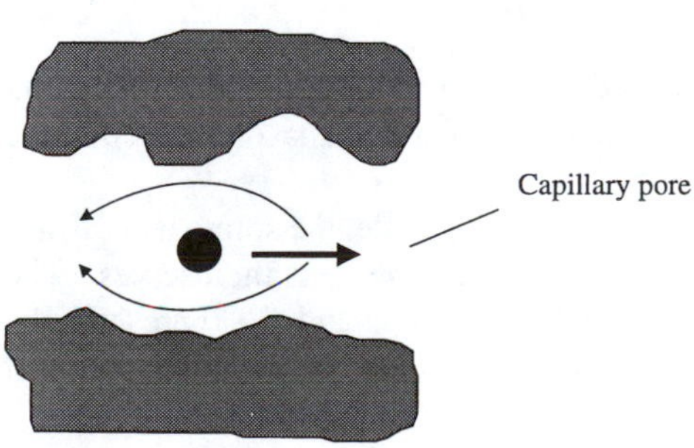

Figure 2.4 Solvent flows across solute in small pores increases drag.

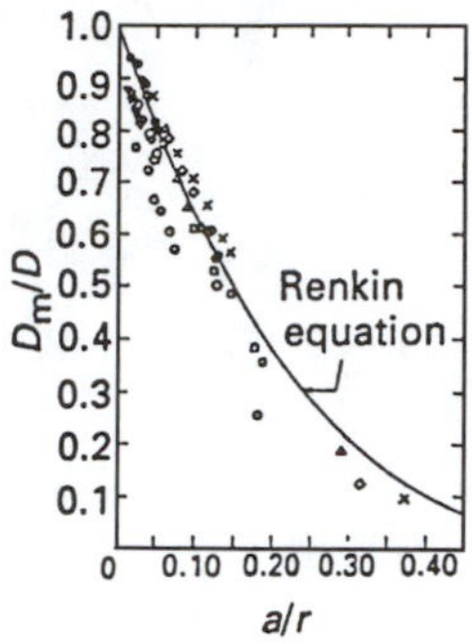

Figure 2.5 Apparent diffusivity in the pores of a membrane (from Beck and Schultz 1970, by permission).

with a narrow pore size distribution. Their data for a number of solutes is shown in Figure 2.5, where the ratio of the pore diffusivity (D_m) and the bulk diffusivity (D) is plotted as a function of the ratio of the solute radius to pore radius (a/r). Note that when the solute size is only 1/10 of the pore size, the solute diffusivity is reduced by 30%. When a/r is equal to 0.40, the solute diffusivity has been reduced by nearly 90%.

The solid line in Figure 2.5 represents the fit of the Renkin equation (1954), given by the following expression, to their data.

$$\frac{D_m}{D} = \left(1 - \frac{a}{r}\right)^2 \left[1 - 2.1\left(\frac{a}{r}\right) + 2.09\left(\frac{a}{r}\right)^3 - 0.95\left(\frac{a}{r}\right)^5\right]$$
$$= K\left(\frac{a}{r}\right) \times \omega_r\left(\frac{a}{r}\right) \tag{2.11}$$

We see that the Renkin equation describes their data rather well. The first term in this equation is the partition coefficient, $K(\frac{a}{r})$, and represents the effect of steric exclusion or the reduction in the solute concentration at the pore mouth compared to its value in the bulk solution. The bracketed term, $\omega_r(\frac{a}{r})$, accounts for the increase in hydrodynamic drag as the solute diffuses through the pore. If the solute is already within the porous structure, then $K = 1$, and $\frac{D_m}{D} = \omega_r(\frac{a}{r})$. Additional expressions for describing the hindered diffusion of large molecules in liquid-filled pores may be found in the review paper by Deen (1987).

With these corrections for the diffusion of a solute in pores, we can now write Fick's law as follows.

$$J_s = -DA_P\left(\frac{K\omega_r}{\tau}\right)\frac{dC}{dx} \tag{2.12}$$

In this equation, τ is an experimentally defined parameter called the *tortuosity* and accounts for the fact that the solute diffusion distance is greater than that of the membrane thickness.

In heterogenous regions like blood or that in the tissue space, the solute can diffuse through both the continuous fluid space as well as through the cells themselves. The transport mechanism in blood or the tissue space is by diffusion and D_T represents the "effective" diffusivity of the solute through the heterogeneous region. The simplest approach for estimating D_T is based on a model developed by Maxwell (1873).

$$\frac{D_T}{D_0} = \frac{2D_0 + D_{\text{cell}} - 2\phi(D_0 - D_{\text{cell}})}{2D_0 + D_{\text{cell}} + \phi(D_0 - D_{\text{cell}})} \tag{2.13}$$

In this equation, D_0 is the diffusivity of the solute through the interstitial fluid space or in the case of blood, $D_0 = D_{\text{plasma}} \cdot D_{\text{cell}}$ is the diffusivity of the solute in the cells. The volume fraction of the cells is represented by ϕ. Interestingly, this equation does not depend on the size of the discrete particles, i.e., the cells, but only on their volume fraction.

More recently, Riley, Buettner et al. (1995a) and Riley, Muzzio et al. (1994, 1995a, 1996) have developed an empirical relation based on Monte Carlo simulations for D_T/D_0 that shows good agreement with available data throughout a wide range of cell volume fractions, $0.04 < \phi < 0.95$.

$$\frac{D_T}{D_0} = 1 - \left(1 - \frac{D_{\text{cell}}}{D_0}\right)(1.727\phi - 0.8177\phi^2 + 0.09075\phi^3) \tag{2.14}$$

The interstitial fluid has a gel-like consistency due to the presence of a variety of macromolecules. Therefore, the solute must diffuse around this random network of macromolecular obstacles. The reduction in the solute diffusivity due to the network of macromolecules within the interstitial fluid gel may be described by the following equation developed by Brinkman (1947).

$$\frac{D_0}{D} = \omega_r = \frac{1}{1 + \kappa a + 1/3(\kappa a)^2} \tag{2.15}$$

In this equation, the ratio of D_0 to D depends on only one parameter, κ, which is a function of the gel's microstructure and may be fitted to experimental data for solute diffusion in the gel.

The steric exclusion, or equilibrium partitioning, of a solute upon entrance from a bulk solution into a gel-like material such as the interstitial fluid may be described by the following equation developed by Ogston (1958).

$$K = \exp\left[-\phi\left(1 + \frac{a}{a_f}\right)^2\right] \tag{2.16}$$

In this equation, a_f represents the radius of the macromolecules which are assumed to form very long cylindrical fibers. If the length of these fibers per volume of gel is given by L, then the volume fraction of these macromolecular fibers is given by $\phi = \pi a_f^2 L$ (Tong and Anderson 1996). Tong and Anderson (1996) showed that Equations 2.15 and 2.16 provided excellent representation of the partitioning and diffusion of two representative globular proteins (albumin and ribonuclease-A) in a polyacrylamide gel.

Nugent and Jain (1984a, 1984b) examined the effective diffusion of various sized solutes through normal and tumor tissue. Their results are shown in Figure 2.6, where we see the ratio of the effective diffusivity of the solute in the tissue (D_T) to its bulk diffusivity (D) plotted as a function of the solute radius. This figure may be used to provide reasonable estimates of the solute diffusivity within tissue. We see that normal tissue results in a greater reduction in the solute diffusivity than that observed for tumor

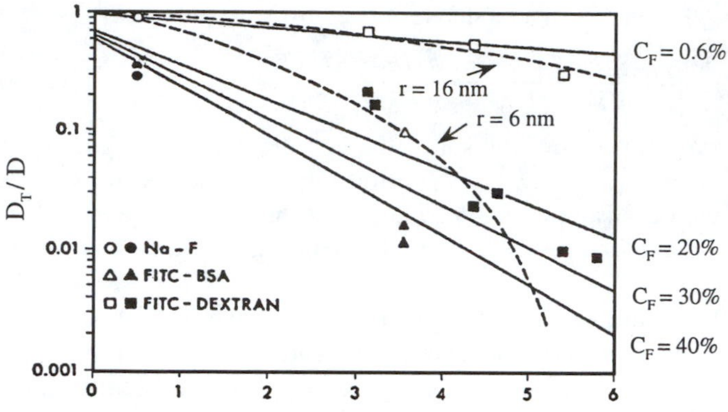

Figure 2.6 Ratios of effective diffusivity in tissue to the aqueous diffusivity as a function of the solute radius. Open and closed symbols represent data for tumor and normal tissue, respectively. Solid lines represent fiber matrix model and dashed lines represent the Renkin pore model (from Nugent and Jain 1984, with permission).

tissue. This reflects the smaller interstitial volume of normal tissue in comparison to tumor tissue.

The solid and dashed lines in the figure represent the corresponding fits of two diffusion models to their data. The solid lines represent a fiber matrix model (Curry and Michel 1980) wherein the diffusivity ratio is given by this equation

$$\frac{D_T}{D} = \exp\left[-(1 + a/a_f)\nu^{1/2}C_F^{1/2}\right] \tag{2.17}$$

The tissue interstitial space is modeled as a random matrix of straight fibers of radius a_f and the solute is considered to be a rigid sphere of radius a. C_F is the fiber concentration in the interstitial fluid and ν the specific volume of the fibers. The fibers once again are assumed to be the macromolecules that form the gel-like consistency of the interstitial fluid. The quantity $C_F\nu$ is related to the fiber volume fraction in the gel by the following expression $\phi/(1-\phi)$. The dashed lines in Figure 2.6 represent the predictions of the diffusivity ratio (D_T/D) based on the hydrodynamic pore model as given by $\omega_r(\frac{a}{r})$ in Equation 2.11.

2.7 SOLUTE PERMEABILITY

Sometimes the term *permeability* (P_m) is used to describe solute transport (N_S, for example in gmoles/sec) across a membrane. Permeability is defined by the following expression using the total membrane surface area S.

$$N_S = P_m S(C_{\text{high}} - C_{\text{low}}) \tag{2.18}$$

$(C_{\text{high}} - C_{\text{low}})$ is the difference in the surface concentrations of the solute on either side of the membrane. In the absence of external diffusion effects, this concentration difference would equal the difference in the bulk concentration of the solution on either side of the membrane. The permeability is also simply equal to the effective diffusivity of the solute in the membrane divided by the thickness of the membrane (D_m/t_m). The permeability may be shown to be given by the following relationship.

$$P_m = \frac{D}{t_m}\left(\frac{A_P}{S}\right)\left(\frac{K\omega_r}{\tau}\right) \tag{2.19}$$

For a membrane with cylindrically-shaped pores, the quantity $K\omega_r$ would be given by Equation 2.11. The tortuosity (τ) would need to be estimated separately, perhaps by comparison to experimental permeabilty data. On the otherhand, for diffusion through a gel-like membrane, $(\frac{A_P}{S})(\frac{K\omega_r}{\tau})$ would be given by the product of Equations 2.15 and 2.16. It is important to note that in both cases, if the solute is already within the membrane, $K = 1$.

In Equation 2.18, we see the product of the solute permeability and the membrane surface area, $P_m S$. Although we just developed a framework for describing P_m in Equation 2.19, there is still a lot of uncertainty in the values that go into estimating the permeability for a given solute by this equation. In some cases, for example for solute diffusion through the capillaries in a large region of tissue, it is also difficult to estimate the surface areas of the capillaries involved. In cases like these, it is more convenient to work with the product of P_m and S, rather than specific values of P_m or S. This is because of the uncertainty in accurately defining either the permeability or the capillary surface area in biological systems.

Renkin and Curry (1979) summarized the results of a variety of experiments that defined continuous capillary $P_m S$ values for solutes of various sizes. These data are shown in Figure 2.7. Note that for a given size solute, the value of $P_m S$ can vary by an order of magnitude. The $P_m S$ value is highly dependent on the nature of the capillaries that exist in the tissue of interest. In addition, fenestrated and discontinuous capillaries can have $P_m S$ values that are several order of magnitudes higher than those given in Figure 2.7.

The solid lines in Figure 2.7 represent a linear least-square regression of the data. The following empirical equations may be used to estimate capillary $P_m S$ values for a

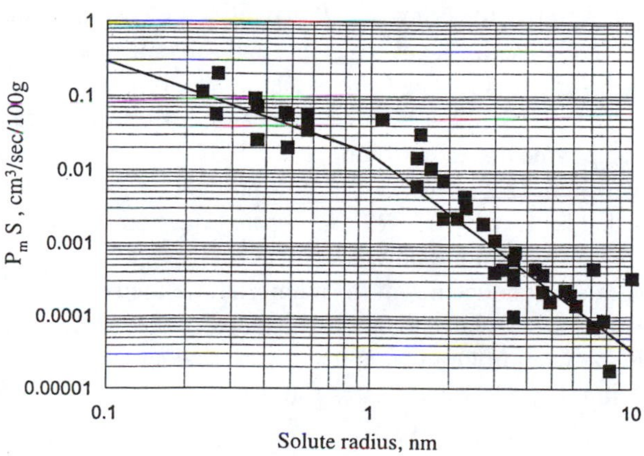

Figure 2.7 Solute permeability of capillaries (based on data from Renkin and Curry [1979]).

given solute of radius a. Note that the units on $P_m S$ are cm^3/sec/100 g of tissue.

$$P_m S = 0.0184a^{-1.223} \qquad a < 1\,\text{nm} \tag{2.20a}$$

$$P_m S = 0.0287a^{-2.92} \qquad a > 1\,\text{nm} \tag{2.20b}$$

2.8 THE IRREVERSIBLE THERMODYNAMICS OF MEMBRANE TRANSPORT

The transport of solute across membranes can occur as discussed in the previous sections by a combination of bulk flow and diffusion. In some cases, the relative magnitudes of these two transport rates for a given solute are comparable and we must consider the fact that these processes are therefore interdependent. This interaction makes the description of the solute transport rate much more complicated and to arrive at the proper result requires an understanding of the thermodynamics of irreversible processes.

The application of irreversible thermodynamics to membrane processes was developed by Staverman (1948) and Kedem and Katchalsky (1958). For dilute solutions, the theory of irreversible thermodynamics states that the filtration rate of the solvent, J, and the filtration rate due to the solute relative to that of the solvent, J_V, both depend in a linear manner on the driving forces ΔP (pressure for flow) and $RT\Delta C$ (concentration for diffusion). This linear combination of the driving forces is then given by the following equations:

$$J = SL_P\Delta P + SL_{PS}RT\Delta C \tag{2.21a}$$

$$J_V = SL_{SP}\Delta P + SL_S RT\Delta C \tag{2.21b}$$

J and J_V are vectors and must have the correct sign sense for proper interpretaton. This is easily accomplished if we simply assume a vertical orientation for the membrane. Flow of solvent and solute from left to right is considered positive. The "Δ" also represents the value of the property (P or C) on the left side of the membrane minus the corresponding value on the right side of the membrane.

The cross coefficients (L_{PS} and L_{SP}) represent secondary effects that are caused by the primary driving forces. For example, the primary ΔP driving force generates the filtration flow (J) and also induces a relative flow between solute and solvent represented by $L_{SP}\Delta P$. This relative flow between the solute and solvent is capable of producing a separation of solute and solvent, by the sieving mechanism discussed earlier, and is referred to as *ultrafiltration*. Along the same line of reasoning, the $RT\Delta C$ primary driving force is responsible for solute diffusion. This diffusive transport produces an additional contribution to the filtration flow that is given by $L_{PS}RT\Delta C$. From Equation 1.4 (Van't Hoff's equation), we recognize this as an osmotic flow.

The coefficient L_P is the hydraulic conductance of the membrane defined earlier by Equation 2.1. We will soon show that L_S is related to the permeability of the membrane.

Another basic theorem from the work of Onsager is that the cross-coefficients L_{SP} and L_{PS} are equal. We can then simplify Equation 2.21 as follows:

$$J = SL_P[\Delta P - \sigma RT\Delta C] \tag{2.22a}$$

$$J_V = SL_P\left[-\sigma\Delta P + \frac{L_S}{L_P}RT\Delta C\right] \tag{2.22b}$$

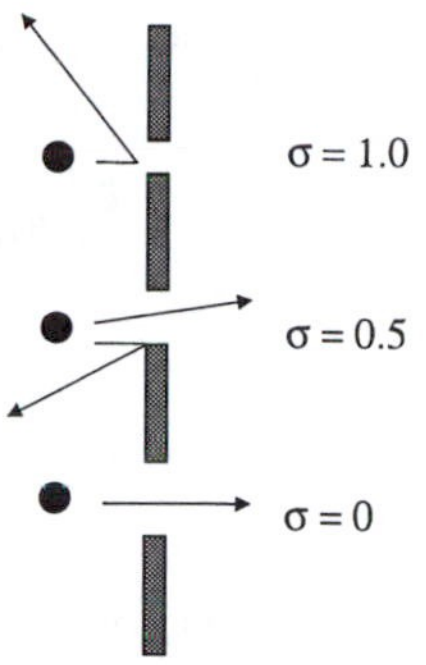

Figure 2.8 The reflection coefficient.

The parameter σ, defined as $-L_{SP}/L_P$ or $-L_{PS}/L_P$, is called the Staverman *reflection coefficient.*

Note the similarity between the filtration flow rate (J) in Equation 2.22a and that given by Equation 1.7, recognizing that $RT\Delta C$ is equivalent to $(\pi_C - \pi_{IF})$. These equations are identical when $\sigma = 1$. This implies that the pores of the membrane are impermeable to the solute as required by the derivation of Equation 1.7, the solute is therefore completely "reflected" by the membrane and the $RT\Delta C$ term is in fact the osmotic pressure difference. We then recover Starling's equation. When $\sigma = 0$ in Equation 2.22a, then from our previous discussion, we have no secondary effect of osmotic flow caused by the primary concentration difference driving force. This means that the solute flows through the pores completely unimpeded, just as easily as the solvent, hence the solute is not "reflected". The concept of the reflection coefficient is illustrated in Figure 2.8. Most solutes have a reflection coefficient that lies between these two extremes.

The total rate of transfer of solute through the pores of the capillary wall would be given by the product of the solute concentration (C) and the combined flow rate of the solution due to the applied pressure and concentration differences, i.e., $C(J + J_V)$. Combining Equations 2.22a and 2.22b results in the following equation for the transport rate of the solute.

$$N_S = CL_PS\left[(1-\sigma)\Delta P + \left(\frac{L_S}{L_P} - \sigma\right)RT\Delta C\right] \tag{2.23}$$

This equation can be rewritten by using Equation 2.22a to express ΔP as a function of J and ΔC.

$$N = C(1-\sigma)J + CSL_P\left(\frac{L_S}{L_P} - \sigma^2\right)RT\Delta C \tag{2.24}$$

The second term in the above equation has the factor $(L_S/L_P - \sigma^2)$ in it that must be determined. Recognizing that when $J = 0$, no filtration flow, diffusion is the only solute transport mechanism, then by Equation 2.22a we have:

$$(\Delta P)_{J=0} = (\sigma RT\Delta C)_{J=0} \tag{2.25}$$

Substituting this expression into Equation 2.23 for $J = 0$ gives the following equation, where we must recognize that in the absence of J, the flow rate of the solute across the membrane must be equal to our previous result given by Equation 2.18.

$$N_{J=0} = CL_P S(RT\Delta C)_{J=0}\left(\frac{L_S}{L_P} - \sigma^2\right) = P_m S(\Delta C)_{J=0} \tag{2.26}$$

We then find that $(L_S/L_P - \sigma^2)$ is given by the following equation:

$$\left(\frac{L_S}{L_P} - \sigma^2\right) = \frac{P_m}{L_P RTC} \tag{2.27}$$

where P_m is the membrane permeability defined earlier by Equation 2.18. Equation 2.24 can now be simplified using Equation 2.27 to give our final result for the solute transport rate through the pores of the capillary wall, recognizing that J is the filtration flow rate defined earlier by Equation 2.22a.

$$N_S = C(1-\sigma)J + P_m S\Delta C \tag{2.28}$$

The first term in the above equation represents the amount of solute transported by bulk flow (convection) of the solvent and the second term represents the contribution of diffusion. C represents the solute concentration on the side of the membrane where the filtration flow (J) originates.

2.9 FINDING L_P, P_m, AND σ

We must know the three parameters, L_P, P_m, and σ, in Equations 2.22a and 2.28 to calculate the transport rate of a solute. L_P is the hydraulic conductance and it can be estimated using Equation 2.1, or it can be measured using a solute whose value of $\sigma = 0$, and measuring the flow rate J across the membrane for a given ΔP. Then by Equation 2.22a, $L_P = \frac{J}{S\Delta P}$.

Recall that P_m is the permeability of the solute in the membrane and may be estimated by Equation 2.21. Alternatively, in the absence of any bulk flow across the membrane ($J=0$), the permeability can be obtained by measuring the solute transport rate (N) for a given concentration difference. Then from Equation 2.28, the permeability would be given by $P_m = \frac{N_S}{S\Delta C}$.

The reflection coefficient, σ, can be measured by one of two ways. In the first method, one employs a pressure drop across the membrane and measures the filtration flowrate J, as well as the transport rate of the solute, N_S, across the membrane in the absence of a solute concentration difference across the membrane. Then from Equation 2.28, we obtain the fact that $(1-\sigma) = \frac{N_S}{CJ}$. Alternatively, we could measure the filtration flowrate J, in the absence of a pressure drop across the membrane, but under the control of a concentration difference across the membrane (osmotic flow). Then from Equation 2.22a, the reflection coefficient would be given by $\sigma = \frac{-J}{SL_P RT\Delta C}$.

Durban (1960) has obtained the data shown in Figure 2.9 for the reflection coefficient of a variety of solutes in dialysis tubing, cellophane, and a wet gel. One could use this figure to directly estimate the reflection coefficient for a solute of known (a/r).

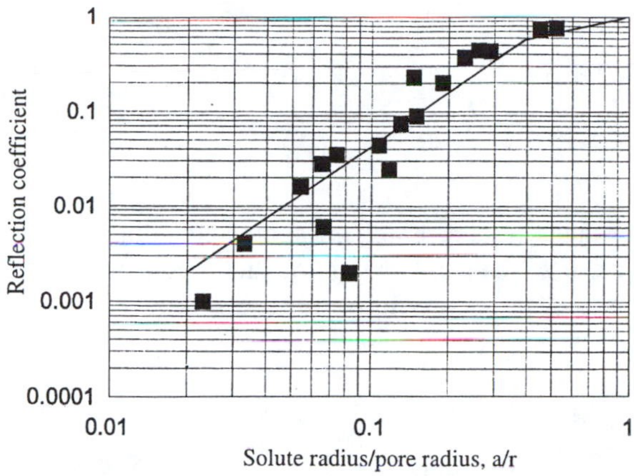

Figure 2.9 Reflection coefficient (based on data from Durbin [1960]).

The reflection coefficient can also be estimated using theories developed to describe the motion of particles in small pores. Anderson and Quinn (1974) reexamined the basic hydrodynamic equations that describe hindered particle motion in small pores. Under the assumptions of a rigid and spherical solute, and no electrostatic interactions with the pore wall, which is assumed to be inert, they showed that the sieving coefficient, $S_a \equiv N/CJ$, is the same as $(1 - \sigma)$. In this expression, C is the concentration of the solute on the nonfiltrate side of the membrane. Recall that their expression for S_a is given by Equation 2.2. This equation can be rewritten to give the following predictive equation for the reflection coefficient.

$$\sigma = 1 - \left(1 - \frac{a}{r}\right)^2 \left[2 - \left(1 - \frac{a}{r}\right)^2\right] \times \left[1 - \frac{2}{3}\left(\frac{a}{r}\right)^2 - 0.163\left(\frac{a}{r}\right)^3\right] \qquad (2.29)$$

The solid line in Figure 2.9 represents a comparison of Equation 2.29 to the experimental data obtained by Durbin (1960) for the reflection coefficient. Equation 2.29 provides a reasonable comparison to Durbin's data at higher values of a/r, where the a/r ratio would be expected to be the dominant factor controlling the value of the reflection coefficient. At lower values of a/r, other effects not accounted for in the derivation of Equation 2.29 dominate. In the absence of experimental data, Equation 2.29 can be used to provide an estimate of the reflection coefficient.

2.10 MULTICOMPONENT MEMBRANE TRANSPORT

Equations 2.22a and 2.28 describe the filtration flow rate and the solute transport rate in a system comprised of only a single solute and a solvent. However, similar equations can be derived for a multicomponent solute and solvent system by recognizing that now

Equations 2.21a and 2.21b become:

$$J = SL_P\Delta P + SRT\sum_i L_{PS_i}\Delta C_i \tag{2.30a}$$

$$J_V = S\Delta P\sum_i L_{SP_i} + SRT\sum_i L_{S_i}\Delta C_i \tag{2.30b}$$

The summations are over each solute in the system. Using the approach outlined earlier for the derivation of Equations 2.22a and 2.28, the following Equations are obtained for the filtration rate and the transport rate of solute i across the membrane in multicomponent systems:

$$J = SL_P\left[\Delta P - RT\sum_i \sigma_i\Delta C_i\right] \tag{2.31a}$$

$$N_i = C_i\left(1 - \sum_i \sigma_i\right)J + P_{m_i}S\Delta C_i \tag{2.31b}$$

2.11 TRANSPORT OF SOLUTES ACROSS THE CAPILLARY WALL

We will continue to use the capillary wall as a representative membrane to illustrate in the following examples the calculation of solute transport rates using the relationships developed in the previous sections. We will find it convenient in our discussion to classify solutes into three categories. The first category concerns the transport of small water-soluble but lipid-insoluble substances such as Na^+ and K^+ ions, glucose, amino acids, and other metabolites. The second category of solutes are lipid-soluble substances such as oxygen and carbon dioxide, and the third category we will discuss is the transport of large lipid insoluble substances such as proteins (macromolecules).

Example 2.2 Calculate the rate of transport by filtration and diffusion of a small water-soluble, but lipid-insoluble solute, such as glucose. The *diffusivity of glucose* (D) is 0.91×10^{-5} cm^2/sec and its molecular radius 0.36 nm. The average concentration of glucose in plasma is about 5 μmole/ml and it is assumed that all the glucose transported to the extracapillary space is consumed instantly by the cells. Assume that all of the plasma proteins are retained by the capillary wall. Also, assume that the tortuosity of the capillary wall is equal to 2.

SOLUTION The filtration flow rate of plasma (J) across the capillary wall is given by Equation 2.22a with $\sigma = 1.0$ and was shown earlier in Example 2.1 to be equal to 5.75×10^{-9} cm^3/hr. Equation 2.28 may be used to calculate the solute transport rate of glucose. We first need to estimate the reflection coefficient for glucose. Equation 2.29 may be used to obtain an estimate assuming the diameter of the capillary pores is 7 nm.

$$\sigma = 1-\left(1-\frac{0.36}{3.5}\right)^2\left[2-\left(1-\frac{0.36}{3.5}\right)^2\right]\times\left[1-\frac{2}{3}\left(\frac{0.36}{3.5}\right)^2-0.163\left(\frac{0.36}{3.5}\right)^3\right]$$

$$\sigma = 0.05$$

The permeability of glucose in the capillary wall may be estimated using Equation 2.21. The quantity $K\omega_r$ may be estimated from Equation 2.11, assuming the pores in the capillary wall are cylindrical in shape.

$$P_m = \frac{0.91 \times 10^{-5}\,\text{cm}^2/\text{sec}}{2 \times 0.5 \times 10^{-6}\,\text{m}} \times \frac{1\,\text{m}}{100\,\text{cm}} \times 0.001 \times \left(1 - \frac{0.36}{3.5}\right)^2$$

$$\times \left[1 - 2.1\left(\frac{0.36}{3.5}\right) + 2.09\left(\frac{0.36}{3.5}\right)^3 - 0.95\left(\frac{0.36}{3.5}\right)^5\right]$$

$$= 5.76 \times 10^{-5}\,\text{cm/sec}$$

Renkin (1977) reports values of the glucose permeability ranging from 1.6 $\times 10^{-6}$ cm/sec for capillaries in the dog brain to as high as 9.6×10^{-5} cm/sec for cat leg capillaries. On average, the capillary glucose permeability is about 2.65×10^{-5} cm/sec, thus the value calculated above is of the correct order of magnitude.

We may now calculate by Equation 2.28 the transport rate of glucose across the capillary wall.

$$N_S = 5\frac{\mu\text{mole}}{\text{cm}^3}(1 - 0.05) \times 5.75 \times 10^{-9}\frac{\text{cm}^3}{\text{hr}} + 5.76 \times 10^{-5}\frac{\text{cm}}{\text{sec}} \times \pi$$

$$\times 10 \times 10^{-6}\,\text{m} \times 0.001\,\text{m} \times (5 - 0)\frac{\mu\text{mole}}{\text{cm}^3} \times \frac{(100\,\text{cm})^2}{\text{m}^2} \times \frac{3600\,\text{sec}}{\text{hr}}$$

$$N_S = 2.73 \times 10^{-8}\frac{\mu\text{mole}}{\text{hr}} + 3.26 \times 10^{-4}\frac{\mu\text{mole}}{\text{hr}}$$

$$= 3.26 \times 10^{-4}\frac{\mu\text{mole}}{\text{hr}}$$

The first term in the calculation of N_S by Equation 2.28 represents the transport rate of glucose by filtration. The second term represents the transport rate by diffusion. Comparing the above result, we find that the transport rate of small water-soluble but lipid-insoluble solutes such as glucose across the capillary wall is several thousand times higher by diffusion than by convection! We may therefore neglect the convective transport of these substances across the capillary wall.

Example 2.3 Now consider the transport rate of lipid soluble solutes such as oxygen. In this case the entire surface area of the capillary wall is available for transport. Oxygen has considerable solubility in the cell membrane and will readily diffuse through the endothelial cells of the capillary wall. Oxygen can also diffuse through the pores in the capillary wall. The *diffusivity of oxygen* (D) is 2.11×10^{-5} cm^2/sec and its molecular radius is 0.16 nm. The average concentration of oxygen in the capillary is 0.09 μmole/cm^3. Once again assume that the tortuosity of the capillary wall is 2.

SOLUTION The filtration flow rate of plasma (J) across the capillary wall was shown earlier to be equal to 5.75×10^{-9} cm^3/hr. Equation 2.28 may be used to calculate the solute transport rate of oxygen through the pores in the capillary wall. We first need to estimate the reflection coefficient for oxygen. Equation 2.29 may be used to obtain an estimate assuming the diameter of the capillary pores is 7 nm.

$$\sigma = 1-\left(1-\frac{0.16}{3.5}\right)^2\left[2-\left(1-\frac{0.16}{3.5}\right)^2\right]\times\left[1-\frac{2}{3}\left(\frac{0.16}{3.5}\right)^2-0.163\left(\frac{0.16}{3.5}\right)^3\right]$$

$$\sigma = 0.01$$

The permeability of oxygen in the capillary pores may now be determined.

$$P_m = \frac{2.11\times10^{-5}\mathrm{cm^2/sec}}{2\times0.5\times10^{-6}\,\mathrm{m}}\times\frac{1\,\mathrm{m}}{100\,\mathrm{cm}}\times0.001\times\left(1-\frac{0.16}{3.5}\right)^2$$
$$\times\left[1-2.1\left(\frac{0.16}{3.5}\right)+2.09\left(\frac{0.16}{3.5}\right)^3-0.95\left(\frac{0.16}{3.5}\right)^5\right]$$
$$= 1.74\times10^{-4}\,\mathrm{cm/sec}$$

We may now calculate by Equation 2.28 the transport rate of oxygen through the pores of the capillary wall.

$$N_S = 0.09\,\frac{\mu\mathrm{mole}}{\mathrm{cm^3}}(1-0.01)\times5.75\times10^{-9}\frac{\mathrm{cm^3}}{\mathrm{hr}}+1.74\times10^{-4}\frac{\mathrm{cm}}{\mathrm{sec}}\times\pi$$
$$\times10\times10^{-6}\,\mathrm{m}\times0.001\,\mathrm{m}\times(0.09-0)\,\frac{\mu\mathrm{mole}}{\mathrm{cm^3}}\times\frac{(100\,\mathrm{cm})^2}{\mathrm{m^2}}\times\frac{3600\,\mathrm{sec}}{\mathrm{hr}}$$

$$N_S = 5.12\times10^{-10}\frac{\mu\mathrm{mole}}{\mathrm{hr}}+1.77\times10^{-5}\frac{\mu\mathrm{mole}}{\mathrm{hr}}$$
$$= 1.77\times10^{-5}\,\frac{\mu\mathrm{mole}}{\mathrm{hr}}$$

Comparing the above result, we again find that the transport rate of small molecules like oxygen through the capillary pores is considerably higher by diffusion than by convection.

The permeability of oxygen through the capillary wall itself is given by the following equation.

$$P_{\mathrm{oxygen}} = \frac{K_{\mathrm{oxygen}}D}{t_m} = \frac{0.25\times2.11\times10^{-5}\frac{\mathrm{cm^2}}{\mathrm{sec}}}{0.5\times10^{-6}\,\mathrm{m}}\times\frac{1\,\mathrm{m}}{100\,\mathrm{cm}} = 0.11\frac{\mathrm{cm}}{\mathrm{sec}}$$

K_{oxygen} is the oxygen partition coefficient and represents the ratio of the oxygen concentration in the capillary wall to that in the bulk solution. The value of 0.25 was used in order to match the oxygen permeability of the capillary wall with the accepted value of about 0.1 cm/sec. The oxygen transport rate from the capillary is then equal to:

$$N_{\mathrm{oxygen}} = 0.11\frac{\mathrm{cm}}{\mathrm{sec}}\times\pi\times10\times10^{-6}\,\mathrm{m}\times0.001\,\mathrm{m}\times(0.09-0)\frac{\mu\mathrm{mole}}{\mathrm{cm^3}}$$
$$\times\frac{(100\,\mathrm{cm})^2}{\mathrm{m^2}}\times\frac{3600\,\mathrm{sec}}{\mathrm{hr}} = 0.011\frac{\mu\mathrm{mole}}{\mathrm{hr}}$$

This is essentially the total oxygen transport rate from the capillary, since the contribution due to the capillary pores is negligible in comparison. The smaller value of the oxygen transport rate through the pores (N_S) in the pores in comparison to the value given above for the entire capillary wall (N_{oxygen}) represents the fact that the pore surface area is 1/1000 that of the capillary wall. Since the permeability of oxygen through the pores of the capillary and the capillary wall itself needs to be multiplied by the same values of S and ΔC to calculate the transport rate of oxygen, it is seen that regardless of these values, the diffusive transport of oxygen is 1000-fold higher through the endothelial cells lining the capillary walls than what is transported by diffusion through the capillary pores. Thus we can conclude that the dominant mode for transport of oxygen is simple diffusion across the entire surface area of the capillary walls.

Example 2.4 Our final category of solute to consider is lipid-insoluble macromolecules such as proteins. These solutes are characterized by a molecular radius that is comparable to the pores in the capillary wall. Calculate the rate of transport by filtration and diffusion through the capillary wall of a protein of molecular weight equal to 50,000 daltons. Assume the concentration of the protein in the plasma is 0.05 micromoles/ml.

SOLUTION The filtration flowrate of plasma (J) across the capillary wall is equal to 5.75×10^{-9} cm^3/hr. Since we only know the molecular weight of the protein, we need to first estimate its radius and its diffusivity. From Equation 2.10 we find that the radius of the solute is equal to 2.71 nm. We can also use Equation 2.8 to estimate the diffusivity. Therefore, we find that $D = 7 \times 10^{-7}$ cm^2/sec. Equation 2.28 may be used to calculate the solute transport rate of the protein. We first need to estimate the reflection coefficient for the protein. Equation 2.29 may be used to obtain an estimate, assuming the diameter of the capillary pores is 7 nm.

$$\sigma = 1 - \left(1 - \frac{2.71}{3.5}\right)^2 \left[2 - \left(1 - \frac{2.71}{3.5}\right)^2\right] \times \left[1 - \frac{2}{3}\left(\frac{2.71}{3.5}\right)^2 - 0.163\left(\frac{2.71}{3.5}\right)^3\right]$$

$$\sigma = 0.95$$

The permeability of the protein in the capillary wall may be estimated using Equation 2.19. The quantity $K\omega_r$ may be estimated from Equation 2.11 assuming the pores in the capillary wall are cylindrical in shape.

$$\begin{aligned} P_m &= \frac{7.0 \times 10^{-7}\ \text{cm}^2/\text{sec}}{2 \times 0.5 \times 10^{-6}\ \text{m}} \times \frac{1\ \text{m}}{100\ \text{cm}} \times 0.001 \times \left(1 - \frac{2.71}{3.5}\right)^2 \\ &\quad \times \left[1 - 2.1\left(\frac{2.71}{3.5}\right) + 2.09\left(\frac{2.71}{3.5}\right)^3 - 0.95\left(\frac{2.71}{3.5}\right)^5\right] \\ &= 2.85 \times 10^{-8}\ \text{cm/sec} \end{aligned}$$

This value of the permeability is representative of the experimental values for macromolecules of this size reported by Renkin (1977).

We may now calculate by Equation 2.28 the transport rate of the protein across the capillary wall.

$$N_S = 0.05\,\frac{\mu\text{mole}}{\text{cm}^3}(1-0.95)\times 5.75\times 10^{-9}\frac{\text{cm}^3}{\text{hr}} + 2.85\times 10^{-8}\frac{\text{cm}}{\text{sec}}$$
$$\times\,\pi\times 10\times 10^{-6}\,\text{m}\times 0.001\,\text{m}\times(0.05-0)\frac{\mu\text{mole}}{\text{cm}^3}\times\frac{(100\,\text{cm})^2}{\text{m}^2}\times\frac{3600\,\text{sec}}{\text{hr}}$$
$$N_S = 1.44\times 10^{-11}\frac{\mu\text{mole}}{\text{hr}} + 1.61\times 10^{-9}\frac{\mu\text{mole}}{\text{hr}}$$
$$= 1.61\times 10^{-9}\frac{\mu\text{mole}}{\text{hr}}$$

Once again we observe that even for a macromolecular solute with a molecular weight of 50,000 daltons, diffusive transport across the capillary wall is still significantly higher than the amount of solute transported by convection.

Example 2.5 Repeat the previous example, assuming that the mean capillary pressure is increased by 20 mm Hg.

SOLUTION The filtration flow rate of plasma (J) across the capillary wall will be significantly increased as a result of the increase in mean capillary pressure. The net filtration pressure will increase from 0.3 to 20.3 mm Hg. We can calculate the filtration rate of the plasma following the procedure outlined in Example 2.1. We can then show that the filtration rate equals 3.89×10^{-7} cm^3/hr. This represents a nearly 68-fold increase in the plasma filtration rate and would lead to edema in the interstitium. The reflection coefficient and solute permeability would still have the same values as calculated in the previous example, i.e., $\sigma = 0.95$ and $P_m = 2.85\times 10^{-8}$ cm/sec.

We may now calculate by Equation 2.28 the transport rate of the protein across the capillary wall for the special case of increased mean capillary pressure.

$$N_S = 0.05\,\frac{\mu\text{mole}}{\text{cm}^3}(1-0.95)\times 3.89\times 10^{-7}\frac{\text{cm}^3}{\text{hr}} + 2.85\times 10^{-8}\frac{\text{cm}}{\text{sec}}\times\pi$$
$$\times\,10\times 10^{-6}\,\text{m}\times 0.001\,\text{m}\times(0.05-0)\frac{\mu\text{mole}}{\text{cm}^3}\times\frac{(100\,\text{cm})^2}{\text{m}^2}\times\frac{3600\,\text{sec}}{\text{hr}}$$
$$N_S = 9.73\times 10^{-10}\frac{\mu\text{mole}}{\text{hr}} + 1.61\times 10^{-9}\frac{\mu\text{mole}}{\text{hr}}$$
$$= 2.58\times 10^{-9}\frac{\mu\text{mole}}{\text{hr}}$$

The first term in the calculation of N_S by Equation 2.28 represents the transport rate of the protein by filtration. The second term represents the transport rate by diffusion. Unlike the calculations for solute transport in the previous examples, we now observe that the increased filtration pressure has significantly increased the transport rate of the solute by convection. Convection of the solute now accounts for 38% of the total solute transport rate and is responsible for the 60% increase in the solute transport rate as a result of the increased mean capillary pressure.

2.12 TRANSPORT OF A SOLUTE BETWEEN A CAPILLARY AND THE SURROUNDING TISSUE SPACE

Now that we have developed an understanding of solute transport across a semipermeable membrane like the capillary wall, it is possible to develop a model for the spatial distribution and consumption or production of a particular species within the tissue space surrounding a given capillary. In the case of oxygen, the situation is somewhat more complex since the red blood cell provides an additional source or sink of oxygen through interactions with hemoglobin. We will discuss oxygen transport later in Chapter 4.

2.12.1 Shell Balances

The concept of a shell balance is an important technique for developing mathematical models involving solute transport in biological systems. The shell balance is conceptually simple to use and is based on application of the following generalized balance equation for a given finite volume of interest.

$$\text{Accumulation} = \text{In} - \text{Out} + \text{Generation} \tag{2.32}$$

This equation may be used to account for changes in such quantities as mass, moles, energy, momentum, and even money. The "Accumulation" term accounts for the change with time of the quantity of interest within the volume of interest. Under steady-state conditions, where the quantity of interest in a given volume is not changing with time, then this term is equal to zero. The "In" term accounts for the entry of the quantity of interest into the system by all routes. For example, as carried in by flow of fluid, by diffusion, or in the case of energy, by conduction. The "Out" term is similar to the "In" term but accounts for loss of the quantity of interest. The "Generation" term accounts for the production or consumption of the quantity of interest within the volume of interest. In the case of a solute, this term could represent consumption or production of the solute by cells.

Example 2.6 To illustrate a simple application of Equation 2.32, consider the filtration of plasma in either a capillary or in a hollow fiber used in the cross-flow membrane plasmapheresis of blood. Assume the length of the flow channel is given by L and the radius of the capillary or fiber is R. Develop the differential equation that describes the change with axial position (z) of the volumetric flowrate of the blood. Assume steady-state flow.

SOLUTION Perform a shell balance across a finite axial control volume of the cylindrical flow channel equal to $\pi R^2 \Delta z$. The "Accumulation" term represents the change in the volumetric flow rate with time at any particular axial location. Since the system is at steady-state, this term will equal zero. The "In" term represents the volumetric flow rate into the control volume and may be written as $Q|_z$. The "Out" term represents the volumetric flow rate out of the control volume and may be expressed as $Q|_{z+\Delta z}$. The "generation" term would equal the loss of plasma by filtration across the capillary wall or the membrane surface. This is equal to $-2\pi R \Delta z j$, where $2\pi R \Delta z$ is the circumferential surface area of the control volume.

Equation 2.32 may then be written as:

$$0 = Q|_z - Q|_{z+\Delta z} - 2\pi R \Delta z j$$

Dividing by Δz, and taking the limit as $\Delta z \to 0$, results in the following differential equation that describes the axial variation of Q due to filtration.

$$\frac{dQ(z)}{dz} = -2\pi R j(z)$$

Integration of this equation for the axial change in the volumetric flow rate requires another relationship between $j(z)$ and $Q(z)$ (see problem 2.11).

2.12.2 The Krogh Tissue Cylinder

Microscopic studies support somewhat the notion that a bed of capillaries in tissue can be represented as a repetitive arrangement of capillaries surrounded by a cylindrical layer of tissue. Figure 2.10 shows an idealized sketch of the capillary bed and the corresponding cylindrical layer of tissue of thickness equal to r_T. As mentioned at the beginning of this chapter, the residence time for blood in the capillary is on the order of 1 second. The diffusivity of a solute in tissue is on the order of 10^{-5} cm^2/sec. The characteristic time for diffusion in the Krogh tissue cylinder can be expressed by the following relation: $t_{\text{diffusion}} = \frac{r_T^2}{2D} \ln(\frac{r_T}{r_C})$ (Lightfoot 1995). Since the blood residence time and the diffusion time must be of the same order of magnitude, we can use this simple relationship to estimate that the Krogh tissue cylinder radius is on the order of 30 microns. Krogh (1919) used this cylindrical capillary-tissue model to study the supply of oxygen to muscle. This approach is now known as the *Krogh tissue model.*

A steady-state mathematical model for the transport of a solute can be developed using the Krogh tissue model. We will treat the tissue space surrounding the capillary as a continuous phase and ignore the fact that it is comprised of discrete cells. Therefore, the diffusion of the solute in the tissue space may be described by an effective diffusivity (D_T) as defined earlier by Equations 2.14 or 2.17. The driving force for diffusion of the solute is the consumption (or production) of the solute by the cells within the tissue space.

We need to have an expression that relates the consumption or production of the solute to the concentration of the solute in the tissue space ($\bar{C}$). The Michaelis-Menten

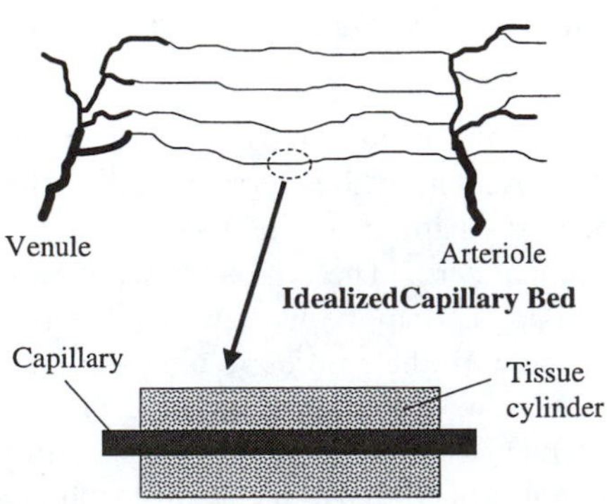

Figure 2.10 The Krogh tissue cylinder.

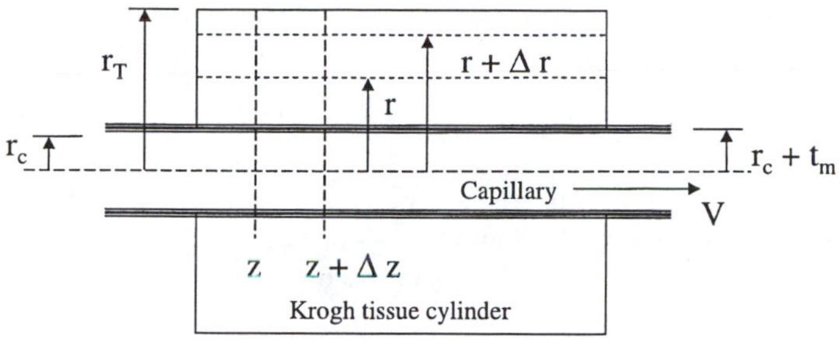

Figure 2.11 Geometry of the Krogh tissue cylinder.

equation can be used to describe the metabolic consumption (or production) of the solute in the tissue space. The Michaelis-Menten equation may be written as follows:

$$R(\bar{C}) = \frac{V_{\max}\bar{C}}{K_m + \bar{C}} \tag{2.33}$$

For consumption of the solute, $R(\bar{C})$ will be assumed to have a positive value, whereas if the solute is produced, then $R(\bar{C})$ will have a negative value. In this equation, $V_{\max}$ represents the maximum reaction rate. The maximum reaction rate occurs when $\bar{C} \gg K_m$, and the reaction rate is then said to be zero-order in the solute concentration (i.e., $\bar{C}^0$). K_m is called the Michaelis constant, and represents the value of $\bar{C}$ for which the reaction rate is one-half the maximal value. When $\bar{C} \ll K_m$, then $R(\bar{C}) = \frac{V_{\max}}{K_m}\bar{C}$, and the reaction is first-order in the solute concentration (i.e., $\bar{C}^1$). In many cases for biological reactions, $\bar{C} \gg K_m$, and we may reasonably assume that $R(\bar{C}) = V_{\max} = R_0$. This will considerably simplify the following mathematical analysis.

Figure 2.11 illustrates a finite volume section of the Krogh tissue cylinder extending from z to $z + \Delta z$. Within the capillary, the solute is transported primarily by convection in the axial direction and by diffusion in the radial direction. Diffusion of the solute from the blood and through the capillary wall is proportional to the concentration difference of the solute and is represented by an overall mass transfer coefficient (K_0). The overall mass transfer coefficient represents the combined resistance of the fluid flowing through the capillary (i.e., k_m) and the permeability of the solute in the capillary wall (P_m). We also treat the blood as a continuous phase and ignore the finite size of the red blood cells.

We will assume that the blood flows through the capillary with an average velocity of V. With these assumptions, a steady-state shell balance on the solute in the blood from z to $z + \Delta z$ provides the following equation:

$$V\pi r_C^2 C\big|_z - V\pi r_C^2 C\big|_{z+z\Delta z} = 2\pi r_C \Delta z K_0 \left(C - \bar{C}\big|_{r_C+t_m}\right) \tag{2.34}$$

Dividing by Δz, and taking the limit as $\Delta z \to 0$, results in the following differential equation for the solute concentration in the blood.

$$-V\frac{dC}{dz} = \frac{2}{r_C}K_0\left(C - \bar{C}\big|_{r_C+t_m}\right) \tag{2.35}$$

A steady-state shell balance at a given value of z from r to $r + \Delta r$ may also be written for the solute concentration in the tissue space.

$$-D_T \frac{d\bar{C}}{dr} 2\pi r \Delta z \bigg|_r + D_T \frac{d\bar{C}}{dr} 2\pi r \Delta z \bigg|_{r+\Delta r} = R(\bar{C}) 2\pi r \Delta r \Delta z \tag{2.36}$$

This equation neglects axial diffusion in comparison to radial diffusion within the tissue space. After dividing by $2\pi r \Delta r$, and taking the limit as $\Delta r \to 0$, the following differential equation results for the solute concentration in the tissue space.

$$\frac{D_T}{r} \frac{d}{dr}\left(r \frac{d\bar{C}}{dr} \right) + R(\bar{C}) = 0 \tag{2.37}$$

The boundary conditions for differential Equations 2.35 and 2.37 are:

$$\begin{aligned} &BC1: && z = 0, && C = C_o \\ &BC2: && r = r_C + t_m, && \bar{C} = \bar{C}\big|_{r_C+t_m} \\ &BC3: && r = r_T, && \frac{d\bar{C}}{dr} = 0 \end{aligned} \tag{2.38}$$

It should be noted that even though we neglect axial diffusion in the tissue space in comparison to radial diffusion, the solute concentration in the tissue space will still have an axial dependence. This axial dependence of the tissue space solute concentration results from the boundary condition at $(r_C + t_m)$ and the solute flux term, $K_0(C - \bar{C}|_{r_C+t_m})$, which couples the solutions for the solute concentration within the capillary and the tissue space. In this way, the axial variation of the solute concentration in the capillary is impressed on that of the tissue space.

The solute concentration in the tissue space is then easily found by solving equation 2.37 subject to boundary conditions 2 and 3 in Equation 2.38. Here we assume the consumption of solute is zero-order as discussed earlier, hence $R(\bar{C}) = R_0$ is a constant. The following equation is then obtained.

$$\bar{C}(r, z) = \bar{C}(z)\big|_{r_C+t_m} + \frac{R_0}{4D_T}[(r^2 - (r_C + t_m)^2)] - \frac{R_0 r_T^2}{2D_T} \ln\left(\frac{r}{r_C + t_m} \right) \tag{2.39}$$

This equation provides no information on the value of $\bar{C}(z)|_{r_C+t_m}$. However, at any axial position z from the capillary entrance, at steady state, the change in solute concentration within the blood must equal the consumption of solute in the tissue space. Therefore we can express this requirement by writing the following equation:

$$V\pi r_C^2 C_0 - V\pi r_C^2 C\big|_z = \pi \left[r_T^2 - (r_C + t_m)^2 \right] z R_0 \tag{2.40}$$

This equation may be rearranged to give the following equation that provides for the axial variation of the solute concentration in the capillary.

$$C(z) = C_0 - \frac{R_0}{V r_C^2} \left[r_T^2 - (r_C + t_m)^2 \right] z \tag{2.41}$$

Equation 2.41 can now be used to find $\frac{dC}{dz}$ in Equation 2.35 with the result that $\bar{C}(z)|_{r_C+t_m}$ is given by:

$$\bar{C}(z)\Big|_{r_C+t_m} = C(z) - \frac{R_0}{2r_C K_0}\left(r_T^2 - (r_C + t_m)^2\right) \tag{2.42}$$

Note that $C(z)$ in Equation 2.42 makes $\bar{C}(z)|_{r_C+t_m}$ depend on z, and by Equation 2.39, the tissue space solute concentration then depends on both r and z as discussed earlier. Equations 2.39, 2.41, and 2.42 can be combined to give the following equation for the solute concentration in the tissue space.

$$\begin{aligned}\bar{C}(r,z) = C_0 &- \frac{R_0}{V r_C^2}\left[r_T^2 - (r_C + t_m)^2\right]z - \frac{R_0}{2r_C K_0}\left[r_T^2 - (r_C + t_m)^2\right]\\ &+ \frac{R_0}{4D_T}[(r^2 - (r_C + t_m)^2)] - \frac{R_0 r_T^2}{2D_T}\ln\left(\frac{r}{r_C + t_m}\right)\end{aligned} \tag{2.43}$$

Under some conditions either the delivery of the solute to the capillary may be limited by the capillary flow rate, or the transport rate of the solute across the capillary wall is limited, or the consumption of the solute by the tissue is very rapid. Any one of these conditions may lead to regions of the tissue that have no solute. We can then define a *critical radius* in the tissue, $r_{\text{critical}}(z)$, defined as the distance beyond which no solute is present in the tissue. For this situation we need to modify boundary condition 3 in equation 2.38 to the following:

$$\text{critical radius } BC3': r = r_{\text{critical}}(z), \quad \frac{d\bar{C}}{dr} = 0 \quad \text{and} \quad \bar{C} = 0 \tag{2.44}$$

Under these conditions, the solute concentrations in the capillary, i.e., $C(z)$, at the interface between the capillary and the tissue space, i.e., $\bar{C}(z)|_{r_C+t_m}$, and in the tissue space itself, i.e., $\bar{C}(r,z)$, would still be given respectively by Equations 2.41, 2.42, and 2.43; however, the Krogh tissue cylinder radius, r_T, is replaced with $r_{\text{critical}}(z)$ once the solute concentration in the tissue at a particular location has reached zero. The critical radius may be obtained by recognizing that at $r_{\text{critical}}(z)$, $\bar{C}(r,z) = 0$. Thus we may use Equation 2.43, with $r_T = r_{\text{critical}}(z)$ and $\bar{C}(r,z) = 0$, to obtain the following expression for the critical radius.

$$\begin{aligned}\left(\frac{r_{\text{critical}}(z)}{r_C + t_m}\right)^2 \ln\left(\frac{r_{\text{critical}}(z)}{r_C + t_m}\right)^2 - \left(\frac{r_{\text{critical}}(z)}{r_C + t_m}\right)^2 + 1 = \left(\frac{4D_T C_0}{R_0(r_C + t_m)^2}\right)\\ - \frac{4D_T}{V r_C^2}\left[\left(\frac{r_{\text{critical}}(z)}{r_C + t_m}\right)^2 - 1\right]z - \frac{2D_T}{r_C K_0}\left[\left(\frac{r_{\text{critical}}(z)}{r_C + t_m}\right)^2 - 1\right]\end{aligned} \tag{2.45}$$

This is a nonlinear algebraic equation that may be solved for the critical radius. This is illustrated in Example 2.8 below.

Of particular interest would be the conditions under which the solute just begins to be depleted within the tissue space. The depletion of the solute will start at the downstream corner of the Krogh tissue cylinder represented by the coordinates $z = L$ and $r = r_T$.

Equation 2.43 may be used to explore the conditions when the solute depletion just begins by recognizing that $\bar{C}(r_T, L) = 0$.

The overall mass transfer coefficient K_0 can be evaluated by recognizing that the solute flux at any value of z is constant (neglecting the curvature of the capillary wall) from the blood to the capillary wall at r_C, and across the capillary wall to the beginning of the tissue space at $r_C + t_m$. The transport rate across the capillary wall would be given by Equation 2.28, recognizing that in most cases the convective term $C(1-\sigma)J$ is negligible in comparison to the amount of solute transported by diffusion. We may then write at a particular axial location along the capillary that:

$$N = 2\pi r_C \Delta z k_m \left(C - \hat{C}\big|_{r_C}\right) = 2\pi r_C \Delta z P_m \left(\hat{C}\big|_{r_C} - \bar{C}\big|_{r_C+t_m}\right)$$
$$= 2\pi r_C \Delta z K_0 \left(C - \bar{C}\big|_{r_C+t_m}\right) \tag{2.46}$$

Where $\hat{C}$ now denotes the solute concentration at the surface of the capillary wall on the blood side. Equation 2.46 may then be rearranged to obtain the following expression for the overall mass transfer coefficient.

$$K_0 = \frac{1}{\frac{1}{k_m} + \frac{1}{P_m}} \tag{2.47}$$

P_m is the solute permeability across the capillary and k_m is the blood side film mass transfer coefficient defined earlier by Equation 2.4.

The transport properties for the blood, as represented by the viscosity and the solute diffusivity, are based on their respective plasma values. This occurs because for blood the Schmidt number, $Sc = \mu/\rho D$, is much greater than unity. Laminar boundary layer theory (Bird, Stewart, and Lightfoot 1960) tells us that the concentration boundary layer, the distance from the capillary wall over which the solute concentration changes from its bulk value to its value at the capillary wall, is much thinner for $Sc \gg 1$ than the momentum boundary layer, the distance from the capillary wall over which the axial velocity changes from its bulk value to zero at the capillary wall. The red blood cells also have a tendency, as we shall discuss in Chapter 3, to accumulate along their axis of flow. This creates a cell-free plasma layer adjacent to the capillary wall. Because of the thinness of the concentration boundary layer, it lies well within the cell-free layer, making it appropriate to base the transport properties on the plasma values.

The calculation of the solute concentration within the capillary and the tissue space surrounding the capillary is illustrated in the following examples.

Example 2.7 Calculate the glucose concentration profiles for an exercising muscle capillary. Assume the capillary properties summarized in Table 2.1 apply and that the glucose consumption rate, R_0, is equal to 0.01 μmole/sec/cm^3. Also assume that the Krogh tissue cylinder radius, r_T, is 40 microns. All other properties are listed as part of the solution to the problem shown in Chart 2.1.

SOLUTION Mathcad® is used to obtain the solution shown in Chart 2.1. Shown in the chart figure are the axial solute concentrations within the capillary and in the tissue space at the Krogh tissue cylinder radius (r_T).

Chart 2.1 for **Example 2.7:** Glucose concentration profiles for an exercising muscle capillary.

$z := 0, .001 \ldots 1$	axial position in capillary, cm
$D_T := 8 \times 10^{-6}$	glucose tissue diffusivity, cm^2/sec, from Figure 2.6
$D := .91 \times 10^{-5}$	plasma solute diffusivity, cm^2/sec
$V := .05$	average blood velocity, cm/sec
$\mu := .012$	plasma viscosity, g/cmsec
$\rho := 1.024$	plasma density, g/cm^3
$D_C := .001$	capillary diameter, cm; $r_C = .5 D_C$
$t_m := 5 \times 10^{-5}$	capillary wall thickness, cm
$R_e := \frac{\rho D_C V}{\mu}$	Reynolds number; $R_e = 0.004$
$Sc := \frac{\mu}{(\rho D)}$	Schmidt number; $Sc = 128$
$C_0 := 5$	entering glucose concentration, $\mu mole/cm^3$
$r_T := 4 \times 10^{-3}$	Krogh tissue cylinder radius, cm
$R_0 := .01$	tissue glucose consumption rate, $\mu mole/cm^3 sec$
$Sh := 3.66$	Sherwood number, assume fully-developed flow
$k_m := \frac{D Sh}{D_C}$	value of mass-transfer coefficient, cm/sec; $k_m := 0.033$ cm/sec
$P_m := 5.76 \times 10^{-5}$	glucose permeability, cm/sec; from Example 2.2
$K_0 := \frac{1}{\left(\frac{1}{k_m} + \frac{1}{P_m}\right)}$	overall mass transfer coefficient; $K_0 = 5.75 \times 10^{-5}$ cm/sec

Now calculate the axial glucose profiles.

$$C(z) := C_0 - \left[\frac{R_0}{(V r_C^2)}\right]\left[r_T^2 - (r_C + t_m)^2\right] z$$

solute axial profile in blood

$$C_W(z) := C(z) - \left[\frac{R_0}{(2 r_C K_0)}\right]\left[r_T^2 - (r_C + t_m)^2\right]$$

solute axial profile on the outside wall of the capillary

$$C_T(z) := C_W(z) + \left[\frac{R_0}{(4 D_T)}\right]\left[r_T^2 - (r_C + t_m)^2\right] - \left[\frac{R_0 r_T^2}{(2 D_T)}\right] \ln\left[\frac{r_T}{(r_C + t_m)}\right]$$

solute axial profile at the Krogh tissue radius

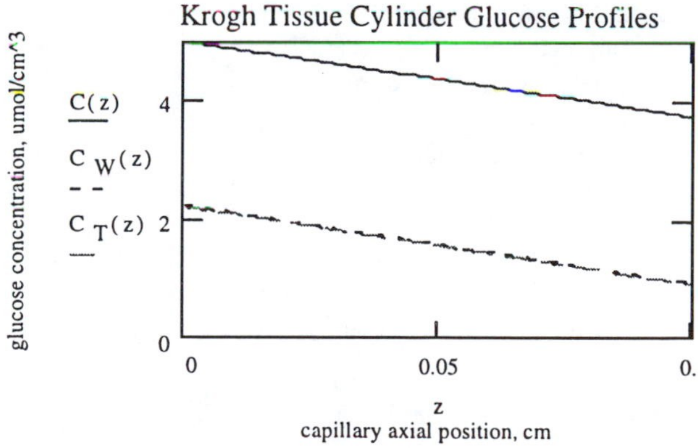

As shown in Chart 2.1, we observe that the axial solute concentration within the capillary and in the tissue space decreases in a linear fashion. We also observe very little difference between the glucose concentration at the outer surface of the capillary wall and at the Krogh tissue cylinder. This indicates that there is little mass transfer resistance to glucose within the tissue space. It is also important to note for glucose transport, that the bulk of the mass transfer resistance is the capillary wall as reflected by the value of the glucose permeability, P_m. We see that the glucose permeability is about 2 orders of magnitude smaller than the film mass transfer coefficient k_m, and for practical purposes we can ignore the blood side mass transfer effect since it offers little resistance in comparison to that of the capillary wall. We also see that because of the large mass transfer resistance of the capillary wall, a significant concentration difference must exist between the capillary and the tissue space to obtain the required mass transfer rate. Transport of glucose for these conditions is therefore said to be diffusion limited.

Example 2.8 Calculate the critical radius for the capillary conditions used in Example 2.7. Assume that the flow rate in the capillary has been reduced by a factor of ten to $V = 0.005$ cm/sec.

SOLUTION Mathcad® is used to obtain the solution shown in Chart 2.2. This example also illustrates the use of the functions "root" and "find" to solve nonlinear algebraic equations.

Chart 2.2 for **Example 2.8:** Critical radius calculation for an exercising muscle capillary under conditions of reduced blood flow.

$z := 0, .001 \ldots 1$	axial position in capillary, cm
$D_T := 8 \times 10^{-6}$	glucose tissue diffusivity, cm^2/sec
$D := .91 \times 10^{-5}$	plasma solute diffusivity, cm^2/sec
$V := .005$	average blood velocity, cm/sec
$\mu := .012$	plasma viscosity, g/cm sec
$\rho := 1.024$	plasma density, g/cm^3
$D_C := .001$	capillary diameter, cm
$r_C := .5 \times D_C$	
$t_m := 5 \times 10^{-5}$	capillary wall thickness, cm
$R_e := \frac{\rho \cdot D_C \cdot V}{\mu}$	Reynolds number; $R_e = 4.267 \times 10^{-4}$
$Sc := \frac{\mu}{(\rho D)}$	Schmidt number; $Sc = 1288$
$C_0 := 5$	entering glucose concentration, μmole/cm^3
$r_T := 4 \times 10^{-3}$	Krogh tissue cylinder radius, cm
$R_0 := .01$	tissue glucose consumption rate, μmole/cm^{3}sec
$Sh := 3.66$	Sherwood number, assume fully-developed flow
$k_m := \frac{DSh}{D_C}$	value of mass-transfer coefficient, cm/sec; $k_m = 0.033$ cm/sec
$P_m := 5.76 \times 10^{-5}$	glucose permeability, cm/sec; from Example 2.2
$K_0 := \frac{1}{\left(\frac{1}{k_m} + \frac{1}{P_m}\right)}$	overall mass-transfer coefficient, cm/sec; $K_0 = 5.75 \times 10^{-5}$ cm/sec

We first need to find at what value of z the tissue solute concentration at r_T first becomes zero, let this position be $z = Z$.

$Z := 0.5 \times .1$

an initial guess of Z equal to 1/2 the capillary length

$$C(Z) := C_0 - \left[\frac{R_0}{(Vr_C^2)}\right]\left[r_T^2 - (r_C + t_m)^2\right] Z$$

solute axial profile in blood

$$C_W(Z) := C(Z) - \left[\frac{R_0}{(2r_C K_0)}\right]\left[r_T^2 - (r_C + t_m)^2\right]$$

solute axial profile on the outside wall of the capillary

$$C_T(Z) := \left[(C_W(Z)) + \left[\frac{R_0}{(4D_T)}\right]\left[r_T^2 - (r_C + t_m)^2\right] - \left[\frac{R_0 r_T^2}{(2D_T)}\right]\ln\left[\frac{r_T}{(r_C + t_m)}\right]\right]$$

solute axial profile at the Krogh tissue radius

We want $C_T = 0$ at Z so we use the root finder function to solve the above equation.

$Z := \mathrm{root}(C_T(Z), Z)$

$Z = 0.018$ cm, axial location where tissue at the Krogh tissue cylinder radius is devoid of solute

Now the solute for positions downstream of Z can be caculated.

$r_{\mathrm{critical}} = 0.5(r_C + r_T)$

initial guess of the critical radius

given

Mathcad nonlinear equation solver

$$\left(\frac{r_{\mathrm{critical}}}{r_C + t_m}\right)^2 \ln\left(\frac{r_{\mathrm{critical}}}{r_C + t_m}\right)^2 - \left(\frac{r_{\mathrm{critical}}}{r_C + t_m}\right)^2 + 1 - \left[\frac{4D_T C_0}{R_0(r_C + t_m)^2}\right]$$
$$+ 4\frac{D_T}{Vr_C^2}\left[\left(\frac{r_{\mathrm{critical}}}{r_C + t_m}\right)^2 - 1\right]z + 2 \times \frac{D_T}{r_C K_0} \times \left[\left(\frac{r_{\mathrm{critical}}}{r_C + t_m}\right)^2 - 1\right] = 0$$

$\mathrm{critical}(z) := \mathrm{find}(r_{\mathrm{critical}})$

$N := \frac{0.1 - Z}{0.001}$

$N := \mathrm{floor}(N) + 1$

$i := 0 \cdots N$

$z_i := Z + 0.001 i$

$r_{\mathrm{critical}_i} = 10000\,\mathrm{critical}(z_i)$; the 10000 converts cm to microns

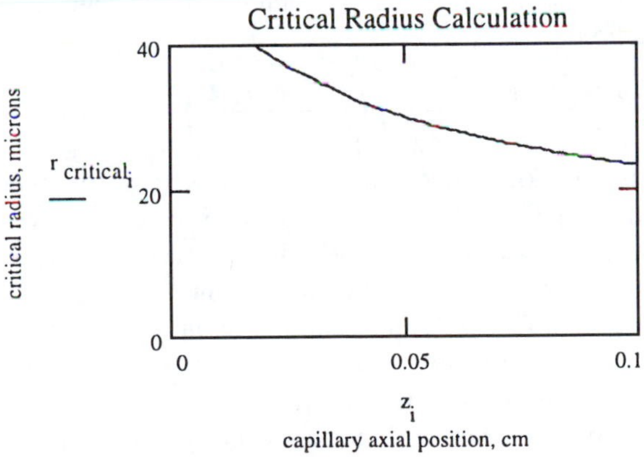

the exiting glucose concentration in the capillary would be given by:

$$C_L := C_0 - \left(\frac{R_0}{Vr_C^2}\right)\left[\left(\frac{r_{\mathrm{critical}_N}}{10000}\right)^2 - (r_C + t_m)^2\right] z_N$$

$C_L = 0.887$ mole/cm^3

The above calculation shows that the glucose concentration at the Krogh tissue cylinder radius is nonzero until 0.018 cm into the capillary of total length 0.1 cm. The region above the curve in the Critical Radius figure represents the region of the tissue where the glucose concentration is zero. We also see that the exiting glucose concentration for the reduced flow conditions in the above example is significantly reduced in comparison to the result obtained in Example 2.7. Although the exiting glucose concentration in the capillary is lower, this does not necessarily mean that the total amount of glucose transferred is higher. In fact the amount of solute transferred is about one-third lower than Example 2.7 because of the reduced velocity of the blood. For solutes with reasonably high capillary permeability, i.e., lipophilic solutes like oxygen or small lipophobic solutes like glucose, the solute transport rate, as in this example, can be flow limited. The transport rate of the solute to the tissue can therefore be increased by increasing the blood flow rate as in Example 2.7. However, for larger lipid insoluble solutes, the capillary solute permeability is significantly reduced and the solute transport rate is diffusion limited. In this case the blood flow rate has little effect on the solute transport rate.

The above analysis can also be simplified considerably for the special case where the solute concentration in the tissue space is zero. We can also assume, as shown in the above example, that $K_0 \approx P_m$. In other words, the blood offers little resistance to the solute transport rate in comparison to that of the capillary wall. For these conditions, Equation 2.35 simplifies to:

$$\frac{\mathrm{d}C}{\mathrm{d}z} = -\frac{2}{V r_C} P_m C = -\frac{2\pi r_C}{Q} P_m C \tag{2.48}$$

Here we have replaced the capillary blood velocity with the volumetric flow rate of blood in the capillary. This equation is easily integrated to obtain the solute concentration in the capillary at any axial position z.

$$E = \frac{[C_0 - C(z)]}{C_0} = 1 - \exp\left(-\frac{2\pi r_C P_m z}{Q}\right) \tag{2.49}$$

This equation is known as the Renkin-Crone equation and describes how the solute concentration varies along the length of the capillary for the special case where the solute concentration in the tissue space is zero. The solute extraction is represented by E. Since the quantity QC_0 represents the maximum amount of solute that can be transported across the capillary, then $E \times QC_0$ represents the actual solute transport rate for the given conditions. The solute extraction is a measure of the removal efficiency of the solute from the capillary. This equation readily shows that solute transport from the capillary is strongly dependent on the ratio of the permeability of the capillary wall ($2\pi r_C P_m z$) to the blood flow rate (Q). This ratio can be used to define conditions under which the solute transport is either flow limited or diffusion limited as follows:

$$\text{if } \frac{2\pi r_C P_m z}{Q} \gg 1, \text{ then flow limited, if } \frac{2\pi r_C P_m z}{Q} \ll 1, \text{ then diffusion limited}$$

For regions of tissue containing many capillaries, we can replace the quantity $2\pi r_C P_m z$

with the group $P_m S$, where S represents the total surface area of the capillaries within the tissue region of interest.

The multiple tracer indicator diffusion technique may be used to obtain a test solute's permeability ($P_m S$) across the capillary wall in organs and large tissue regions. In this technique, a solution containing equal concentrations of a permeable test solute and a nonpermeable reference solute (typically labelled albumin), is rapidly injected into a main artery leading to the region of interest (Levick 1991). Venous blood samples are then collected over the next few seconds and analyzed for the concentrations of the test and reference solutes. Because the test solute is permeable, its concentration in the venous blood will fall below that of the reference solute due to its diffusion into the surrounding interstitium. Because of the transient nature of this technique, the concentration of the test solute in the interstitium may be assumed to be close to zero. The concentration of the reference solute in the venous blood provides an estimate of what the test solute concentration would have been if no transport of the test solute out of the capillaries had occurred. After several seconds, the reference solute concentration in the blood will fall below that of the test solute since the reference solute has been nearly washed out of the capillary and the test solute concentration in the interstitium is now larger than that in the blood. Hence, the test solute starts to diffuse back into the blood. The concentrations of the test (C) and reference solutes (C_0) in the venous blood may then be used to calculate, using Equation 2.49, the test solute extraction (E) and the value of $P_m S$ for a given blood flowrate (Q). It is important to note that accurate results require that the test be repeated at increasing blood flow rates (Q) to insure that the solute transport is not flow limited.

The term *clearance* often times is used to describe the removal of a solute from a flowing fluid such as in a capillary. Clearance, given the symbol CL, is the volumetric flow rate of the fluid that has been totally cleared of the solute. Clearance is then defined by the following expression:

$$CL(z) \equiv \frac{Q(C_0 - C(z))}{C_0} = QE \qquad (2.50)$$

This expression shows that capillary solute clearance is linearly dependent on the flow rate under flow limited conditions, i.e., $E \approx 1$. However, under diffusion limited conditions, the clearance becomes independent of the flowrate Q and is limited by the capillary wall permeability.

2.12.3 A Comparison of Convection and Diffusion Effects

Recall that in developing Equations 2.41 and 2.43 that we ignored axial diffusion of the solute within the capillary in comparison to axial convection of solute, the solute being carried along by the flowing blood. We can easily check the validity of this assumption through the following arguments. The amount of solute carried by convection is on the order of:

$$\text{solute transport by axial convection} = \frac{\pi}{4} D_C^2 V C_0 \qquad (2.51)$$

and the amount of solute carried by axial diffusion is on the order of:

$$\text{solute transport by axial diffusion} = \frac{\pi}{4} D_C^2 \frac{DC_0}{L} \tag{2.52}$$

By taking the ratio of Equations 2.51 and 2.52 we obtain what is called the *Peclet (Pe) number*, whose magnitude represents the importance of axial convection in comparison to axial diffusion. The criterion for ignoring axial diffusion is then given by the fact that $Pe = VL/D \gg 1$. For the example problems presented above:

$$Pe = VL/D = (0.05\ \text{cm/sec} \times 0.1\ \text{cm})/0.91 \times 10^{-15}\ \text{cm}^2/\text{sec} = 550 \gg 1$$

and we conclude that we can ignore axial diffusion.

A similar line of reasoning may be applied to the tissue space to support the neglect of axial diffusion in comparison to radial diffusion. The solute transport by radial diffusion would be on the order of:

$$\text{solute transport by radial diffusion} = \left(\frac{D_T C_0}{r_T - r_C}\right) 2\pi r_C L \tag{2.53}$$

and the amount of solute transport by axial diffusion would be on the order of:

$$\text{solute transport by axial diffusion} = D_T \frac{C_0}{L} \pi \left(r_T^2 - r_C^2\right) \tag{2.54}$$

The ratio of Equation 2.53 and 2.54 would be a measure of the relative importance of radial diffusion in comparison to axial diffusion. If this ratio is $\gg 1$, then axial diffusion can be neglected. For the example problem we then have:

$$\frac{2r_C L^2}{(r_T - r_C)\left(r_T^2 - r_C^2\right)} = \frac{2 \times 0.5 \times 10^{-3}\ \text{cm} \times (0.1\ \text{cm})^2}{(4 \times 10^{-3} - 0.5 \times 10^{-3})\left[(4 \times 10^{-3})^2 - (0.5 \times 10^{-3})^2\right]\ \text{cm}^3} = 181 \gg 1$$

and we conclude that we can ignore axial diffusion in the tissue space.

2.13 SOLUTE TRANSPORT IN A VASCULAR BED

The Krogh tissue cylinder model provides a description of solute transport within a single capillary. In many cases a model is needed to describe the solute transport in a much larger region containing numerous capillaries as illustrated for the capillary bed shown in Figure 2.10. Equation 2.49 with $2\pi r_C P_m z$ replaced with $P_m S$ applies to vascular beds but assumes the solute concentration is zero in the tissue space.

One simple approach to describe solute transport in this situation would be to treat the blood in the capillary bed and the tissue space as separate well-mixed regions. To facilitate the analysis, we introduce two parameters that characterize the degree of vascularization in the region of interest. The first parameter would be the *surface area capillary density* (s), which is the capillary surface area per unit volume of tissue. Assuming the geometry of the Krogh tissue cylinder shown in Figure 2.11, the following equation provides an estimate of (s) based on given values of the capillary radius, r_C, and the Krogh tissue cylinder radius, r_T.

$$s = \frac{2r_C}{r_T^2} \tag{2.55}$$

The second parameter is the *volume of capillaries per unit volume of tissue* (v). In terms of the Krogh tissue model, this may also be expressed in terms of r_C and r_T.

$$v = \left(\frac{r_C}{r_T}\right)^2 \tag{2.56}$$

A steady-state solute balance, assuming only diffusion is relevant for solute transport across the capillary wall, may be written for the capillary bed illustrated in Figure 2.10. We may therefore write that the solute carried into the capillary bed by blood flow must equal the amount removed by diffusion across the capillary walls and the amount that leaves with the blood flow.

$$QC_0\big|_{\text{in}} = P_m S(C - \bar{C}) + QC\big|_{\text{out}} \tag{2.57}$$

Q is the volumetric flow rate of blood to the region of interest. If V_T is the volume of the tissue region considered, then $q_b = \frac{Q}{V_T}$, is defined as the *tissue blood perfusion rate*. Assuming a capillary of length L, and blood velocity within the capillary of V, it can be easily shown that:

$$q_b = \left(\frac{r_C}{r_T}\right)^2 \left(\frac{V}{L}\right) = \frac{v}{\tau} \tag{2.58}$$

Where τ is the blood residence time ($\tau = \frac{L}{V} = 0.1$ cm/0.05 cm/sec $= 2$ sec).

The diffusion term, $P_m S(C - \bar{C})$, for a zero-order reaction occurring within the tissue space must equal $V_T R_0$. Equation 2.57 is then readily solved to provide the solute concentration in the blood exiting the tissue region of interest.

$$C = C_0 - \frac{V_T R_0}{Q} = C_0 - \frac{R_0}{q_b} \tag{2.59}$$

The solute concentration within the tissue space is then given by the following equation.

$$\bar{C} = C_0 - R_0\left[\frac{1}{q_b} + \frac{V_T}{P_m S}\right] \tag{2.60}$$

The following example illustrates the use of these equations.

Example 2.9 Consider the transport of glucose from capillary blood to exercising muscle tissue. As a basis, consider 1 gram of tissue. The glucose consumption of the tissue is 60 μmole/min/100 g or 0.01 μmole/sec/g. Blood flow to the region is 60 ml/min/100 g or 0.01 ml/sec/g. The arterial glucose concentration is 5 mM (5 μmol/cm^3). The value of $P_m S$, based on capillary recruitment during exercise, is 0.33 cm^3/sec/100 g or for the 1 g of tissue considered here, 0.0033 cm^3/sec.

Calculate the glucose concentration in the tissue space and in the exiting blood. (Based on data provided in Levick 1991.)

SOLUTION

$$C = 5\,\frac{\mu\text{mole}}{\text{cm}^3} - \frac{0.01\,\frac{\mu\text{mole}}{\text{g sec}}}{0.01\ \text{ml/g sec}} = 4.0\,\frac{\mu\text{mole}}{\text{cm}^3} = 4\ \text{mM}$$

(by Equation 2.59)

$$\bar{C} = 5\,\frac{\mu\text{mole}}{\text{cm}^3} - 0.01\,\frac{\mu\text{mole}}{\text{g sec}}\left[\frac{1}{0.01\ \text{ml/g sec}} + \frac{1\ \text{g}}{0.0033 \times \frac{\text{cm}^3}{\text{sec}}}\right]$$

(by Equation 2.60)

$$= 5\,\frac{\mu\text{mole}}{\text{cm}^3} - 4.03\,\frac{\mu\text{mole}}{\text{cm}^3} = 0.97\,\frac{\mu\text{mole}}{\text{cm}^3} = 0.97\ \text{mM}$$

PROBLEMS

1. Consider a polymeric membrane that is 30 microns thick, with a total surface area of 1 cm^2. The membrane has pores equivalent in size to a spherical molecule with a molecular weight of 100,000, a porosity of 80%, and a tortuosity of 2.5. On one side of the membrane we have a protein solution that is to be concentrated (protein properties, a = 3 nm and $D = 6.0 \times 10^{-7}$ cm^2/sec) at a concentration of 80 g/L. Assume the solution viscosity is 1 cP. On the other side of the membrane is pure water. The pressure on the protein side of the membrane is 20 psi higher than on the filtrate side of the membrane. Determine the convective flow rate of the solution and the rate at which the protein crosses the membrane.

2. Hemoglobin has a molecular weight of 64,460. Estimate its molecular radius and its diffusivity at 37°C. Compare the predicted hemoglobin diffusivity to the data shown in Figure 2.1.

3. Show graphically that Equation 2.9 provides a reasonable estimation for the data shown in Figure 2.1.

4. Tong and Anderson (1996) obtained the followng data in a polyacrylamide gel for the partition coefficient (K), as a function of gel volume fraction (ϕ), for bovine serum album (BSA). The BSA they used had a molecular weight of 67,000 daltons, a molecular radius of 3.6 nm, and a diffusivity of 6×10^{-7} cm^2/sec. Compare the Ogston equation to their data and obtain an estimate for the radius of the cylindrical fibers (a_f) that comprise the gel.

Gel volume fraction (ϕ)	K_{BSA}
0.00	1.0
0.025	0.35
0.05	0.09
0.06	0.05
0.075	0.017
0.085	0.02
0.105	0.03

5. Tong and Anderson (1996) also obtained the following data for the diffusive permeability of bovine serum albumin (BSA) ($K D_0$) in polyacrylamide gels as a function of the gel volume fraction (ϕ). Show that the combined Brinkman and Ogston equations may be used to represent this data.

Gel volume fraction (ϕ)	Diffusive Permeability ($K D_0$), cm^2/sec
0.00	5.5×10^{-7}
0.02	1.0×10^{-7}
0.025	6.5×10^{-8}
0.05	8.0×10^{-9}
0.065	4.5×10^{-9}
0.075	1.0×10^{-9}

6. Show for a protein with a molecular weight of 50,000 daltons that $S_0 = S_a$ for blood flowing through a capillary.

7. Investigate the fit of the Maxwell (1873) equation and the equation developed by Riley, Buettner et al. (1995) and Riley, Muzzio et al. (1994, 1995a, 1995b, 1995, 1996) to the data presented in Figure 2.6 (Nugent and Jain 1984a, 1984b).

8. The following data was obtained by Iwata, Morikawa, and Ikada. (1996) for the effective diffusivity of several solutes through a 5% agarose hydrogel. Show how well the Brinkman (1947) equation and the Renkin (1954) equation represent this data. What is the effective pore size of the gel based on the Renkin equation?

Solute	Molecular weight	Gel diffusivity, cm^2/sec
glucose	180	4.5×10^{-6}
vitamin B_{12}	1,200	1.7×10^{-6}
myoglobin	17,000	4.0×10^{-7}
BSA	69,000	1.0×10^{-7}
IgG	150,000	1.3×10^{-7}

9. Iwata, Morikawa, and Ikada (1996) also measured the sieving coefficient for a variety of solutes using an XM-50 ultrafilter reported to have a nominal molecular weight cutoff of 50,000 daltons. Compare their measured seiving coefficients to predictions for the sieving coefficient calculated by Equation 2.2. What pore diameter is needed to get a reasonable representation of the data?

Solute molecular weight	Sieving coefficient
180	0.98
1,200	0.94
14,300	0.55
17,000	0.12
69,000	0.01

10. Derive Equation 2.39

11. Membrane plasmapheresis (Zydney 1995; Zydney and Colton 1986) is a technique used to separate plasma from the cellular components of blood. The plasma that is collected may be processed further to yield a variety of substances that are useful for the treatment of a variety of blood disorders. Membrane-based systems employ blood flow that is parallel to the membrane.

The filtrate of plasma is therefore perpendicular to the membrane. A cell-free filtrate with minimal retention of plasma proteins is obtained since the membrane pores are typically 0.2 to 1.0 microns. The filtrate flux is limited, however, by the accumulation (concentration polarization) of cells along the surface of the membrane that provides an additional hydraulic resistance.

At any axial position (z) from the device inlet, this layer of cells at steady state represents a balance between the flux of cells carried to the membrane by the filtration flux, i.e., $j(z)C(z)$, and the diffusive flux of cells carried away from the membrane, i.e., $D\frac{dC(z)}{dy}$. Here we let $j(z)$ represent the filtration flux ($J(z)/S$), $C(z)$ is the concentration of the cells in the blood, D is the diffusivity of the red blood cells, and y is the distance from the membrane surface. Show that after equating these flux expressions, and integrating over the thickness of the cell boundary layer (δ), that the following expression may be obtained for the filtrate flux.

$$j(z) = \frac{D}{\delta}\ln\left(\frac{C_w}{C_b}\right) = k_m \ln\left(\frac{C_w}{C_b}\right)$$

In this equation, k_m represents the local film mass transfer coefficient (i.e., D/δ) and C_w and C_b represent the cell concentration at the membrane surface and in the bulk blood, respectively. Here the cell boundary layer is much thinner than that of the flow channel. For these conditions, the axial velocity profile will vary linearly across the cell boundary layer and the mass transfer coefficient may be evaluated from Equation 2.5 with the lead constant of 1.86 replaced by the value of 1.03. The shear-induced diffusivity of the red blood cells is given by the following equation (Zydney and Colton 1986).

$$D = 0.03a^2\gamma_W$$

Here (a) represents the radius of the red blood cells, about 4.2 microns, and γ_W is the wall shear rate, given by the following equation for flow of blood in a single cylindrical hollow fiber of radius R (discussed in Chapter 3).

$$\gamma_W = \frac{4Q(z)}{\pi R^3}$$

Show that Equation 2.5 (with 1.86 replaced with 1.03) may be rewritten as follows for the flow of blood in the hollow fiber membrane.

$$k_m = 0.05\left(\frac{a^4}{z}\right)^{1/3}\gamma_W$$

The above expressions may be substituted into the following equation that was obtained in Example 2.6 for the axial change of the volumetric flow rate as a result of plasma filtration.

$$\frac{dQ(z)}{dz} = -2\pi R j(z)$$

This equation may be integrated to find the fractional filtrate yield, defined as the ratio of the total plasma filtrate formed (Q_f) and the inlet blood flow rate $Q_B(0)$. However, to obtain an analytical expression that is useful for design calculations, several assumptions are needed.

To obtain an analytical expression, Zydney and Colton (1986) assumed that the bulk cell concentration remained constant at its inlet value ($C_b(0)$) and that the wall shear rate (γ_W) varied linearly with $Q(z)$. Using these assumptions, show that the following equation is obtained for $Q_f/Q_B(0)$.

$$\frac{Q_f}{Q_B(0)} = 1 - \exp\left[-0.90\beta \ln\frac{C_W}{C_b(0)}\right] \qquad \text{where } \beta = \frac{2}{3}\left(\frac{a^2L}{R^3}\right)^{2/3}$$

β is a dimensonless length. Zydney and Colton (1986) provided the following performance data for a variety of membrane plasmapheresis units. Show that the above equation provides a remarkable prediction of the fractional filtration yield for these units. Use 0.35 and 0.95 for the concentration of the red blood cells entering the blood and at the wall, respectively.

β	Fractional filtrate yield
0.20	0.12
0.27	0.20
0.27	0.23
0.32	0.22
0.32	0.25
0.35	0.25
0.40	0.30
0.42	0.32
0.43	0.30
0.44	0.39
0.45	0.38
0.49	0.36
0.51	0.35
0.58	0.40
0.60	0.35
0.62	0.37
0.70	0.41

12. If the surface area of capillaries is 7,000 cm^2/100 g of tissue, estimate the Krogh tissue cylinder radius.

13. In a discussion of glucose transport through exercising muscle capillaries, Renkin (1977) reports the capillary glucose permeability (P_m) to be on the order of 12×10^{-6} cm/sec and a capillary surface area (S) of 7,000 cm^2/100 g of tissue. How does the value of $P_m \times S$ compare to the data shown in Figure 2.7?

14. Using the Renkin-Crone equation, make a graph of capillary solute extraction (E) as a function of capillary position (z) for various values of the ratio $(2\pi r_C P_m z)/Q$. Identify conditions on your plot for which the solute transport is flow limited and diffusion limited.

15. Levick (1991) presented the following data obtained by Renkin (1977) for the capillary clearance of antipyrene and urea as a function of blood flow in skeletal muscle. Based on this data, which of these solutes transport rate is flow limited and which is diffusion limited? Calculate the value of $P_m S$ for urea. How does this value of $P_m S$ for urea compare to the results presented in Figure 2.7?

Solute	Blood flow (Q, ml/min/100 g)	Clearance (CL, ml/min/100 g)
antipyrene	4.5	4
	4.5	4.5
	6.5	6.5
	10	10
urea	5	3.5
	5.5	3
	7	4
	8	2.5
	8.5	2.6
	8.5	4
	10	2.6
	10.5	2.5
	12	4.5
	13.5	2.6

16. Mann, Smaje, and Yudilevich (1979) obtained the following concentration-versus-time data for several solutes in cat fenestrated salivary glands using the multiple tracer indicator diffusion technique. Fenestrated capillaries have solute permeabilities that are significantly higher than the permeabilities of continuous capillaries. The test solutes were cyanocobalamin (vitamin B_{12}, MW = 1,353) and insulin (MW = 5,807). Albumin (MW = 69,000) served as the nonpermeable reference solute. The perfusion rate for the gland was 8 ml/min/g. The following table summarizes the concentration of these solutes (% dose/0.2 ml of perfusate sample) for various periods of time following their injection into the gland. From this data calculate the value of $P_m S$ for B_{12} and insulin. How do the $P_m S$ values for B_{12} and insulin compare to the results shown in Figure 2.7?

Time, seconds	C, B_{12}	C, Insulin	C, Albumin
0.3	1.5	2.2	2.3
0.66	4.4	6.3	7.3
1.0	7.3	10.0	11.9
1.5	8.6	11.0	12.9
1.9	7.6	10.2	10.7
2.1	6.0	7.3	7.8
2.5	5.0	5.9	5.7
2.8	4.1	4.2	3.9
3.1	3.8	3.7	3.0
3.5	3.0	2.6	2.2
3.8	2.7	2.2	2.0

CHAPTER

THREE

THE PHYSICAL AND FLOW PROPERTIES OF BLOOD

3.1 PHYSICAL PROPERTIES OF BLOOD

Blood is a viscous fluid mixture consisting of plasma and cells. Table 3.1 summarizes the most important physical properties of blood. Recall that the chemical composition of the plasma was previously shown in Table 1.2. Proteins represent about 7–8 wt% of the plasma. The major proteins found in the plasma are albumin (MW = 69,000; 4.5 g/100 ml), the globulins (MW = 35,000–1,000,000; 2.5 g/100 ml), and fibrinogen (MW = 400,000; 0.3 g/100 ml). *Albumin* has a major role in regulating the pH and the colloid osmotic pressure. The so-called *alpha* and *beta globulins* are involved in solute transport, whereas the *gamma globulins* are the antibodies that fight infection and form the basis of the humoral component of the immune system. *Fibrinogen*, through its conversion to long strands of fibrin, has a major role in the process of blood clotting. *Serum* is simply the fluid remaining after blood is allowed to clot. For the most part, the composition of serum is the same as that of plasma, with the exception that the clotting proteins, primarily fibrinogen, and the cells, have been removed.

3.2 CELLULAR COMPONENTS

The cellular component of blood consists of three main cell types. The most abundant cells are the red blood cells (RBCs) or *erythrocytes* comprising about 95% of the cellular component of blood. Their major role is the transport of oxygen by hemoglobin contained within the red blood cells. Note from Table 3.1 that the density of a red blood cell is higher than that of plasma. Therefore, in a quiescent fluid, the red blood cells will tend to settle. The red blood cell volume fraction is called the *hematocrit* and typically

Table 3.1 Physical properties of adult human blood

Property	Value
Whole blood	
pH	7.35–7.40
Viscosity (37°C)	3.0 cP (at high shear rates)
Specific gravity (25/4°C)	1.056
Venous hematocrit	
Male	0.47
Female	0.42
Whole blood volume	~78 ml/kg body weight
Plasma or serum	
Colloid osmotic pressure	~330 mm H_2O
pH	7.3–7.5
Viscocity (37°C)	1.2 cP
Specific gravity (25/4°C)	1.0239
Formed elements	
Erythrocytes (RBCs)	
pH	7.396
Specific gravity (25/4°C)	1.098
Count	
Male	5.4×10^9/ml whole blood
Female	4.8×10^9/ml whole blood
Average life span	120 days
Production rate	4.5×10^7/ml whole blood/day
Hemoglobin concentration	0.335 g/ml of erythrocyte
Leukocytes	
Count	$\sim 7.4 \times 10^6$/ml whole blood
Diameter	7–20 microns
Platelets	
Count	$\sim 2.8 \times 10^8$/ml whole blood
Diameter	~2–5 microns

Data from Cooney, D. O. 1976. *Biomedical engineering principles*, 39. New York: Marcel Dekker.

varies between 40–50%. The true hematocrit (H) is about 96% of the measured hematocrat (Hct).

The red blood cell has a unique shape described as a *biconcave discoid*. Figure 3-1 illustrates the shape of the red blood cell and Table 3.2 summarizes typical dimensions. Red blood cells can form stacked coin-like structures called *rouleaux*. Rouleaux can also clump together to form larger RBC structures called aggregates. Both rouleaux and aggregates break apart under conditions of increased blood flow or higher shear rates.

The next most abundant cell type are the *platelets*, comprising about 4.9% of the cell volume. The platelets are major players in blood coagulation and *hemostasis*, which is prevention of blood loss. The remaining 0.1% of the cellular component of blood consists of the white blood cells (WBCs) or *leukocytes* which form the basis of the cellular component of the immune system. Since the white blood cells and platelets only comprise about 5% of the cellular component of blood, their effect on the macroscopic flow characteristics of blood is negligible.

Table 3.2 Dimensions of the normal red blood cell

Property	Value
Diameter	8.1 ± 0.43 microns
Greatest thickness	2.7 ± 0.15 microns
Least thickness	1.0 ± 0.3 microns
Surface area	138 ± 17 microns^2
Volume	95 ± 17 microns^3

Data from Burton, A. C. 1972. *Physiology and biophysics of the circulation*, 2nd ed. Chicago: Year Book Medical Publishers.

3.3 RHEOLOGY

The field of rheology concerns the deformation and flow behavior of fluids, the prefix *rheo-* is Greek and refers to something that flows. Because of the particulate nature of blood, we would expect the rheological behavior of blood to be somewhat more complex than a simple fluid such as water.

To understand the flow behavior of blood, we first must define the relationship between *shear stress* (τ) and the *shear rate* ($\dot{\gamma}$). To develop this relationship, consider the situation shown in Figure 3.2. A fluid is contained between two large parallel plates both of area A. The plates are separated by a small distance equal to h. Initially the system is at rest. At time $t = 0$, the lower plate is set into motion in the x direction at a constant velocity V. As time proceeds, momentum is transferred in the y direction to successive layers of fluid from the plate that is in motion in the x direction. Momentum therefore "flows" from a region of high velocity to a region of low velocity. After a sufficient length of time, a steady-state velocity profile is obtained that is a linear function of y, i.e., $v_x(y) = \frac{V}{h}(h - y)$. Recall that velocity is a vector and v_x represents the component of the fluid velocity in the x direction. For the situation shown in Figure 3.2, the other components of the velocity vector, i.e., v_y and v_z are equal to zero.

At steady state, a constant force (F) must be applied to maintain the motion of the lower plate. For the situation shown in Figure 3.2, the *shear stress* on the lower plate is

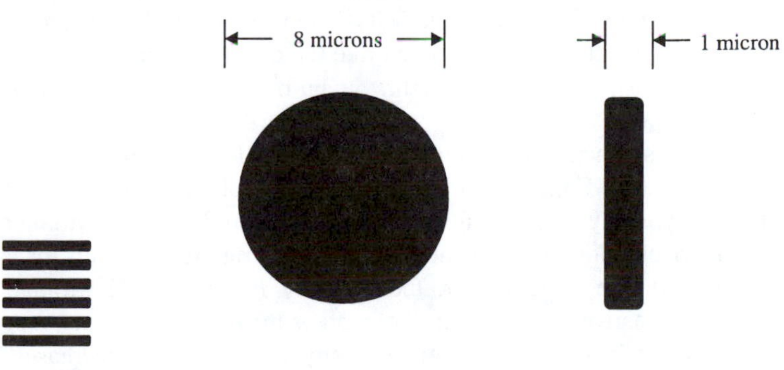

Figure 3.1 The shape and dimensions of the red blood cell.

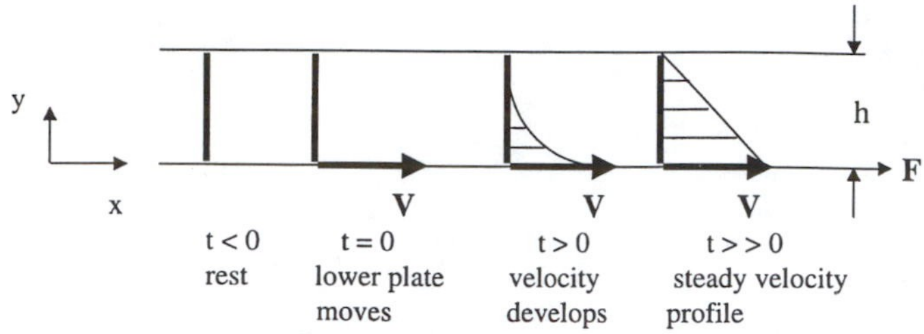

Figure 3.2 Velocity profile development for flow of a fluid between two parallel plates.

defined as F/A and is given the symbol τ_{yx}. The subscript yx on τ denotes the viscous flux of x momentum in the y direction (Bird, Stewart, and Lightfoot 1960). Note that the momentum flux (momentum/area/time) of the fluid adjacent to the plate that is in motion would be given by ρVV, where ρ is the fluid density. The *shear rate* at any position y in the fluid is defined as $-\frac{dv_x(y)}{dy} = \frac{V}{h} = \dot{\gamma}$. Shear rate is given the symbol $\dot{\gamma}$. Notice that shear rate has units of reciprocal time. *Newton's law of viscosity* states that for laminar (nonturbulent) flow the shear stress is proportional to the shear rate. The proportionality constant is called the *viscosity*, μ, which is a property of the fluid and is a measure of the flow resistance of the fluid. For the situation shown in Figure 3.2, this may be stated as follows.

$$\frac{F}{A} = \tau_{yx} = \mu\dot{\gamma} = -\mu\frac{dv_x}{dy} = \mu\frac{V}{h} \tag{3.1}$$

Most simple homogeneous liquids and gases obey this law and are called *Newtonian fluids*.

For more complicated geometries, the steady-state velocity profile is not linear. However, Newton's law of viscosity may be stated at any point in the laminar flow field as:

$$\tau_{yx} = \mu\dot{\gamma} = -\mu\frac{dv_x}{dy} \tag{3.2}$$

where v_x is the velocity in the x direction at position y. This equation tells us that the momentum flows in the direction of decreasing velocity. The velocity gradient is therefore the driving force for momentum transport much like the concentration gradient is the driving force for mass transport. Recall Ficks law given by Equation 2.7 and note the similarity to Newton's law of viscosity.

Fluids whose shear stress–shear rate relationship do not follow Equation 3.2 are known as non-Newtonian fluids. Figure 3.3 illustrates the type of shear stress–shear rate relationships that are typically observed for *non-Newtonian fluids*. The Newtonian fluid is shown for comparison. Note that the shear stress–shear rate relationship for a Newtonian fluid is linear with the slope equal to the viscosity. For the non-Newtonian fluids, the slope of the line at a given value of the shear rate is the *apparent viscosity*. A *dilatant* fluid thickens or has an increase in apparent viscosity as the shear rate increases. A *pseudoplastic* fluid, on the other hand, tends to thin out, or its apparent viscosity decreases, with an increase in shear rate.

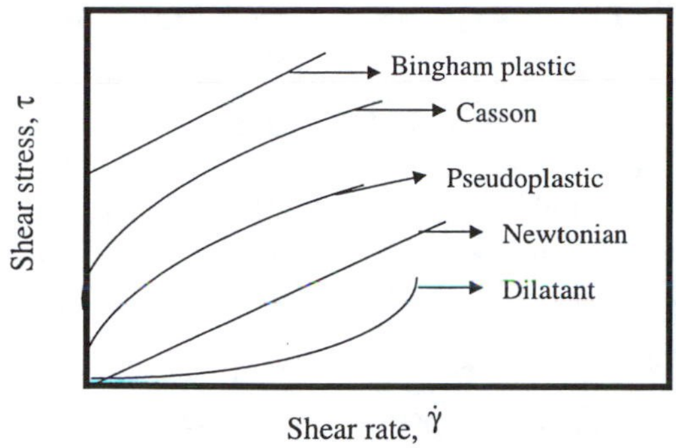

Figure 3.3 Types of shear stress–shear rate relationship.

Heterogeneous fluids that contain a particulate phase that forms aggregates at low rates of shear exhibit a yield stress, τ_y. Two types of fluids that exhibit this behavior are the *Bingham plastic* and the *Casson fluid*. The yield stress must be exceeded in order to get the material to flow. In the case of the Bingham plastic, once the yield stress is exceeded, the fluid behaves as if it were Newtonian. For the Casson fluid, we see that as the shear rate increases, the apparent viscosity decreases indicating that the particulate aggregates are getting smaller and smaller and at some point, the fluid behaves as a Newtonian fluid. Blood is a heterogeneous fluid with the particulates consisting primarily of the red blood cells. As mentioned earlier and shown in Figure 3.1, the red blood cells form rouleaux and aggregates at low shear rates. We will see in the discussion to follow that blood follows the curve shown for the Casson fluid.

The apparent viscosity of blood as a function of shear rate is shown in Figure 3.4 at a temperature of 37°C. At low shear rates, the apparent viscosity of blood is quite high due to the presence of rouleaux and aggregates. However, at shear rates above about 100/sec, only individual cells exist, and blood behaves as if it were a Newtonian fluid. We then approach the asymptotic limit for the apparent viscosity of blood, about 3 cP (1 P [poise] = 100 cP [centipoise] = 1 g/cm sec = 1 dyne sec/cm^2 = 0.1 Newton sec/m^2).

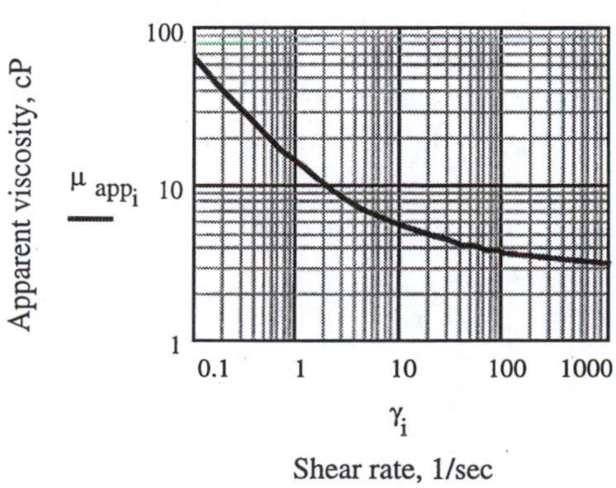

Figure 3.4 Apparent viscosity of blood at 37°C.

3.4 RELATIONSHIP BETWEEN SHEAR STRESS AND SHEAR RATE

The simplest approach for examining the shear stress–shear rate behavior of blood is through the use of the capillary viscometer shown in Figure 3.5. The diameter of the tube is typically on the order of 500 microns. To eliminate entrance effects, the ratio of the capillary length to its radius should be greater than 100 (Rosen 1993). For a given flow rate of blood, Q, the pressure drop across the viscometer, ΔP, is measured. From this information it is possible to deduce an analytical expression for the shear stress–shear rate relationship, i.e., $\dot{\gamma} = \dot{\gamma}(\tau_{rz})$.

Analysis of the capillary flow illustrated in Figure 3.5 requires the use of cylindrical coordinates (r, θ, and z) and the following simplifying assumptions: the length of the tube (L) is much greater than the tube radius (R) (i.e., $L/R \gg 100$) to eliminate entrance effects; steady incompressible and isothermal flow; no external forces acting on the fluid; no holes in the tube so that there is no radial velocity component v_r; axisymmetric flow or no swirls so that the tangential velocity v_θ is also zero; and the no-slip condition at the wall ($r = R$) that requires $v_z = 0$. Continuity or conservation of mass for an incompressible fluid with these assumptions therefore requires that only an axial velocity component exists and it will be a function of the tube radius only, therefore $v_z = v_z(r)$.

Now consider in Figure 3.5 a cylindrical volume of fluid of radius r and length Δz. For steady flow, the viscous force acting to retard the fluid motion, $\tau_{rz} 2\pi r \Delta z$, must be balanced by the force developed by the pressure drop acting on the volume of fluid of length L, $-\pi r^2(P|_{z+\Delta z} - P|_z)$. Equating these forces, dividing by Δz, and taking the limit as $\Delta z \to 0$, the following equation is obtained for the shear stress distribution for the fluid flowing within the capillary tube.

$$\tau_{rz}(r) = -\frac{r}{2}\frac{\mathrm{d}P}{\mathrm{d}z} = \frac{(P_0 - P_L)r}{2L} \tag{3.3}$$

P_0 is the entering fluid pressure and P_L is the pressure of the fluid exiting the capillary tube.

We note that the shear stress vanishes at the centerline of the capillary and achieves its maximum value, τ_w, at the wall. The *wall shear stress* is important since it is easily calculated by the following equation in terms of the measured pressure drop $(P_0 - P_L)$ and the tube dimensions (R and L).

$$\tau_w = -\frac{R}{2}\frac{\mathrm{d}P}{\mathrm{d}z} = \frac{(P_0 - P_L)R}{2L} \tag{3.4}$$

It should also be pointed out that Equations 3.3 and 3.4 hold whether the fluid is Newtonian or non-Newtonian. Nothing has been said so far about the relationship

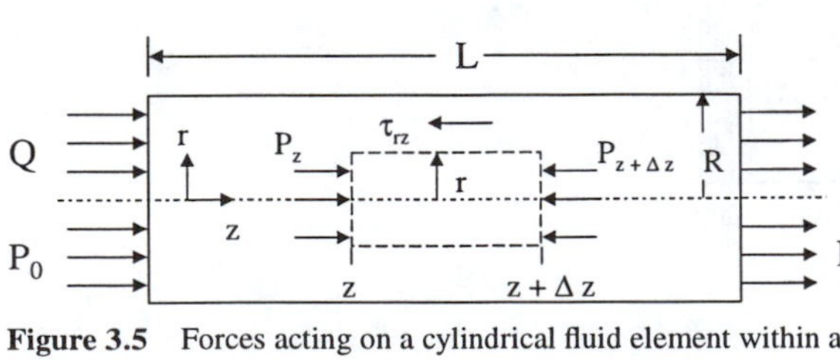

Figure 3.5 Forces acting on a cylindrical fluid element within a capillary viscometer.

between a particular fluid's shear stress and shear rate. Using Equation 3.4, one may rewrite Equation 3.3 to give the shear stress in terms of the wall shear stress and the fractional distance from the center line of the capillary, r/R.

$$\tau_{rz}(r) = \tau_w \frac{r}{R} = \frac{(P_0 - P_L)r}{2L} \tag{3.5}$$

What we want to find through the use of data obtained from the capillary viscometer is a general relationship between the shear rate and some function of the shear stress in terms of the measurable quantities Q, $P_0 - P_L$, L, and R. That is:

$$\dot{\gamma} = -\frac{\mathrm{d}v_z}{\mathrm{d}r} = \dot{\gamma}(\tau_{rz}) \tag{3.6}$$

For the special case of a Newtonian fluid, we have:

$$\dot{\gamma}(\tau_{rz}) = \frac{\tau_{rz}}{\mu} \tag{3.7}$$

A general relationship between $\dot{\gamma}$ and τ_{rz}, and the measurable quantities of the capillary viscometer, can be obtained by the following development. First we write the total flow rate Q in terms of the axial velocity profile as follows.

$$Q = 2\pi \int_0^R v_z(r) r \, \mathrm{d}r \tag{3.8}$$

Next, we integrate this equation by parts to obtain this equation.

$$Q = -\pi \int_0^R r^2 \left(\frac{\mathrm{d}v_z(r)}{\mathrm{d}r} \right) \mathrm{d}r \tag{3.9}$$

Using Equations 3.5 and 3.6, one can show that the following equation then results. This is called the *Rabinowitsch equation.*

$$Q = \frac{\pi R^3}{\tau_w^3} \int_0^{\tau_w} \dot{\gamma}(\tau_{rz}) \tau_{rz}^2 \, \mathrm{d}\tau_{rz} \tag{3.10}$$

For data obtained from a given capillary viscometer, the experiments would provide Q as a function of the wall shear stress, τ_w, which is related by Equation 3.4 to the observed pressure drop, $P_0 - P_L$, and the capillary dimensions R and L. The appropriate shear rate and shear stress relationship, Equation 3.6, would be the one that best fits the data according to Equation 3.10. This will be illustrated for blood in the discussion to follow.

3.5 THE HAGAN-POISEUILLE EQUATION

But first consider the case of a Newtonian fluid. Equation 3.7 provides the relationship for $\dot{\gamma}(\tau_{rz})$. Incorporating this equation into Equation 3.10, one readily obtains the following result for the volumetric flow rate.

$$Q = \frac{\pi R^4 (P_0 - P_L)}{8\mu L} \tag{3.11}$$

This is the *Hagan-Poiseuille law* for laminar flow of a Newtonian fluid in a cylindrical tube. It provides a simple relationship between the volumetric flowrate in the tube given by Q, and the pressure drop, $P_0 - P_L$, for the given dimensions of the tube R and L, and the fluid viscosity μ.

Using the analogy that flow is proportional to a driving force divided by a resistance, we see that the driving force is the pressure drop, $P_0 - P_L$, and the resistance is given by, $\frac{8\mu L}{\pi R^4}$. For laminar flow within blood vessels, the radius of the vessel is the determining factor that controls the blood flow rate for a given pressure drop. The arterioles, the smallest elements of the arterial system, with diameters less than 100 microns, consist of an endothelial cell lining that is surrounded by a layer of vascular smooth muscle cells. Contraction of the smooth muscle layer provides a reactive method for controlling arteriole diameter and hence the blood flow rate within organs and tissues.

Using Equations 3.3, 3.6, and 3.7, we can also write for a Newtonian fluid that:

$$\tau_{rz} = -\mu \frac{\mathrm{d}v_z}{\mathrm{d}r} = \frac{(P_0 - P_L)r}{2L} \tag{3.12}$$

This Equation can be easily integrated recognizing that at $r = R$, $v_z = 0$ and that at $r = 0$, $\frac{\mathrm{d}v_z}{\mathrm{d}r} = 0$. The result of this integration provides an equation that describes the velocity profile for laminar tube flow of a Newtonian fluid. The following equation predicts that the velocity profile will have a parabolic shape.

$$v_z(r) = \frac{(P_0 - P_L)R^2}{4\mu L}\left[1 - \left(\frac{r}{R}\right)^2\right] \tag{3.13}$$

Note that the maximum velocity occurs at the centerline of the capillary, $r = 0$.

3.6 OTHER USEFUL FLOW RELATIONSHIPS

Some other useful relationships for Newtonian flow may be obtained by combining Equations 3.4 and 3.11. First we obtain the following expression that relates the wall shear stress to the volumetric flow rate:

$$\tau_w = \left(\frac{4\mu}{\pi}\right)\frac{Q}{R^3} \tag{3.14}$$

Also, by writing Equation 3.7 at the tube wall and using the above result, it is found that the wall or apparent shear rate is given by:

$$|\dot{\gamma}_w| = \frac{4Q}{\pi R^3} = \frac{4V_{\text{average}}}{R} \tag{3.15}$$

where V_{average} is the average velocity in the tube.

The reduced average velocity $\bar{\mathbf{U}}$ is related to the wall shear rate and is defined as the ratio of the average velocity and the tube diameter. It is given by the following equation:

$$\bar{\mathbf{U}} = \frac{V_{\text{average}}}{D} = \frac{4Q}{\pi D^3} \tag{3.16}$$

Example 3.1 A large artery may have a diameter of about 0.5 cm whereas a large vein may have a diameter of about 0.8 cm. The average blood velocity in these arteries and veins would respectively be on the order of 40 cm/sec and 20 cm/sec. Calculate for these blood flows the wall shear rate, $|\dot{\gamma}_w|$. What would the shear rate be at the center line of these blood vessels?

SOLUTION We may use Equation 3.15 to calculate the wall shear rates.
For the large artery:

$$|\dot{\gamma}_w| = \frac{4 \times 40\,\frac{\text{cm}}{\text{sec}}}{0.25\,\text{cm}} = 640\,\frac{1}{\text{sec}}$$

and for the large vein:

$$|\dot{\gamma}_w| = \frac{4 \times 20\,\frac{\text{cm}}{\text{sec}}}{0.40\,\text{cm}} = 200\,\frac{1}{\text{sec}}$$

From Equation 3.12, at the center line, $r = 0$, therefore the shear rate in both cases is zero.

3.7 RHEOLOGY OF BLOOD

As shown in Figure 3.4, blood is a non-Newtonian fluid except at shear rates above about 100/sec where the viscosity is seen to approach its asymptotic limit of about 3 cP. Therefore, as shown in Example 3.1, blood flow in arteries and veins as well as smaller vessels is Newtonian near the vessel wall, since the wall shear rate is significantly higher than 100/sec. However, as one gets closer to the center line of the vessel, the shear rate approaches zero, and blood exhibits non-Newtonian behavior. Therefore, blood cannot in general be expected to follow the relationships summarized in Table 3.3 for Newtonian fluids.

Figure 3.6 provides the shear stress–shear rate relationship for blood under a variety of conditions as obtained by Repogle, Meiselman, and Merrill (1967) using a capillary viscometer at ambient temperatures. This log-log plot clearly shows that below a shear

Table 3.3 Summary of key flow equations

Relationship	Non-Newtonian	Newtonian
Shear stress, $\tau_{rz} =$	$\frac{(P_0-P_L)r}{2L} = \tau_w \frac{r}{R}$	$\frac{(P_0-P_L)r}{2L} = \tau$
Wall shear stress, $\tau_w =$	$\frac{(P_0-P_L)R}{2L}$	$\frac{(P_0-P_L)R}{2L}$
Shear rate, $\dot{\gamma} =$	$-\frac{dv_z}{dr} = \dot{\gamma}(\tau_{rz})$	$-\frac{dv_z}{dr} = \frac{\tau_{rz}}{\mu}$
Volumetric flow rate, $Q =$	$\pi \frac{R^3}{\tau_w^3} \int_0^{\tau_w} \dot{\gamma}(\tau_{rz})\tau_{rz}^2\, d\tau_{rz}$	$\frac{\pi R^4 \Delta P}{8\mu L}$
Wall shear stress, $\tau_w =$	—	$\frac{4\mu Q}{\pi R^3}$
Wall shear rate, $|\dot{\gamma}_w| =$	—	$\frac{4Q}{\pi R^3}$
Reduced average velocity, $\bar{\text{U}} =$	$\frac{4Q}{\pi D^3}$	$\frac{4Q}{\pi D^3}$

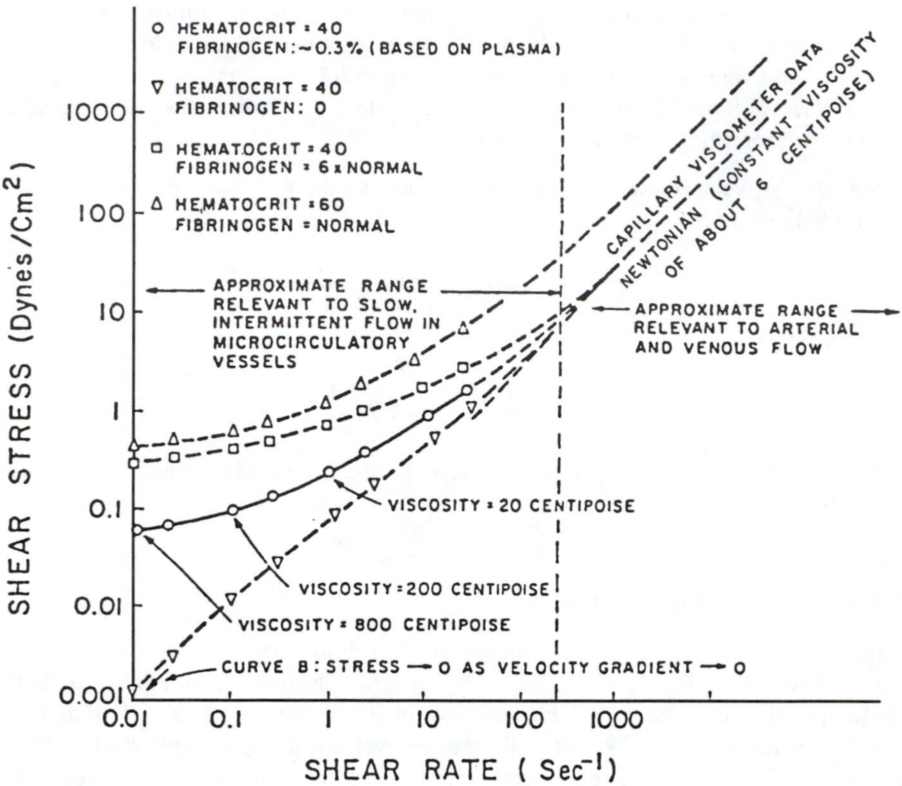

Figure 3.6 Shear stress versus shear rate for blood [from Replogle, Meiselman, and Merrill (1967), with permission].

stress of about 100/sec, the flow of blood (open circle) becomes non-Newtonian. Note in this figure the rapid rise in the viscosity of blood as the shear rate decreases. Above a shear rate of 100/sec, the fluid is Newtonian with a constant viscosity. If the clotting protein fibrinogen (typically about 0.3 wt% in plasma) is removed, keeping the hematocrit constant, the resulting red blood cell suspension (upside down triangle) behaves nearly like a Newtonian fluid for shear rates down to as low as 0.01/sec. Fibrinogen and its effect of making the red blood cells stick together is therefore primarily responsible for the non-Newtonian flow behavior of blood at low shear rates. Thus the other plasma proteins, such as albumin and the globulins, do not contribute to the non-Newtonian behavior of blood, although their concentration will affect the viscosity of the plasma. It should also be pointed out that serum and plasma alone are Newtonian fluids.

Increasing the hematocrit has the effect of increasing the flow resistance of blood (compare open circles, $H = 40\%$, with open triangles, $H = 60\%$). The same effect is also seen when the fibrinogen concentration is increased (compare open circles with open squares). Therefore, the rheology of blood can be significantly affected by the hematocrit and the plasma protein concentrations.

3.8 THE CASSON EQUATION

The shear stress–shear rate relationship for blood may be described by the following empirical equation known as the *Casson equation*.

$$\tau^{1/2} = \tau_y^{1/2} + s\dot{\gamma}^{1/2} \tag{3.17}$$

In this equation, τ_y is the yield stress and s is a constant, both of which must be determined from viscometer data. The yield stress represents the fact that a minimum force must be applied to stagnant blood before it will flow. This was illustrated earlier in Figure 3.3. The yield stress for normal blood at 37°C is about 0.04 dynes/cm^2. It is important to point out, however, that the effect of the yield stress on the flow of blood is small, as the following example will show.

Example 3.2 Estimate the pressure drop required to just overcome the yield stress.

SOLUTION One may use Equation 3.4 to solve for the pressure drop needed to overcome the yield stress.

$$(P_0 - P_L)_{\min} = \frac{2L\tau_y}{R}$$

Using the yield stress of 0.04 dynes/cm^2 and an L/R of 200 for a blood vessel, the pressure drop required to just initiate flow in the capillary is on the order of 0.01 mm Hg. This is considerably less than the mean capillary pressure of about 17 mm Hg.

At large values of the shear rate, the Casson equation should approach the asymptotic viscosity of the Newtonian fluid as shown in Figure 3.4. Equation 3.17 at large shear rates thus provides the interpretation that the parameter s is the square root of this asymptotic Newtonian viscosity. Table 3.1 shows the asymptotic viscosity of blood to be 3 cP. Therefore, the parameter s is 1.732 (cP)$^{1/2}$ or 0.173 (dyne sec/cm^2)$^{1/2}$. It should be stressed that the values of the yield stress and the parameter s will also depend on plasma protein concentrations, hematocrit, and temperature. Therefore, one must be careful to properly calibrate the Casson equation to the actual conditions used in the viscometer or the situation under study.

3.9 USING THE CASSON EQUATION

We are now ready to use the Casson equation to represent the shear stress–shear rate relationship for blood. Consider the data shown in Figure 3.7 and obtained by Merrill et al. (1965) for the flow of blood in various diameter tubes. The figure provides data for the pressure drop as a function of the flow rate for five tube diameters ranging from 288 microns to 850 microns. The hematocrit is 39.3 and the temperature is about 20°C.

If the flow were Newtonian, we could use the Hagan-Pouiselle law, Equation 3.11, to express the relationship between the pressure drop and the volumetric flow rate. In a log-log system of coordinates, the Hagan-Pouiselle equation becomes:

$$\log \Delta P = \log Q + \log\left(\frac{8\mu L}{\pi R^4}\right) \tag{3.18}$$

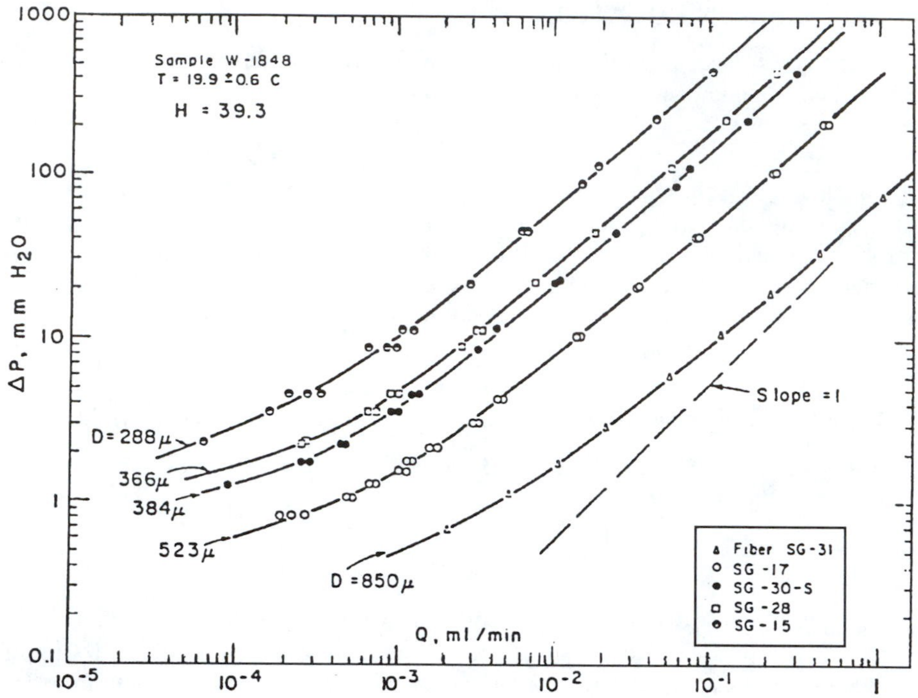

Figure 3.7 Pressure-flow curves for blood [from Merrill et al. (1965), with permission].

The data shown in the log-log plot of ΔP versus Q in Figure 3.7 should therefore have a slope of unity. However, at the lower flow rates we see significant deviation from this requirement as we would expect, since for a given tube radius, Equation 3.15 predicts a decrease in the wall shear rate with decreasing volumetric flow. Thus there is a transition from Newtonian to non-Newtonian flow at the lower flow rates.

To facilitate analysis of the data shown in Figure 3.7, it is convenient to express the pressure drop in terms of the wall shear stress according to Equation 3.4, and to express the volumetric flow rate in terms of the reduced average velocity using Equation 3.16. Figure 3.8 provides a plot of the data shown in Figure 3.7 as a plot of τ_w versus $\bar{\mathbf{U}}$. We observe that all of the data shown in Figure 3.7 now fall nearly along a single line. This result is expected when the Rabinowitsch Equation (3.10) is recast as $\bar{\mathbf{U}}$ versus τ_w.

$$\bar{\mathbf{U}} = \frac{4Q}{\pi D^3} = \frac{1}{1\tau_w^3} \int_0^{\tau_W} \dot{\gamma}(\tau_{rz})\tau_{rz}^2 \, d\tau_{rz} \tag{3.19}$$

This equation predicts that the reduced average velocity should only be a function of the wall shear stress and this result is generally supported by the data presented in Figure 3.8. This result also applies whether or not the fluid is Newtonian or non-Newtonian.

The shear rate or $\dot{\gamma}(\tau_{rz})$ may be written as follows for the Casson Equation (3.17):

$$\dot{\gamma}(\tau_{rz}) = \frac{\left[\tau_{rz}^{1/2} - \tau_y^{1/2}\right]^2}{s^2} \tag{3.20}$$

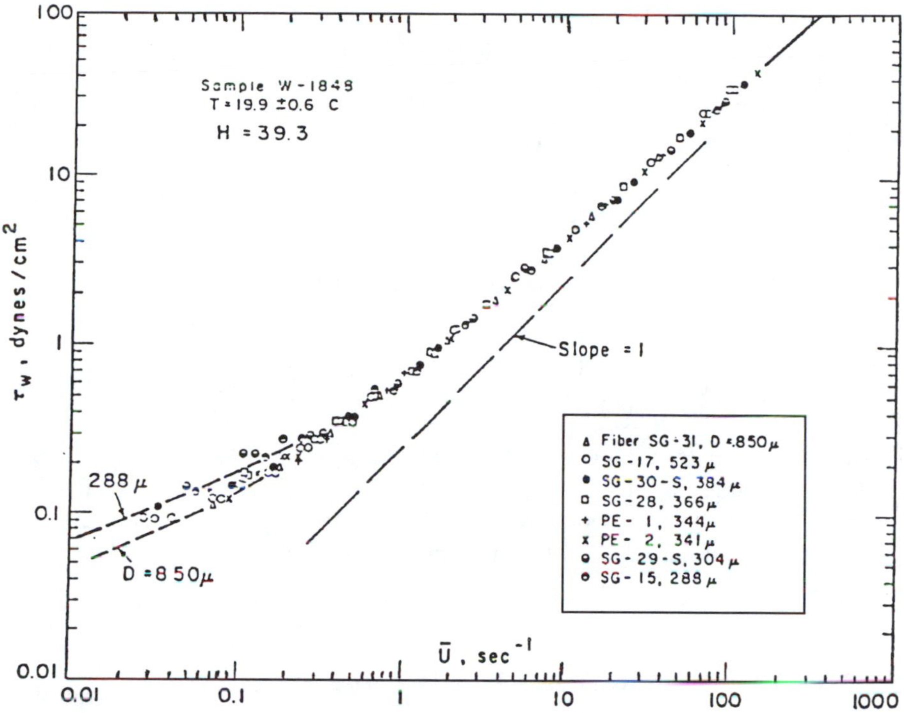

Figure 3.8 Wall shear stress versus the reduced average velocity [from Merrill et al. (1965), with permission].

This equation may be substituted into Equation 3.19 and the resulting equation intergrated to obtain the following fundamental Equation that describes tube flow of the Casson fluid. This equation depends on only two parameters, τ_y and s, that may be determined from experimental data.

$$\bar{U} = \frac{1}{2s^2}\left[\frac{\tau_w}{4} - \frac{4}{7}\sqrt{\tau_y}\sqrt{\tau_w} - \frac{1}{84}\frac{\tau_y^4}{\tau_w^3} + \frac{\tau_y}{3}\right] \tag{3.21}$$

Merrill et al. (1965) provides the following values of these parameters at about 20°C: $\tau_y = 0.0289$ dynes/cm^2 and $s = 0.229$ (dynes-sec/cm^2)$^{1/2}$. It is left as an exercise to show that the Casson equation with these values of the parameters provides an excellent fit to the data shown in Figure 3.8.

3.10 THE VELOCITY PROFILE FOR TUBE FLOW OF A CASSON FLUID

We can also use the Casson equation to obtain an expression for the axial velocity profile for flow of blood in a cylindrical tube or vessel. The maximum shear stress, τ_w, is at the wall. If the yield stress, τ_y, is greater than τ_w, then there will be no flow. On the other hand, if τ_w is greater than τ_y, there will be flow. However, there will be a critical radius (r_{critical}) at which the local shear stress will equal τ_y. From the tube center line to this critical radius, the core fluid will have a flat velocity profile, i.e., $v_z(r) = v_{\text{core}}$, and will move as if it were a solid body or in what is known as "plug flow." For the region from the critical radius to the tube wall ($r_{\text{critical}} \le r \le R$), Equation 3.3 describes once again the shear stress distribution as a function of radial position. We can set this equation equal to the shear stress relation provided by the Casson Equation (3.17), and rearrange to obtain the following equation for the shear rate.

$$\dot{\gamma} = -\frac{\mathrm{d}v_z(r)}{\mathrm{d}r} = \left\{ \frac{P_0 - P_L}{2L} r - 2\tau_y^{1/2} \left[\frac{(P_0 - P_L)}{2L} r \right]^{1/2} + \tau_y \right\} \frac{1}{s^2} \tag{3.22}$$

This equation may be integrated, using the boundary condition that at $r = R$, $v_z(R) = 0$, to obtain the following result.

$$v_z(r) = \frac{R\tau_w}{2s^2} \left\{ \left[1 - \left(\frac{r}{R} \right)^2 \right] - \frac{8}{3} \sqrt{\frac{\tau_y}{\tau_w}} \left[1 - \left(\frac{r}{R} \right)^{3/2} \right] + 2 \left(\frac{\tau_y}{\tau_w} \right) \left(1 - \frac{r}{R} \right) \right\} \tag{3.23}$$

This equation applies for the region defined by $r_{\text{critical}} \le r \le R$. At this point however, we need to determine the value of r_{critical}. Since the shear stress at r_{critical} must equal the yield stress, it is easy to then show that $r_{\text{critical}} = R(\frac{\tau_y}{\tau_w})$. For locations from the center line to the critical radius, the velocity of the core would be given by Equation 3.23 after setting $\frac{r}{R} = \frac{r_{\text{critical}}}{R} = \frac{\tau_y}{\tau_w}$.

3.11 TUBE FLOW OF BLOOD AT LOW SHEAR RATES

Note in Figure 3.8 the scatter in the data at lower shear rates, i.e., $\bar{\text{U}}$ less than about 1/sec. The data start to show significant deviations from Equation 3.19, which predicts that τ_w versus $\bar{\text{U}}$ should be independent of the tube diameter. Clearly at these lower shear rates, RBC aggregates are starting to form, and their characteristic size is becoming comparable to that of the tube diameter. The assumption of blood being a homogeneous fluid is no longer correct. However, for practical purposes, blood flow in the body and through medical devices will be at significantly higher shear rates. Therefore, we really need not concern ourselves with the limiting case of blood flow at low shear rates.

3.12 THE EFFECT OF THE DIAMETER AT HIGH SHEAR RATES

Tube flow of blood at high shear rates (>100/sec) shows two anomolous effects that involve the tube diameter. These are the *Fahraeus effect* and the *Fahraeus-Lindquist effect* (Barbee and Cokelet 1971a, 1971b; Gaehtgens 1980). To understand the effect of tube diameter on blood flow, consider the conceptual model illustrated in Figure 3.9. In this figure, blood flows from a larger vessel, such as an artery, into progressively smaller vessels ultimately reaching the size of a capillary, after which the vessels progress in size to a vein. We make the distinction between the hematocrit found in the feed to a smaller vessel represented by H_F, the hematocrit in the vessel of interest H_T, and the discharge hematocrit H_D. In tubes with diameters from about 15 to 500 microns, it is

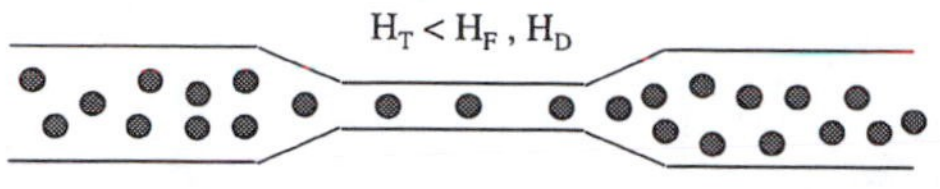

Figure 3.9 The Fahraeus effect.

found that the tube hematocrit (H_T) is actually less than that of the discharge hematocrit ($H_D = H_F$, at steady state). This is called the Fahraeus effect and is shown in Figure 3.10 where we see H_T/H_D plotted as a function of tube diameter, D, in microns. We observe a pronounced decrease in the tube hematocrit in comparison to the discharge hematocrit (or feed hematocrit). For tubes less than about 15 microns in diameter, H_D is no longer equal to H_F, and the ratio of H_T to H_D starts to increase as shown in Figure 3.10. One reason for this is plasma skimming. When the tube size is smaller than about 15 microns, the small vessel draws blood from a much larger vessel and it becomes physically very difficult for the red blood cells to even enter the vessel. Hence the orientation of the branch as well as that of the red blood cells will affect the tube hematocrit as well as the resulting discharge hematocrit.

To explain the Fahraeus effect, it has been shown both by in vivo and in vitro experiments that the cells do not distribute themselves evenly across the tube cross section. Instead, the RBCs tend to accumulate along the tube axis, forming, in a statistical sense, a thin cell-free layer along the tube wall. This is illustrated in Figure 3.11. Recall from Equation 3.13 that the fluid velocity is maximal along the tube axis. The axial accumulation of RBCs, in combination with the higher fluid velocity, maintains the RBC balance ($H_F = H_D$) even though the hematocrit in the tube is reduced. The thin cell-free layer along the tube wall is called the *plasma layer*. The thickness of the plasma

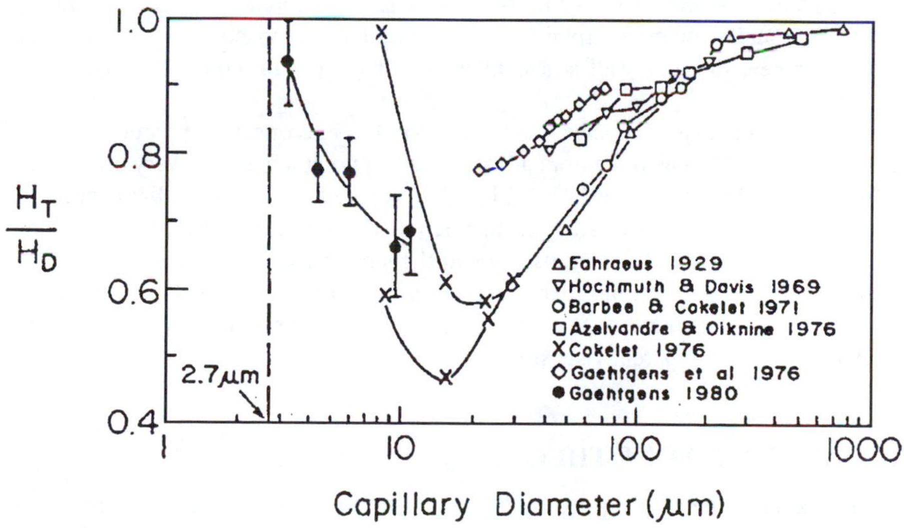

Figure 3.10 The Fahaeus effect [from Gaehtgens (1980), with permission.] Data source identity may be found in this reference.

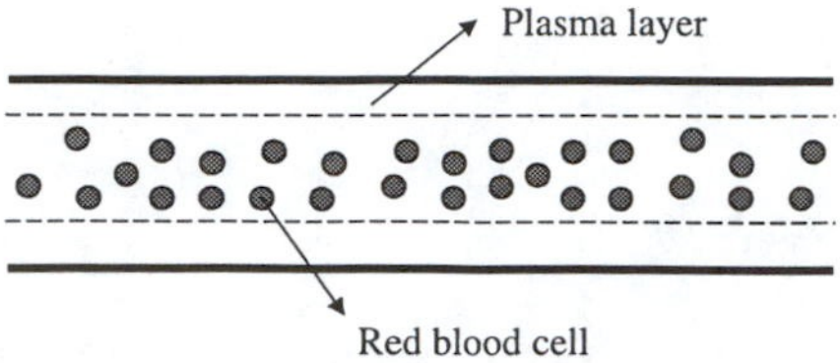

Figure 3.11 Axial accumulation of red blood cells.

layer depends on the tube diameter and the hematocrit and is typically on the order of several microns.

Because of the Fahraeus effect, it is found that as the tube diameter decreases below about 500 microns, the viscosity of blood also decreases. The decreased viscosity is a direct result of the decrease in the tube hematocrit. The reduction in the viscosity of the blood is known as the Fahraeus-Lindquist effect. Since the rheology of blood now depends on the tube diameter, the simple relationship (Equation 3.11) developed to describe tube flow of blood at high shear rates no longer applies. However, since for tube flow of blood we can still measure both Q and $\Delta P/L$, we can use these measurements to calculate for a given tube, the apparent viscosity of blood. This apparent viscosity can be written in terms of the observed values of Q and $\Delta P/L$ by rearranging Equation 3.11.

$$\mu_{\text{apparent}} = \frac{\pi R^4 \Delta P}{8LQ} \tag{3.24}$$

Figure 3.12 illustrates the Fahraeus-Lindquist effect where we see the ratio of the apparent viscosity to its large tube value (μ_F) plotted as a function of the tube diameter. We observe a significant decrease in the apparent viscosity of blood between tube diameters from about 500 microns to 4–6 microns. The decrease in the apparent viscosity in the smaller tubes is explained by the fact that the viscosity in the marginal plasma layer near the tube wall is less than that of the central core that contains the cells.

As the tube diameter becomes less than about 4–6 microns, the apparent viscosity is seen in Figure 3.12 to increase dramatically . At this point it becomes very difficult for an RBC to enter the tube (Fung 1993). The RBC is about 8 microns in diameter and is deformable to some extent and can enter tubes smaller than this characteristic dimension. A tube diameter of about 2.7 microns is about the smallest size that an RBC can enter. However, by Equation 3.11, in order to maintain the flow rate under these conditions, the pressure drop must increase to overcome the resistance presented by the RBC, that is, the increase in the apparent viscosity.

3.13 MARGINAL ZONE THEORY

The marginal zone theory proposed by Haynes (1960) may be used to characterize the Fahraeus-Lindquist effect in the range from 4–6 microns to 500 microns. Using this theory it is possible to obtain an expression for the apparent viscosity in terms of the plasma layer thickness, tube diameter, and the hematocrit. Development of the marginal

Table 3.4 Calculation of H_c and δ from the data of Bayliss and Gaehtgens

d, microns	μ_{app}/μ_F	δ, microns	H_C	H_T	δ/d
130	1.0	0	0.40	0.40	0
100	0.93	1.15	0.40	0.38	0.012
80	0.90	1.4	0.40	0.38	0.018
60	0.82	2.27	0.41	0.35	0.038
40	0.77	2.22	0.43	0.34	0.055
30	0.69	2.92	0.48	0.31	0.093
25	0.63	3.87	0.59	0.28	0.155

zone theory makes use of the relationships for a Newtonian fluid that we developed earlier. These are summarized in Table 3.3.

As shown in Figure 3.11, the blood flow within a tube or vessel is divided into two regions; a central core that contains the cells with a viscosity μ_c, and the cell-free marginal or plasma layer that consists only of plasma with a thickness of δ, and a viscosity equal to that of the plasma given by the symbol μ_p. In each region the flow is considered to be Newtonian and Equation 3.12 applies to each. For the core region, we may then

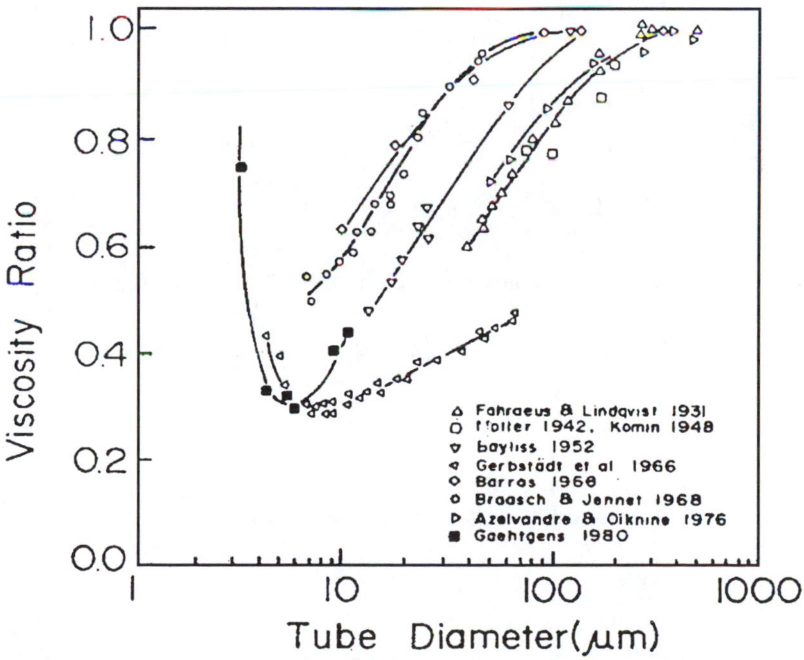

Figure 3.12 The Fahaeus-Lindquist effect [from Gaehtgens (1980), with permission.] Data source identity may be found in this reference.

write:

$$\tau_{rz} = \frac{(P_0 - P_L)r}{2L} = -\mu_c \frac{dv_z^c}{dr}$$
$$\text{BC1: } r = 0, \frac{dv_z^c}{dr} = 0 \tag{3.25}$$
$$\text{BC2: } r = R - \delta, \tau_{rz}\big|_c = \tau_{rz}\big|_p$$

Boundary condition (BC1) expresses the fact that the axial velocity is a maximum at the center line of the tube. The second boundary condition (BC2) derives from the fact that the transport of momentum must be continuous across the interface between the core and the plasma layer. For the plasma layer the following equations apply.

$$\tau_{rz} = \frac{(P_0 - P_L)r}{2L} = -\mu_p \frac{dv_z^p}{dr}$$
$$\text{BC3: } r = R - \delta, v_z^c = v_z^p \tag{3.26}$$
$$\text{BC4: } r = R, v_z^p = 0$$

Boundary condition 3 states the requirement that the velocity in each region must be the same at the interface. The last boundary condition (BC4) simply requires that the axial velocity is zero at the tube wall.

Equations 3.25 and 3.26 may be readily integrated to give the following expressions for the axial velocity profiles in the core and plasma regions.

$$v_z^p(r) = \frac{(P_0 - P_L)R^2}{4\mu_p L}\left[1 - \left(\frac{r}{R}\right)^2\right] \quad \text{for } R - \delta \le r \le R \tag{3.27}$$

$$v_z^c(r) = \frac{(P_0 - P_L)R^2}{4\mu_c L}\left\{1 - \left(\frac{R-\delta}{R}\right)^2 - \frac{\mu_p}{\mu_c}\left(\frac{r}{R}\right)^2 + \frac{\mu_p}{\mu_c}\left(\frac{R-\delta}{R}\right)^2\right\} \quad \text{for } 0 \le r \le R - \delta \tag{3.28}$$

The core and plasma volumetric flow rates are given by the following equations.

$$Q_p = 2\pi \int_{R-\delta}^{R} v_z^p(r) r \, dr$$
$$Q_c = 2\pi \int_0^{R-\delta} v_z^c(r) r \, dr \tag{3.29}$$

Integration of the above equations with the values of v_z^p and v_z^c from Equations 3.27 and 3.28 provides the following result for the volumetric flowrates of the plasma layer and the core:

$$Q_p = \frac{\pi(P_0 - P_L)}{8\mu_p L}[R^2 - (R - \delta)^2]^2 \tag{3.30}$$

$$Q_c = \frac{\pi(P_0 - P_L)R^2}{4\mu_p L}\left[(R-\delta)^2 - \left(1 - \frac{\mu_p}{\mu_c}\right)\frac{(R-\delta)^4}{R^2} - \frac{\mu_p}{\mu_c}\frac{(R-\delta)^4}{2R^2}\right] \tag{3.31}$$

The total flow rate of blood would equal the sum of the flow rates in the core and plasma regions. After adding the above two equations, we obtain the following expression for the total flow rate:

$$Q = \frac{\pi R^4 (P_0 - P_L)}{8\mu_p L}\left[1 - \left(1 - \frac{\delta}{R}\right)^4 \left(1 - \frac{\mu_p}{\mu_c}\right)\right] \tag{3.32}$$

Comparison of this equation with Equation 3.24 allows one to develop a relationship for the apparent viscosity based on the marginal zone theory. We then arrive at the following expression for the apparent viscosity in terms of δ, R, μ_c, and μ_p.

$$\mu_{\text{apparent}} = \frac{\mu_p}{1 - \left(1 - \frac{\delta}{R}\right)^4 \left(1 - \frac{\mu_p}{\mu_c}\right)} \tag{3.33}$$

As $\delta / R \to 0$, then $\mu_{\text{apparent}} \to \mu_c \to \mu$, the bulk viscosity of blood in a large tube at high shear rates, i.e., a Newtonian fluid, as one would expect. We can use Equations 3.32 and 3.33 for blood flow calculations in tubes less than 500 microns in diameter, provided that we know the thickness of the plasma layer, δ, and the viscosity of the core, μ_c. Alternatively, one could simply obtain the apparent viscosity from Figure 3.12. However, often times for design calculations, it is necessary and more convenient to have an equation to describe the flow behavior of blood.

3.14 USING THE MARGINAL ZONE THEORY

We can now use Equation 3.32, based on the marginal zone theory, to fit the apparent viscosity data shown in Figure 3.12. This will allow us to obtain values of the plasma layer thickness and the core hematocrit as a function of the tube diameter. In order to do this, we must develop the relationship for how the core hematocrit (H_c) depends on the feed hematocrit (H_F) and the thickness of the plasma layer.

We will also need an equation to describe the dependence of the blood viscosity on the hematocrit, since the value of H_c will be larger than H_F because of the axial accumulation of the RBCs. This relative increase in the core hematocrit will make the blood in the core have a higher viscosity than the blood in the feed. The following equation may be used to express the dependence of the viscosity of blood at high shear rates on the hematocrit and temperature (Charm and Kurland, 1974).

$$\mu = \mu_p \frac{1}{1 - \alpha H}, \quad \text{for } H \leq 0.6$$

$$\text{where } \alpha = 0.070 \exp\left[2.49H + \frac{1107}{T}\exp(-1.69H)\right] \tag{3.34}$$

In the above equations, the temperature (T) is in K. These equations may be used to a hematocrit of 0.60, the stated accuracy is within 10%. If a subscript occurs on the viscosity, then the corresponding values of H and α in the above equations will carry the same subscript. For example, if we are talking about the core region, then $\mu = \mu_c$, $H = H_c$, and $\alpha = \alpha_c$.

Going back to Equation 3.32, we can rewrite the total volumetric flow rate Q using Equation 3.34 in place of the viscosity ratio $\frac{\mu_p}{\mu_c}$ to obtain the following equation:

$$Q = \frac{\pi R^4 (P_0 - P_L)}{8L\mu_p}\left[1 - \left(1 - \frac{\delta}{R}\right)^4 \alpha_c H_c\right] \tag{3.35}$$

Continuity or conservation of the cells allows us to write the following expression that relates the feed hematocrit to the core hematocrit: $Q_F H_F = Q_c H_c$. Recall that the core volumetric flow rate Q_c can also be written as $Q_F - Q_p$, where Q_p is the volumetric flow rate of the marginal plasma layer. We can now use Equations 3.30, 3.31, and 3.34 to obtain an expression for the hematocrit of the core, $\frac{H_c}{H_F} = \frac{Q_F}{Q_c} = 1 + \frac{Q_P}{Q_c}$, and with $\sigma \equiv 1 - \frac{\delta}{R}$.

$$\frac{\mathrm{H}_c}{\mathrm{H}_F} = 1 + \frac{(1-\sigma^2)^2}{\sigma^2\left[2(1-\sigma^2) + \sigma^2 \frac{\mu_p}{\mu_c}\right]} \tag{3.36}$$

The tube hematocrit would also be given by the expression $H_T = \sigma^2 H_c$.

Equation 3.36 is nonlinear in H_c (since $\mu_c = \mu_c(H_c)$ by Equation 3.34) and also depends for a given tube diameter, D, on the thickness of the plasma layer, δ, which at this point is not known. Therefore, we have two unknowns, H_c and δ. The viscosity ratio for a given tube diameter provides the other relationship that may be used, along with Equation 3.36, to determine H_c and δ. The viscosity ratio, which is also the ordinate of Figure 3.12, may be determined by dividing both sides of Equation 3.33 by the feed or large tube viscosity, μ_F and using Equation 3.34 to eliminate μ_p/μ_F in favor of the term $(1 - \alpha_F H_F)$.

$$\frac{\mu_{\text{apparent}}}{\mu_F} = \frac{1 - \alpha_F H_F}{[1 - \sigma^4 \alpha_c H_c]} \tag{3.37}$$

For a given tube diameter and corresponding value of $\frac{\mu_{\text{apparent}}}{\mu_F}$, Equations 3.36 and 3.37 may be solved to find δ and H_c. This is illustrated in the following example for some of the data shown in Figure 3.12.

Example 3.3 Find values of δ and H_c as a function of tube diameter for the data of Bayliss and Gaehtgens shown in Figure 3.12.

SOLUTION Chart 3.1 illustrates a Mathcad® program that can be used to find for a given value of D and $\mu_{\text{apparent}}/\mu_F$, the corresponding values of δ and H_c. Table 3.4 summarizes the results for tube diameters from 130 to 25 microns using the Haynes marginal zone layer theory. Tubes smaller than 25 microns in diameter give a core hematocrit greater than 0.60 which is beyond the range allowed by Equation 3.34. Knowing how δ and H_c vary with tube diameter for a particular data set, allows one to calculate the apparent viscosity from Equation 3.37, and the flow rate, for a given pressure drop, from Equation 3.24.

Chart 3.1 for **Example 3.3:** Calculating the core hematocrit and the plasma layer thickness for the data of Bayliss and Gaehtgens, 60-micron tube.

$H_F := .40$ feed hematocrit
$H_c := .45$ core hematocrit, assumed
$\delta := 0$ plasma layer thickness, microns, assumed
ratio := .82 viscosity ratio for the current conditions
$d_t := 60$ tube diameter, microns

$$R := \frac{d_t}{2}$$

$$\sigma(\delta) := 1 - \frac{\delta}{R}$$

$$\alpha_F := .070 \exp\left(2.49 H_F + \frac{1107}{310} \cdot \exp(-1.69 H_F)\right)$$

$$\alpha_C(H_c) := .070 \exp\left(2.49 H_c + \frac{1107}{310} \cdot \exp(-1.69 H_c)\right)$$

given

$$1 + \frac{(1 - \sigma(\delta)^2)^2}{\left[\sigma(\delta)^2 \cdot \left[2 \cdot \left(1 - \sigma(\delta)^2\right) + \sigma(\delta)^2 \cdot \left(1 - \alpha_C(H_c) \cdot H_c\right)\right]\right]} - \frac{H_c}{H_F} = 0$$

$$\frac{1 - \alpha_F \cdot H_F}{(1 - \sigma(\delta)^4 \cdot \alpha_C(H_c) \cdot H_c)} - \text{ratio} = 0$$

$$\begin{pmatrix} \delta \\ H_c \end{pmatrix} := \text{find}(\delta, H_c)$$

$\sigma = 2.2$ plasma layer thickness, microns
$H_c = 0.414$ core hematocrit, $H_T := \left(1 - \frac{\delta}{R}\right)^2 \cdot H_c$
$H_T = 0.353$ tube hematocrit

3.15 GENERALIZED MECHANICAL ENERGY BALANCE EQUATION

Equation 3.11 provides a useful relationship for describing the laminar flow of a fluid in a cylindrical tube of constant cross section. We will often times need a more generalized relationship that can account for not only the effect of the pressure drop on fluid flow, but also the effect of changes in elevation, tube cross section, changes in fluid velocity, sudden contractions or expansions, pumps, as well as the effect of fittings such as valves. This general relationship for the flow of a fluid is called the Bernoulli equation and is valid for laminar or turbulent flow.

The Bernoulli equation accounts for the effect of changes in fluid pressure, potential energy, and kinetic energy on the flow of the fluid. It also accounts for the energy added to the fluid by pumps and accounts for energy losses due to a variety of frictional effects. For steady-state flow of an incompressible fluid (density ρ = constant) such as blood from inlet station "1" to exit station "2", the Bernoulli equation may be written as follows (Bird, Stewart, and Lightfoot 1960; McCabe, Smith, and Harriott 1985):

$$\frac{P_1}{\rho} + gZ_1 + \frac{\alpha_1 V_1^2}{2} + \eta W_p = \frac{P_2}{\rho} + gZ_2 + \frac{\alpha_2 V_2^2}{2} + h_{\text{friction}} \tag{3.38}$$

The units of each term in this equation must be consistent with one another and in general are expressed in terms of energy per unit mass of fluid. In this equation,

P represents the pressure, Z the elevation relative to a reference plane, and V the average fluid velocity. Gravitational acceleration is represented by g and is equal to 9.8 m/sec^2. The α's are kinetic energy correction terms. For laminar flow, i.e., $Re < 2000$, $\alpha = 2.0$; and for turbulent flow, $\alpha = 1.0$. W_p represents work done per unit mass of fluid. The pump efficiency is represented by η and accounts for frictional losses and other inefficiencies of the pump. Therefore, $\eta < 1$. Other frictional effects between positions 1 and 2 due to the tube itself, e.g., contractions, expansions, and fittings, are accounted for by the term h_{friction}.

In addition to the Bernouilli equation, we also need to write a mass balance on the fluid. At steady state, this simply says that the rate of mass leaving the system, or a region of interest, 2, must equal the rate at which mass enters the system at region 1. For steady-state flow, this can be expressed by the continuity equation given below.

$$\dot{m} = (\rho V S)_1 = (\rho V S)_2 = \text{constant} \tag{3.39}$$

In this equation, $\dot{m}$ is the mass flow rate and S represents the cross- sectional area of the tube.

The frictional effects are described by the following equation.

$$h_{\text{friction}} = \left(4 \sum_i f_i \frac{L_i V_i^2}{2D_i} + \sum_j \frac{V_j^2}{2} K_{\text{fitting}_j} \right) \tag{3.40}$$

In this equation, f_i is called the friction factor and represents the skin friction contribution in tube segment i with length L_i and diameter D_i. V_i represents the average velocity within tube section i. The friction factor (also known as the Fanning friction factor) is defined as the ratio of the wall shear stress (τ_w) and the average velocity head $\frac{\rho V^2}{2}$. For laminar flow, it is easy to show that $f = 16/Re$. For turbulent flow ($Re > 2000$) in smooth tubes, the friction factor may be evaluated from the following equation (McCabe, Smith, and Harriott 1985). If the flow is turbulent and the walls are not smooth, then the friction factor charts in McCabe, Smith, and Harriott (1985) should be consulted.

$$\frac{1}{\sqrt{f}} = 4.07 \log(Re\sqrt{f}) - 0.60 \tag{3.41}$$

The term in the second summation of Equation 3.40 represented by K_{fitting} accounts for energy loss in the fluid due to tube contractions, expansions, or fittings such as valves. The velocity in this summation is for the fluid just downstream of the contraction, expansion, or fitting. K_{fitting} for such items as valves is highly dependent on the valve's degree of openness and on the type of valve used. It is generally best to consult the manufacturer's literature for the particular valve being considered. As examples, a gate valve that is wide open has a K_{fitting} of about 0.2 and a value of about 6 when half open, and a globe valve that is wide open may have a value as high as 10. Simple fittings such as a tee have a value of 2.0 and a 90° elbow has a value of about 1.0. For sudden contractions and expansions of the fluid the following equations may be used to estimate

K_{fitting}. Once again S refers to the cross-sectional area of the tube.

$$K_{\text{fitting}} = 0.4\left(1 - \frac{S_{\text{downstream}}}{S_{\text{upstream}}}\right), \quad \text{sudden contraction, turbulent flow}$$
$$K_{\text{fitting}} = \left(\frac{S_{\text{downstream}}}{S_{\text{upstream}}} - 1\right)^2, \quad \text{sudden expansion, turbulent flow} \tag{3.42}$$

For the most part these fitting losses tend to be negligible for laminar flow of a fluid.

If the tube through which a fluid flows is not circular, then the equivalent diameter can be used in the above calculations. The equivalent diameter for flow in tubes of noncircular cross section is defined as 4 times the hydraulic radius (r_H). The hydraulic radius is defined as the ratio of the cross sectional area of the flow channel to the wetted perimeter of the tube.

The following example illustrates the use of the above realtionships.

Example 3.4 The simplest patient infusion system is that of gravity flow from an intravenous (IV) bag. A 500-ml IV bag containing an aqueous solution is connected to a vein in the forearm of a patient. Venous pressure in the forearm is 0 mmHg (gauge pressure)[1]. The IV bag is placed on a stand such that the entrance to the tube leaving the IV bag is exactly 1 m above the vein into which the IV fluid enters. The length of the IV bag is 30 cm. The IV is fed through an 18-gauge tube (internal diameter = 0.953 mm) and the total length of the tube is 2m. Calculate the flow rate of the IV fluid. Estimate the time needed to empty the bag.

SOLUTION We apply Bernouilli's equation from the surface of the fluid in the IV bag ("1") to the entrance to the vein ("2"). We expect the flow of the fluid through the bag and the tube to be laminar and therefore neglect the contraction at the entry to the feed tube and the expansion at the vein. The pressure at the surface of the fluid in the bag will be atmospheric ($P_1 = 760$ mm Hg absolute), since the bag collapses as the fluid leaves the bag. The venous pressure, or P_2, is 0 mm Hg gauge or 760 mm Hg absolute. Because the fluid takes some length of time to leave the bag, we neglect the velocity of the surface of the fluid in the bag in comparison to the fluid velocity at the exit of the tube. Therefore, we assume $V_1 = 0$ and $V_2 \gg V_1$. We also set the reference elevation as the entrance to the patient's arm, hence $Z_2 = 0$. Therefore, Z_1 is equal to the elevation of the bag relative to the position where the fluid enters the patient's arm. This would equal 1 m plus the 30 cm length of the bag. Since there are no pumps in the system, $W_p = 0$. We can now write the Bernouilli equation for this particular problem as:

$$gZ_1 = V_2^2 + \left(4f\frac{L}{D}\right)\frac{V_2^2}{2}$$

Recall that the friction factor for laminar flow is equal to $16/Re$. We may substitute this relationship into the above equation to obtain the following quadratic equation

[1]Gauge pressure is that pressure relative to the local atmospheric pressure. Absolute pressure is gauge pressure plus local atmospheric pressure.

that can be solved for the exiting velocity, V_2.

$$V_2^2 + \left(\frac{32\mu L}{\rho D^2}\right) V_2 - gZ_1 = 0$$

This equation may now be solved for the exit velocity recognizing that this quantity must be positive.

$$V_2 = \frac{-\left(\frac{32\mu L}{\rho D^2}\right) + \left[\left(\frac{32\mu L}{\rho D^2}\right)^2 + 4gZ_1\right]^{1/2}}{2}$$

Assuming the IV fluid has the same properties as water, and substituting the appropriate values for the parameters in the above equation, the exit velocity is calculated as follows:

$$V_2 = \frac{-\left(\dfrac{32\times0.001\frac{\text{kg m sec}}{\text{sec}^2\text{ m}^2}\times2.0\text{ m}}{1000\frac{\text{kg}}{\text{m}^3}\times0.000953^2\text{ m}^2}\right)+\left[\left(\dfrac{32\times0.001\frac{\text{kg m sec}}{\text{sec}^2\text{m}^2}\times2.0\text{ m}}{1000\frac{\text{kg}}{\text{m}^3}\times0.000953^2\text{ m}^2}\right)^2+4\times9.8\frac{\text{m}}{\text{sec}^2}\times1.3\text{ m}\right]^{1/2}}{2}$$

$$V_2 = 0.18\,\frac{\text{m}}{\text{sec}} \times \frac{100\text{ cm}}{\text{m}} \times \frac{60\text{ sec}}{\text{min}} = 1082\,\frac{\text{cm}}{\text{min}}$$

$$Q = 1082\,\frac{\text{cm}}{\text{min}} \times \frac{\pi}{4}(0.0953\text{ cm})^2 \times \frac{\text{ml}}{\text{cm}^3} = 7.72\,\frac{\text{ml}}{\text{min}}$$

$$t_{\text{empty}} \approx \frac{V_{\text{bag}}}{Q} = \frac{500\text{ ml}}{7.72\frac{\text{ml}}{\text{min}}} = 65\text{ minutes}$$

With the exit velocity of the fluid now estimated, we need to check the Reynolds number to see if our assumption of laminar flow is valid.

$$Re = \frac{\rho D V_2}{\mu} = \frac{1000\,\frac{\text{kg}}{\text{m}^3} \times 0.000953\text{ m} \times 0.18\,\frac{\text{m}}{\text{sec}}}{0.001\,\frac{\text{kg m sec}}{\text{sec}^2\text{ m}^2}} = 72$$

Since the $Re < 2000$, our assumption of laminar flow is correct.

PROBLEMS

1. Derive Equations 3.10 and 3.21.

2. Using Equation 3.21 with a yield stress $\tau_y = 0.0289$ dynes/cm^2 and $s = 0.229$ (dynes sec/cm^2)$^{1/2}$, show that this equation provides an excellent fit to the data given in Figure 3.8.

3. Estimate the shear rate at the wall of a capillary.

4. Obtain an expression for the radial dependence of the shear rate for a Newtonian fluid. What is the ratio of the average shear rate to the wall shear rate?

5. In Example 3.4, we obtained a simple estimate of the time needed to drain the IV bag. However, the IV bag has a length of 30 cm, and as the fluid drains from the bag, the potential energy (Z_1) of the remaining fluid that drives the flow will change with time. Therefore, to predict correctly the time to drain the bag, we also need to include how Z_1 changes with time. This can be obtained by combining the expression for the exit velocity (V_2) derived in Example 3.4 with a mass balance on the bag itself. Letting *M(t)* denote the mass of IV fluid remaining in the bag at any time t, we can write the IV bag mass balance as follows:

$$\frac{dM(t)}{dt} = \rho S_{\text{bag}} \frac{dZ_1(t)}{dt} = -\rho\pi \frac{D_{\text{tube}}^2}{4} V_2$$

In this equation S_{bag} is the cross-sectional area of the IV bag and D_{tube} is the diameter of the tube. Use this equation, and the results from Example 3.4, to obtain the time to just drain the bag. How does this time compare to the simple estimate of drain time obtained in Example 3.4?

6. Derive Equation 3.11. Start with Equation 3.10 and the assumption of a Newtonian fluid. Also show that Q may be obtained by integrating $v_z(r)$ in Equation 3.13 using Equation 3.8.

7. The cardiac output is normally about 6 L/min. Blood enters the right side of the heart at a pressure of about 0 mm Hg gauge and flows via the pulmonary arteries to the lungs at a mean pressure of 11 mm Hg gauge. Blood returns to the left side of the heart through the pulmonary veins at a mean pressure of 8 mm Hg gauge. The blood is then ejected from the heart through the aorta at a mean pressure of 90 mm Hg gauge. Use the Bernoulli equation to obtain an estimate of the total work performed by the the heart. Carefully state any assumptions and express your answer in watts (recall that 1 watt = 1 Joule (J)/sec, where a J = Nm = kg m^2/sec^2).

8. Use the Bernoulli equation to describe the expected velocity and pressure changes upstream, within, and downstream of an arterial stenosis. A stenosis is a partial blockage or narrowing of an artery by formation of plaque (athersclerosis).

9. Blood is flowing through a bundle of hollow fiber tubes that are each 50 microns in diameter. There are 10,000 tubes in the bundle. The tube length is 12 cm and the pressure drop across each tube is found to be 250 mm Hg. The hematocrit of the blood is 0.40. The blood flow rate for these conditions is 80 μl/hr in each tube. Estimate the marginal zone thickness.

10. Blood enters a hollow fiber unit that is used as an artifical kidney. The unit consists of 10,000 hollow fibers arranged in a shell and tube configuration. Blood flows from an artery in the patient's arm through a tube and is uniformly distributed to the fibers via an arterial head space region at the entrance of the unit. The blood then leaves each fiber through the venous head space region of the unit and is returned to a vein in the patient's arm. Each hollow fiber has an inside diameter of 250 microns and a length of 15 cm. The fiber void volume with the shell of the unit is 50%. Assuming the maximum available pressure drop across the hollow fiber unit is 90 mm Hg, estimate the total flow rate of blood in ml/min through the hollow fiber unit.

CHAPTER

FOUR

OXYGEN TRANSPORT IN BIOLOGICAL SYSTEMS

4.1 THE DIFFUSION OF OXYGEN IN MULTICELLULAR SYSTEMS

The growth of simple multicellular systems beyond a diameter of a hundred microns or so is limited by the availability of oxygen. The metabolic needs of the cells near the center of the cell aggregate exceed the supply of oxygen available by simple diffusion. This is illustrated in Figure 4.1 for isolated islet organs (Brockman bodies) of the fish, *Osphronemus gorami*. The islet organs, typically 800 microns in diameter, were placed in culture and the oxygen partial pressure in the surrounding media and within the cells was measured radially using an oxygen microelectrode.

For the case with no convection in the media (closed symbols), we see that the oxygen partial pressure drops sharply and has been reduced to nearly zero within a few hundred microns of the islet organ's surface. Those cells that experience reduced oxygen levels suffer from *hypoxia*. If the oxygen level is reduced beyond some critical value the cells may die, a condition called *necrosis*. Note that with increased convection (fluid mixing) in the surrounding media (open symbols), the partial pressure of oxygen at the surface and core regions of the islet is increased significantly in comparison to the case without convection. This is a result of the decrease in the resistance to oxygen transport in the fluid surrounding the islet. However, we still observe a large decrease in the oxygen partial pressure within the islet body.

Clearly the development of such complex cellular systems as vertebrates required the development of oxygen delivery systems that are more efficient than diffusion alone. This has been accomplished by the development of two specialized systems that enhance the delivery of oxygen to cells. The first is the circulatory system, which carries oxygen in blood by convection to tiny blood vessels called capillaries in the neighborhood of the cells. Here oxygen is released and then diffuses over much shorter distances to the cells. Secondly, a specialized oxygen carrier protein, called *hemoglobin*, is contained within the

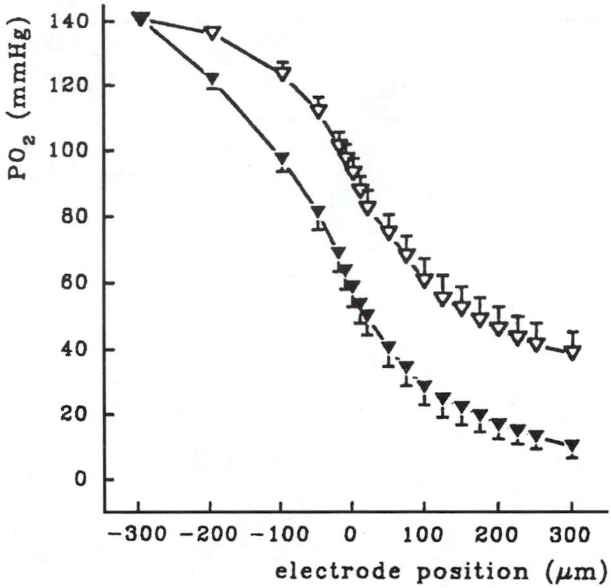

Figure 4.1 pO_2 profiles of 12 Brockman bodies with (open symbols) and without (closed symbols) convection in the medium. Measurements were started 300 microns outside of the tissue (−300). At the surface of the body, the electrode position is 0 microns. (From Schrezenmeir et al. [1994], with permission.)

red blood cells that are suspended within the blood plasma. The presence of hemoglobin overcomes the very low solubility of oxygen in water, the major component of plasma. In this chapter, we will investigate the factors that affect oxygen delivery to cells.

4.2 HEMOGLOBIN

Figure 4.2 shows the basic structural features of the hemoglobin molecule. The hemoglobin found in adult humans is called hemoglobin A. Hemoglobin consists of four polypeptide chains (a polymer of amino acids), two of one kind and two of another kind. The polypeptide chains of hemoglobin A, two α and two β chains, are held together by noncovalent attractions. The hemoglobin molecule is nearly spherical with a diameter of about 5.5 nm and a molecular weight of 68,000 daltons.

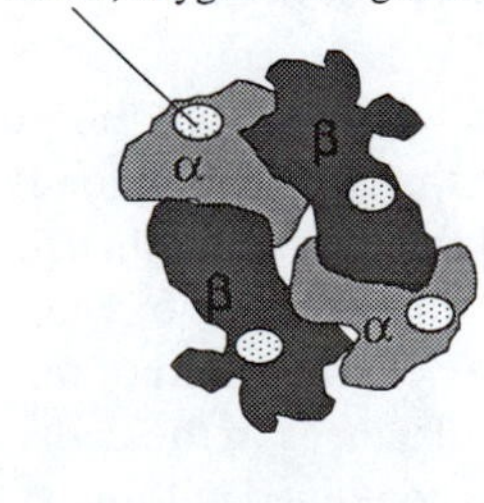

Figure 4.2 Structure of hemoglobin.

The oxygen-binding capacity of hemoglobin (Hb) depends on the presence of a nonpolypeptide unit called the *heme* group. The iron atom within heme gives blood its distinctive red color. The heme group is also called a *prosthetic group* because it gives the hemoglobin protein its overall functional activity. The heme group consists of an organic part that binds through four nitrogen atoms with the iron atom at the center. The organic part is called the *protoporphyrin* and consists of four pyrole rings linked by methene bridges. The iron atom can form up to six bonds, four of which are committed to the pyrole rings. So we have two additional bonds available, one on each side of the heme plane. A heme group can bind one molecule of oxygen.

The four heme groups in a hemoglobin molecule are located near the surface of the molecule and the oxygen-binding sites are about 2.5 nm apart. Each of the four chains contains a single heme group and therefore a single oxygen-binding site. A molecule of hemoglobin therefore has the potential to bind four molecules of oxygen. Hemoglobin is an *allosteric protein* and its oxygen-binding properties are affected by interactions between these separate and nonadjacent oxygen-binding sites. The binding of oxygen to hemoglobin enhances the binding of additional oxygen molecules. The binding of oxygen to hemoglobin is therefore said to be *cooperative*.[1]

4.3 THE HEMOGLOBIN OXYGEN-DISSOCIATION CURVE

The binding of oxygen with hemoglobin is described by the oxygen dissociation curve shown in Figure 4.3. This is an equilibrium curve that expresses, for a given oxygen partial pressure (or pO_2), the fractional occupancy of the hemoglobin oxygen-binding sites (Y). The equilibrium partial pressure of oxygen, or pO_2, above a solution of blood is related to the dissolved oxygen concentration by *Henry's law* which is expressed by the following equation:

$$pO_2 \equiv P y_{\text{oxygen}} = H_{\text{oxygen}} C_{\text{oxygen}} = HC \tag{4.1}$$

where H_{oxygen} is Henry's constant, with a value for normal blood at 37°C of 0.74 mm Hg/μM. C_{oxygen} is the dissolved oxygen concentration (μM) and y_{oxygen} is the mole fraction of oxygen in the gas phase. It is important to realize that the pO_2 is exerted only by the dissolved oxygen. Oxygen bound to hemoglobin does not affect the pO_2 but serves only as a source or sink for oxygen.

4.4 OXYGEN LEVELS IN BLOOD

Table 4.1 summarizes some typical properties of arterial and venous blood. The concentration of hemoglobin in blood (see Table 3.1) is about 150 g/L or 2200 μM. Since

[1]An analogy for cooperative binding would be several people entering a life-raft from the water. Because of the high sides of the life raft it is very difficult for the first person to get aboard. Once one person is aboard he can help the next person. Two people on board make it even easier to load the next person and so on.

Table 41 Oxygen levels in blood

Oxygen property	Arterial	Venous
Partial pressure(tension), pO_2, mm Hg	95	40
Dissolved O_2, μM	130	54
As oxyhemoglobin, μM	8500	5820
Total effective (dissolved + oxyHb), μM	8630	5874

Other useful data: $H_{oxygen} = 0.74$ mm Hg/μM; saturated oxyhemoglobin = 8800 μM; $P_{50} \sim 26$ mm Hg; $n = 2.34$.

each molecule of hemoglobin can bind at most 4 molecules of oxygen, the saturated ($Y = 1$) concentration of oxygen bound to hemoglobin is equal to 8800 μM. Comparing this saturated value, or even the arterial or venous values shown in Table 4.1, to the dissolved oxygen value of 130 μM clearly shows that the bulk of the oxygen that is available in the blood is carried by hemoglobin. The P_{50} represents the value of the oxygen partial pressure (pO_2) at which 50% of the oxygen-binding sites are filled, that is, $Y = 0.5$. From Figure 4.3, we see this occurs when the $pO_2 \approx 26$ mm Hg.

We can get a feeling for the importance of the cooperative binding of oxygen through the following simple calculation. Consider an active capillary bed (see Figure 2.10) for which the entering arterial blood has a pO_{2A} of 95 mm Hg and the exiting venous blood has a pO_{2V} of 20 mm Hg. The amount of oxygen delivered to the tissue surrounding the capillary bed would be proportional to the difference in pO_2 between the arterial and venous blood, ($pO_{2A} - pO_{2V}$). This difference in pO_2 from Figure 4.3 is equivalent to a 60% change in the hemoglobin saturation. If the oxygen binding was noncooperative, then $n = 1$, and assuming the same value for P_{50}, the difference in hemoglobin saturation is now only 35%. Therefore, cooperative binding of oxygen to hemoglobin provides for about a 70% increase in oxygen delivery.

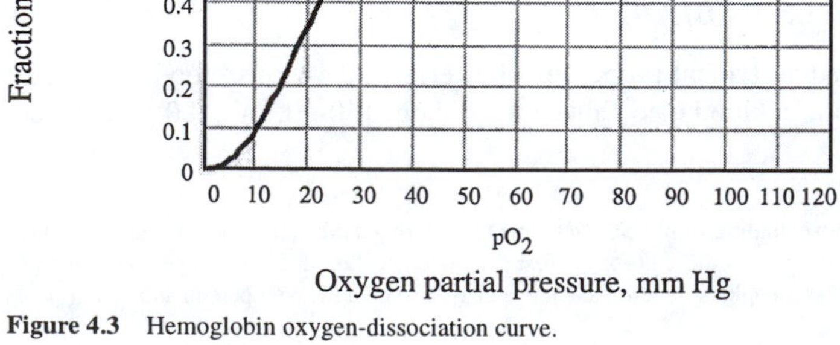

Figure 4.3 Hemoglobin oxygen-dissociation curve.

4.5 THE HILL EQUATION

In 1913, Archibold Hill showed that the sigmoidal shape of the oxygen dissociation curve could be described by the following equilibrium reaction.

$$Hb(O_2)_n \underset{k_1}{\overset{k_{-1}}{\leftrightarrows}} Hb + nO_2 \tag{4.2}$$

In this equation, n represents the number of molecules of oxygen that bind with hemoglobin (Hb) to form oxyhemoglobin $[Hb(O_2)_n]$. Writing an elementary rate expression for the appearance of Hb, and assuming equilibrium, provides the following equation.

$$\frac{d[Hb]}{dt} = k_1[Hb(O_2)_n] - k_{-1}[Hb][O_2]^n = 0 \tag{4.3}$$

Equation 4.3 can then be rearranged to express the concentration of oxygenated hemoglobin in terms of the hemoglobin and dissolved oxygen concentrations.

$$[Hb(O_2)_n] = \kappa[Hb][O_2]^n \tag{4.4}$$

In this equation, κ is defined as k_{-1}/k_1. The fraction of the hemoglobin that is saturated (Y) would be given by the following relationship:

$$Y \equiv \frac{[Hb(O_2)_n]}{[Hb(O_2)_n] + [Hb]} = \frac{\frac{[Hb(O_2)_n]}{[Hb]}}{1 + \frac{[Hb(O_2)_n]}{[Hb]}} \tag{4.5}$$

This equation can be simplified using Equations 4.1 and 4.4 to obtain the following result.

$$Y = \frac{\kappa[O_2]^n}{1 + \kappa[O_2]^n} = \frac{(pO_2)^n}{P_{50}^n + (pO_2)^n} \tag{4.6}$$

The P_{50}^n value was substituted for H_{oxygen}^n/κ. This equation is called the *Hill equation* and can be used to provide a mathematical relationship between Y and pO_2.

Regression of Y and pO_2 data is facilitated by the following rearrangement of the Hill equation.

$$\ln\frac{Y}{1-Y} = n\ \ln(pO_2) - n \ln P_{50} \tag{4.7}$$

A plot of $\ln(Y/1-Y)$ versus $\ln(pO_2)$ is linear with a slope equal to n and a y-intercept equal to $-n \ln P_{50}$. This is called the *Hill plot*. The following example illustrates the use of the Hill equation to represent the oxygen-hemoglobin dissociation curve.

Example 4.1 The following table presents data that represents the oxygen-hemoglobin dissociation. Show that the Hill equation provides excellent representation of this data.

Oxygen partial pressure, pO_2, mm Hg	Fractional hemoglobin saturation
10	0.12
20	0.28
30	0.56
40	0.72
50	0.82
60	0.88
70	0.91
80	0.93
90	0.95
100	0.96

From Guyton, A. C. 1991. *Textbook of medical physiology,* 8th ed., 436. Philadelphia: W. B. Saunders.

SOLUTION Mathcad® Chart 4.1 illustrates the solution using a linear regression of the above data. The P_{50} value obtained from the regression is 26 mm Hg and the value of n is 2.34. We see that the Hill equation provides excellent representation of the oxygen dissociation curve.

Chart 4.1 for **Example 4.1:** Oxygen-hemoglobin dissociation curve fit to the Hill equation.

$i = 0 \cdots 9$ index for data point

oxygen partial pressure, mm Hg, and fractional saturation

pO_{2i}	Y_i
10	.12
20	.28
30	.56
40	.72
50	.82
60	.88
70	.91
80	.93
90	.95
100	.96

$y_i := \ln\left[\frac{Y_i}{(1 - Y_i)}\right]$

$x_1 := \ln(pO_{2i})$

$n := \text{slope}(x, y)$

$n := 2.336$

$b := \text{intercept}(x, y)$

$b := -7.627$

$P_{50} := e^{-\frac{b}{n}}$

$P_{50} := 26.201$ mm Hg

$p := 0 \cdots 100$

$z(p) := \frac{p^n}{(P_{50}^n + p^n)}$ the predicted saturation of Hb

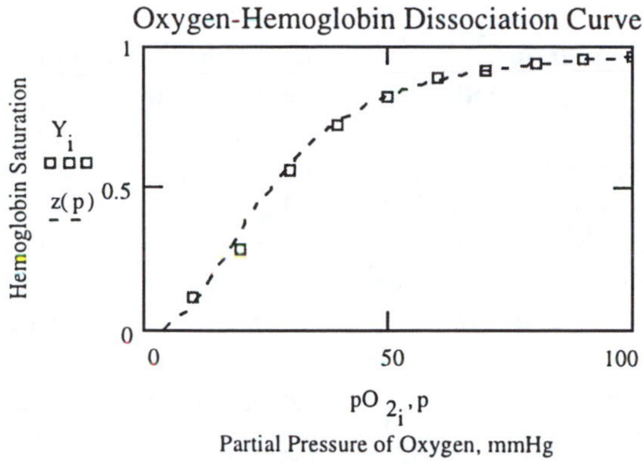

4.6 OTHER FACTORS THAT CAN AFFECT THE OXYGEN-DISSOCIATION CURVE

Other molecules such as H^+, CO_2, and organic phosphates (such as 2,3-diphosphoglycerate, DPG) also bind on specific sites on the hemoglobin molecule and greatly affect its oxygen binding ability through *allosteric interactions.* A lowering of the pH shifts the oxygen dissociation curve to the right as shown in Figure 4.4. This decreases, for a given pO_2, the oxygen affinity of the hemoglobin molecule. CO_2 at increased levels and constant pH also lowers the oxygen affinity of hemoglobin. In metabolically active tissues such as the muscle, the higher levels of CO_2 and H^+ in the capillaries therefore has a beneficial effect by promoting the release of oxygen. This is called the *Bohr effect.*

As shown in Figure 4.4, at high partial pressures of carbon dioxide (pCO_2), CO_2 can reversibly bind to hemoglobin, displacing oxygen to form a compound known as *carbaminohemoglobin.* The reverse process also occurs, that is, oxygen-binding to hemoglobin has the ability to displace CO_2 bound to hemoglobin. This is important in

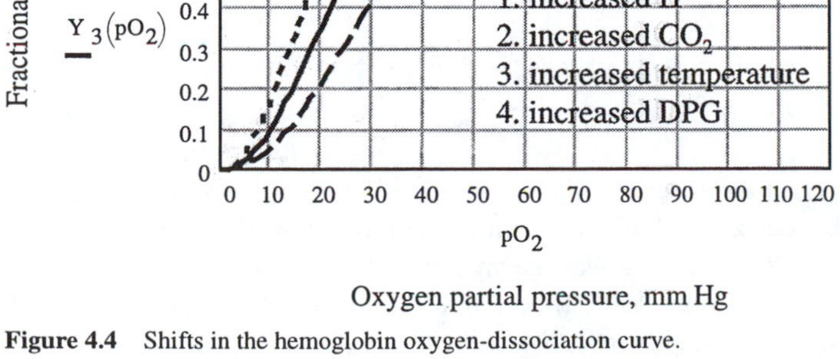

Figure 4.4 Shifts in the hemoglobin oxygen-dissociation curve.

both tissues and in the lungs. In the tissue capillaries, oxygen is released from hemoglobin and the higher pCO_2 in the tissue space results in the formation of carbaminohemoglobin. Within the lungs, where the pO_2 is much higher within the gas space of the alveoli, there is displacement of CO_2 from the hemoglobin by oxygen. This is called the *Haldane effect.*

The affinity of hemoglobin for oxygen within the red blood cell is significantly lower than that of hemoglobin in free solution. Within the red blood cell, there is an organic phosphate called 2,3-diphosphoglycerate (DPG) at about the same concentration as hemoglobin. DPG binds to hemoglobin and reduces the oxygen-binding capacity as shown in Figure 4.4. Without DPG, the value of P_{50} is reduced to about 1 mm Hg. This makes it very difficult for hemoglobin to release the bound oxygen at low pO_2. Note that without DPG, there would be essentially no release of oxygen in the physiological range of pO_2 values from 40–95 mm Hg. Therefore, the presence of DPG within the red blood cell is vital for hemoglobin to perform its oxygen carrying role.

4.7 TISSUE OXYGENATION

With this understanding of the oxygen-binding ability of hemoglobin, we can now focus on tissue oxygenation. Hemoglobin bound oxygen within the red blood cells is transported by the circulatory system to the capillaries where a change in pO_2 levels along the length of the capillary causes a release of oxygen. The released oxygen then diffuses across the capillary wall and into the surrounding tissue space where it is consumed by the cells.

Consider the capillary bed shown earlier in Figure 2.10. What would be the nominal oxygen consumption rate of the tissue lying within this capillary bed? To answer this question, we can develop the following simple mathematical model to describe oxygen usage by the tissue. We assume that the capillary bed and the surrounding tissue space are each well-mixed with respect to oxygen. This assumption allows us to write the following mass balance equations for oxygen. We will let V_T represent the volume of the tissue space (including the capillaries), $\Gamma_{\text{metabolic}}$ represents the tissue metabolic volumetric oxygen consumption rate (μM/sec), and q is the tissue blood perfusion rate (ml of blood/cm^3 tissue/min). C represents the dissolved oxygen concentration in the blood (μM) and C' represents the concentration of oxygen in the blood that is bound to hemoglobin (μM). C_T represents the concentration of oxygen dissolved within the interstitial space of the tissue (μM). P_c (cm/sec) is the permeability of oxygen through the capillary wall, and S_c represents the total surface area of the capillaries (cm^2).

$$\underline{\text{blood}} \qquad 0 = qV_T(C + C')_A - qV_T(C + C')_V - P_cS_c(C_V - C_T) \qquad (4.8)$$

$$\underline{\text{tissue}} \qquad 0 = P_cS_c(C_V - C_T) - V_T\Gamma_{\text{metabolic}} \qquad (4.9)$$

Note that the convective terms (containing q) in Equation 4.8 include oxygen transported in the dissolved states as well as that bound to hemoglobin. However, the mass transfer of oxygen across the capillary wall is only based on the difference between the dissolved oxygen concentration in the blood and that within the tissue interstitial space. The above two equations may be added together and solved to give the following

Table 42 Blood flow to different organs and tissues under basal conditions

Organ	Percent	ml/min	ml/min/100 g
Brain	14	700	50
Heart	4	200	70
Bronchial	2	100	25
Kidneys	22	1100	360
Liver	27	1350	95
Portal	(21)	(1050)	—
Arterial	(6)	(300)	—
Muscle (inactive state)	15	750	4
Bone	5	250	3
Skin (cool weather)	6	300	3
Thryoid gland	1	50	160
Adrenal glands	0.5	25	300
Other tissues	35.5	175	1.3
Total	100.0	5000	—

From Guyton, A.C. 1991. *Textbook of medical physiology*, 8th ed., 186. Philadelphia: W. B. Saunders.

equation for the value of the metabolic oxygen consumption rate.

$$\Gamma_{\text{metabolic}} = q[(C + C')_A - (C + C')_V] \tag{4.10}$$

To solve this equation for the metabolic oxygen consumption rate of tissue requires a value of the tissue blood perfusion rate, q. Table 4.2 provides representative values for the blood flow to various organs and tissues in the body. The blood perfusion is expressed on the basis of flow in ml per min per 100 g of tissue or organ. The kidneys are highly perfused because of their role in the filtration of blood. A nominal tissue perfusion rate is on the order of 0.5 ml/cm^3/min (assuming the density of tissue, $\rho_{\text{tissue}} \approx 1$ g/cm^3). The nominal arterial and venous pO_2 levels are 95 and 40 mm Hg, respectively. We can then use Equation 4.10 to obtain an estimate of the metabolic oxygen consumption rate, $\Gamma_{\text{metabolic}}$ as shown in the following example.

Example 4.2 Calculate the metabolic oxygen consumption rate in μM/sec using the above nominal values for the blood perfusion rate and the arterial and venous pO_2 levels.

SOLUTION

$$\Gamma_{\text{metabolic}} = \frac{0.5\,\text{ml}}{\text{cm}^3\,\text{min}} \times [8630 - 5874]\mu\text{M} \times \frac{1\,\text{min}}{60\,\text{sec}} \times \frac{1000\,\text{cm}^3}{\text{Liter}_{\text{tissue}}}$$
$$\times \frac{\mu\text{mole}}{\text{Liter}_{\text{blood}}\,\mu\text{M}} \times \frac{1\,\text{Liter}_{\text{blood}}}{1000\,\text{ml}} \times \frac{\mu\text{M}\,\text{Liter}_{\text{tissue}}}{\mu\text{mole}} = 22.97\,\frac{\mu\text{M}}{\text{sec}}$$

This calculation shows that the tissue metabolic oxygen consumption rate is on the order of 20 μM/sec. The value for any given tissue will depend on the cells specific requirement for oxygen.

If the value of $\Gamma_{\text{metabolic}}$ were known for a given tissue, then Equation 4.10 could be rearranged to solve for the change in blood oxygenation in order to provide the oxygen demands of the tissue.

$$(C + C')_V = (C + C')_A - \frac{\Gamma_{\text{metabolic}}}{q} \tag{4.11}$$

To illustrate this type of calculation, where $\Gamma_{\text{metabolic}}$ is known, consider the *islet of Langerhans*. The islets of Langerhans are a specialized cluster of cells located within the pancreas. The islets amount to about 1–2% of the pancreatic tissue mass. The β cells that lie within the islet are responsible for the control of glucose metabolism through secretion of the hormone insulin. Loss of the β cells as a result of a defect in the immune system results in diabetes. An islet is only about 150 microns in diameter. The *single islet blood flow* has been determined to be about 7 nl/min (about 4 ml/cm^3/min) (Lifson et al. 1980). This value is about 4 × the blood perfusion rate of the pancreas itself and is probably related to the hormonal function of the islet.

The oxygen consumption rate for tissue like islets may be described by Michaelis-Menten type kinetics. Here we see that the oxygen consumption rate of the tissue is dependent on the pO_2 in the tissue.

$$\Gamma_{\text{metabolic}} = \frac{V_{\max} pO_2}{K_m + pO_2} \tag{4.12}$$

For islets, the value of K_m is 0.44 mm Hg and the value of $V_{\max}$ is 26 μM/sec when the islets are exposed to basal levels of glucose (100 mg/dl) and 46 μM/sec under stimulated glucose levels (300 mg/dl) (Dionne, Colton, and Yarmush 1989, 1991).

Because of the small value of K_m, the tissue oxygen consumption rate is generally independent of the tissue pO_2 until the pO_2 in the tissue reaches a value of just a few mm Hg. Cellular oxygen consumption is therefore relatively insensitive to the local pO_2 level until it reaches a value of 1–2 mm Hg. Therefore, we can approximate the value of $\Gamma_{\text{metabolic}}$ as simply the value of $V_{\max}$. Only at very low pO_2 levels would this approximation be no longer valid.

In Equation 4.11 we know the value of $\Gamma_{\text{metabolic}}$, and the value of $(C + C')_A$ would be the nominal arterial value of 8630 μM (from Table 4.1). We now must solve for the value of $(C + C')_V$ recognizing that C' (amount of oxygen bound to hemoglobin) depends on C (the dissolved oxygen concentration) through the Hill equation (Equation 4.6) where we make use of the fact that $pO_2 = HC$ (i.e., Henry's law).

$$C' = \frac{C'_{\text{SAT}}(HC)^n}{P_{50}^n + (HC)^n} \tag{4.13}$$

Substituting this equation into Equation 4.11 for the value of C', we obtain the equation shown below.

$$\left[C + \frac{C'_{\text{SAT}}(HC)^n}{P_{50}^n + (HC)^n}\right] = (C + C')_A - \frac{\Gamma_{\text{metabolic}}}{q} \tag{4.14}$$

For given values of q, $\Gamma_{\text{metabolic}}$, and $(C + C')_A$, this equation can be solved for the dissolved oxygen concentration (C) in the blood leaving the tissue, which in this case would be an islet. Once we obtained the value of C, then the concentration of oxygen bound to hemoglobin, C', can be found from Equation 4.13. The pO_2 of the exiting blood would be given by Henry's law, Equation 4.1. The following examples illustrate these calculations.

Example 4.3 Calculate the change in blood oxygenation within an islet of Langerhans. Perform the calculations under conditions of normal blood perfusion for basal and stimulated levels of glucose. Find the critical blood perfusion rate to maintain the exiting pO_2 at 20 mm Hg under conditions of basal and stimulated levels of glucose.

SOLUTION Since Equation 4.14 is nonlinear in C, it is convenient to solve this equation using the root finder function in Mathcad®. Chart 4.2 illustrates the solution for the case of basal glucose stimulation.

Chart 4.2 for **Example 4.3:** Illustrate how to calculate the exiting pO_2 from a tissue capillary bed of an islet of Langerhans, basal glucose levels.

$\Gamma_{met} := 25.9$	tissue (islets) basal metabolic O_2 consumption rate, μM/sec
$q := 4$	tissue perfusion rate, ml/cm^3/min
$C_{IN} := 8630$	total O_2 concentration entering the capillary bed, μM
$C_{SAT} := 8800$	Hb saturated, μM
$H := .74$	Henry's constant, mm Hg/μM
$P_{50} := 26$	pO_2 in mm Hg when $Y = .5$
$n := 2.34$	constant in the O_2 dissociation curve
$C := 50$	initial guess of the O_2 concentration in the exiting blood, μM

$$f(C) := \Gamma_{met}\,60 - q\left[C_{in} - \left[C + \frac{C_{sat}(HC)^n}{[P_{50}^n + (HC)^n]}\right]\right]$$

$C := \mathrm{root}(f(C), C)$	
$C = 102.718$	O_2 concentration in exiting blood, μM
$pO_2 := CH$	
$pO_2 = 76.011$	pO_2 of the exiting blood, mm Hg

$$Y := \frac{(HC)^n}{[P_{50}^n + (HC)^n]}$$

$Y = 0.925$	fractional Hb saturation

Chart 4.3 shows the exiting oxygen levels under stimulated glucose conditions.

Chart 4.3 for **Example 4.3:** Illustrate how to calculate the exiting pO_2 from a tissue capillary bed of an islet of Langerhans, stimulated glucose levels.

$\Gamma_{met} := 46.1$	tissue (islets) stimulated metabolic O_2 consumption rate, μM/sec
$q := 4$	tissue perfusion rate, ml/cm^3/min
$C_{IN} := 8630$	total O_2 concentration entering the capillary bed, μM
$C_{SAT} := 8800$	Hb saturated, μM
$H := .74$	Henry's constant, mm Hg/μM
$P_{50} := 26$	pO_2 in mm Hg when $Y = .5$
$n := 2.34$	constant in the O_2 dissociation curve
$C := 50$	initial guess of the O_2 concentration in the exiting blood, μM

$$f(C) := \Gamma_{met}60 - q\left[C_{in} - \left[C + \frac{C_{SAT}(HC)^n}{[P_{50}^n + (HC)^n]}\right]\right]$$

$C := \text{root}(f(C), C)$
$C = 86.711$ O$_2$ concentration in exiting blood, μM
$pO_2 := CH$
$pO_2 = 64.166$ pO_2 of the exiting blood, mm Hg

$$Y := \frac{(HC)^n}{\left[P_{50}^n + (HC)^n\right]}$$

$Y = 0.892$ fractional Hb saturation

In both of the above cases we observe a modest decrease in oxygen levels in the blood exiting the islet. This is because the islets are highly perfused with blood.

The critical perfusion rate is defined as that blood flow rate to the tissue that just maintains a critical pO_2 level, for example 20 mm Hg. Chart 4.4 shows that the critical blood perfusion rate for an islet is 0.28 ml/cm^3/min under basal glucose conditions. A similar calculation shows that the critical blood perfusion rate is 0.50 ml/cm^3/min under stimulated glucose conditions.

Chart 4.4 for **Example 4.3:** Illustrate how to calculate the critical perfusion rate for a given exiting pO_2 from a tissue capillary bed of an islet of Langerhans.

$\Gamma_{met} := 25.9$ tissue (islets) basal metabolic O$_2$ consumption rate, μM/sec
$q := .4$ tissue perfusion rate, ml/cm^3/min, a guess
$C_{IN} := 8630$ total O$_2$ concentration entering the capillary bed, μM
$C_{SAT} := 8800$ Hb saturated, μM
$H := .74$ Henry's constant, mm Hg/μM
$P_{50} := 26$ pO_2 in mm Hg when $Y = .5$
$n := 2.34$ constant in the O$_2$ dissociation curve
$pO_2 := 20$ mm Hg, exiting blood value

$$C := \frac{pO_2}{H}$$

$$f(q) := \Gamma_{met}60 - q\left[C_{IN} - \left[C + \frac{C_{SAT}(HC)^n}{\left[P_{50}^n + (HC)^n\right]}\right]\right]$$

$q :=$ root(f(q),q)
$q = 0.282$ critical blood perfusion rate, ml/cm^3/min
$pO_2 := CH$
$pO_2 = 20$ pO_2 of the exiting blood, mm Hg

$$Y := \frac{(HC)^n}{\left[P_{50}^n + (HC)^n\right]}$$

$Y = 0.351$ fractional Hb saturation

4.8 OXYGEN TRANSPORT IN A BIOARTIFICIAL ORGAN

This discussion on oxygen transport is also of critical importance in the design of bioartificial organs (Colton 1995). In one type of device, therapeutic cells such as the islets of Langerhans for the treatment of diabetes, are sandwiched between two permselective membranes that are adjacent to a polymer matrix material that has been vascularized by the host (Sarver et al., 1995). The permselective membrane has a porous structure at the molecular level such that small molecules like oxygen, key nutrients, and the therapeutic agent are readily permeable, whereas the components of the host immune system cannot

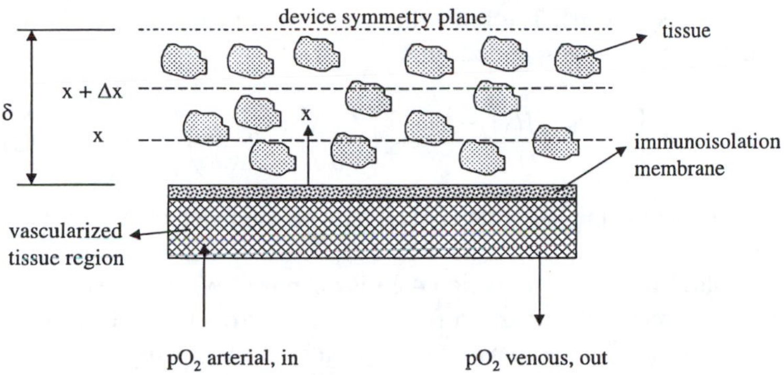

Figure 4.5 Conceptual model for a bioartificial organ.

cross the membrane. The cells are therefore said to be *immunoprotected.* We will discuss this in greater detail in Chapter 8, but here our interest is in providing sufficient oxygen to the cells.

Consider the bioartificial organ shown in Figure 4.5. As a first approximation, we will assume that the major resistance to oxygen transport lies in the immunoprotective membrane and the layer of transplanted cells. We will therefore prescribe the pO_2 level at the outer surface of the membrane to be some average of that of the blood pO_2. A shell balance over a thickness Δx of the cell layer for oxygen may be written as follows:

$$0 = -AD_e \frac{dC}{dx}\bigg|_x + AD_e \frac{dC}{dx}\bigg|_{x+\Delta x} - \Gamma_{\text{metabolic}} A\Delta x(1-\varepsilon) \tag{4.15}$$

Note that the volumetric metabolic oxygen consumption rate of the tissue, $\Gamma_{\text{metabolic}}$ (assumed constant) is multiplied by the islet volume, $A\Delta x(1-\varepsilon)$, to get the amount of oxygen consumed by the tissue in the shell of thickness Δx. The void volume within the tissue space is given by ε. D_e is the effective diffusivity of oxygen within the combined tissue and void space and was defined earlier in Chapter 2. After dividing by Δx, and taking the limit as $\Delta x \to 0$, the following equation is obtained. Henry's law was used to express the dissolved oxygen concentration in terms of the oxygen partial pressure.

$$D_e \frac{d^2 pO_2}{dx^2} = \Gamma_{\text{metabolic}} H(1-\varepsilon) \tag{4.16}$$

Since this equation is a second order differential equation, we require the following two boundary conditions (BC) in order to obtain the solution for the pO_2 profile within the tissue layer.

$$\begin{aligned} &\text{BC1:}\ x = 0,\ pO_2 = pO_2^{x=0} \\ &\text{BC2:}\ x = \delta,\ \frac{dpO_2}{dx} = 0 \end{aligned} \tag{4.17}$$

The first boundary condition (BC1) expresses the pO_2 level at the interface between the membrane and the transplanted cells. The second boundary condition (BC2) results from device symmetry and expresses the fact that there is no net flow of oxygen across the symmetry plane. Equation 4.16 can now be integrated twice and Equation 4.17 can be used to find the integration constants. The following equation is then obtained for the oxygen profile within the layer of transplanted cells.

$$pO_2(x) = pO_2^{x=0} + \left[\frac{\Gamma_{\text{metabolic}} H(1-\varepsilon)\delta^2}{2D_e}\right]\left[\left(\frac{x}{\delta}\right)^2 - 2\left(\frac{x}{\delta}\right)\right] \tag{4.18}$$

The diffusivity of oxygen within the tissue layer (D_e) may be estimated using methods developed in Chapter 2.

At this point the value of $pO_2^{x=0}$ in Equation 4.18 is unknown. We need to develop an additional equation that relates the transport rate of oxygen across the membrane to the total consumption of oxygen by the islets. This is given by the following equation:

$$\frac{AP_m}{H}\left(pO_2^B - pO_2^{x=0}\right) = A\delta(1-\varepsilon)\Gamma_{\text{metabolic}} \tag{4.19}$$

where P_m is the permeability of the membrane and pO_2^B is the average partial pressure of oxygen in the blood in the vascularized region which is assumed to be known. Rearranging the above equation provides the value of $pO_2^{x=0}$.

$$pO_2^{x=0} = pO_2^B - \left[\frac{\Gamma_{\text{metabolic}} H(1-\varepsilon)\delta}{P_m}\right] \tag{4.20}$$

Equations 4.18 and 4.20 can now be used for basic calculations on the effect of oxygen transport on the design of a bioartificial organ. These calculations are illustrated in the following example.

Example 4.4 Consider a bioartificial organ such as that shown in Figure 4.5. Assume the value of pO_2^B is the mean of the arterial and venous values, i.e., 68 mm Hg. Let the transplanted cells be the islets of Langerhans, assumed to be spheres with a diameter of 150 microns, with the basal metabolic oxygen consumption rate of 25.9 μM/sec. The half-thickness (δ) of the islet layer is 200 microns and the void volume fraction within the layer of transplanted cells (ε) is equal to 0.85. The oxygen permeability of the membrane (P_m) is equal to 4×10^{-3} cm/sec. Calculate the pO_2 profile in the tissue space of the bioartificial organ.

SOLUTION Using a value for the bulk diffusivity of oxygen as 2.11×10^{-5} cm^2/sec, the effective diffusivity by Equation 2.13, for the given void fraction of 0.85, is equal to 1.67×10^{-5} cm^2/sec. Chart 4.5 provides a plot of the pO_2 profile within the islet layer of the bioartificial organ. The graph shows that for these conditions the tissue layer is fully oxygenated.

Chart 4.5 for **Example 4.4:** Calculating the pO_2 profile in the tissue space of a bioartificial organ.

basic data

$pO_2^B := 68$	average pO_2 of the blood, mm Hg
$\Gamma_{met} := 25.9$	metabolic oxygen consumption rate, μM/sec
$H := .74$	Henry's constant, mm Hg/μM

$\varepsilon := .85$	void volume within the tissue space
$\delta := 200$	half-thickness of the tissue space, microns
$\delta := \delta \times 10^{-4}$	
$P_m := 4 \times 10^{-3}$	membrane oxygen permeability, cm/sec
$D_e := 1.67 \times 10^{-5}$	effective diffusivity of oxygen of the tissue space, cm^2/sec
$x := 0 \cdots 200$	position within the tissue space, microns

$$pO_2^{x=0} := pO_2^B - \left[\frac{\Gamma_{\text{met}} H(1-\varepsilon)\delta}{P_m}\right]$$

$pO_2^{x=0} = 53.625$ — mm Hg, the pO_2 at the interface between the tissue and the membrane

pO_2 profile within the tissue layer

$$pO_2(x) := pO_2^{x=0} + \left[\frac{\Gamma_{\text{met}} H(1-\varepsilon)\delta^2}{(2D_e)}\right]\left[\left(\frac{x \times 10^{-4}}{\delta}\right)^2 - 2\left(\frac{x \times 10^{-4}}{\delta}\right)\right]$$

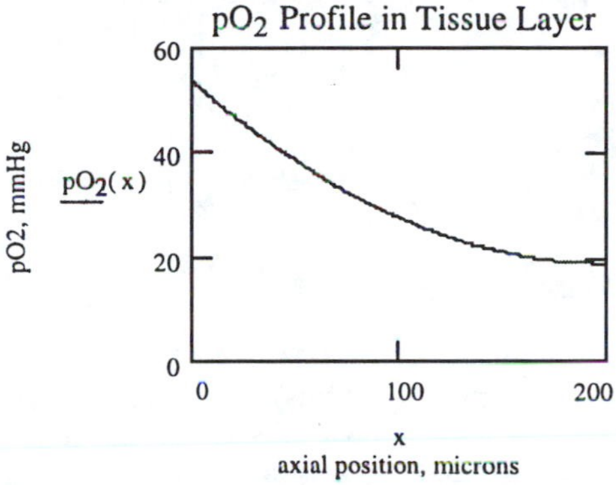

4.9 OXYGEN TRANSPORT IN THE KROGH TISSUE CYLINDER

Our previous analysis for tissue oxygenation (Equation 4.10) treats the tissue and blood as well-mixed regions, i.e., each region is spatially lumped or averaged. This type of analysis provides no information about the oxygen diffusion and consumption within the tissue surrounding a given capillary. Oxygen transport by diffusion to the tissue surrounding a given capillary is usually the rate-limiting step as far as maintaining viable tissue is concerned. The effective oxygenation distance within the tissue outside the capillary is quite small, typically only several capillary diameters, thus requiring an extensive network of capillaries to meet the tissue oxygen demand.

We can use the Krogh tissue cylinder approach (Krogh 1919) developed earlier in Chapter 2 to analyze oxygen transport to the tissue surrounding a given capillary. The Krogh tissue model is a good place to start before considering more advanced models that are described in the recent literature (Mirhashemi, Messmer, and Intaglietta 1987; Tsai, Arfors, and Intaglietta 1990; Lagerlund and Low 1993; Secomb et al. 1993; Intaglietta 1997; Li, Yipintosoi, and Bassingthwaighte 1997).

Figure 2.11 illustrated a concentric shell of the Krogh tissue cylinder. Considering blood first, the shell balance equation for oxygenated hemoglobin may be written as follows[2]:

$$(2\pi r\Delta r\Delta z)\frac{\partial C'}{\partial t} = 2\pi r\Delta r V C'|_z - 2\pi r\Delta r V C'|_{z+\Delta z} + R_{\text{HbO}} 2\pi r\Delta r\Delta z \tag{4.21}$$

In this equation, we ignore the particulate nature of blood as well as the mass transfer resistance of the red blood cell. The blood is assumed to be in plug flow with an average velocity represented by V. Also note that the hemoglobin is carried along by the red blood cell at the average blood velocity (V) and accordingly, there is no diffusive transport of hemoglobin. R_{HbO} represents the volumetric production rate of oxygenated hemoglobin. After dividing by $2\pi r\Delta r\Delta z$, and taking the limit as $\Delta z \to 0$, we obtain the following differential equation that describes the mass balance for oxygenated hemoglobin within the blood flowing through the capillary.

$$\frac{\partial C'}{\partial t} + V\frac{\partial C'}{\partial z} = R_{\text{HbO}} \tag{4.22}$$

A shell balance for the dissolved oxygen in the blood is given by the next equation.

$$\begin{aligned}(2\pi r\Delta r\Delta z)\frac{\partial C}{\partial t} &= 2\pi r\Delta r V C|_z - 2\pi r\Delta r V C|_{z+\Delta z} + (-2\pi r\Delta z) D\left.\frac{\partial C}{\partial r}\right|_r \\ &\quad -(-2\pi r\Delta z) D\left.\frac{\partial C}{\partial r}\right|_{r+\Delta r} + (-2\pi r\Delta r) D\left.\frac{\partial C}{\partial Z}\right|_z \\ &\quad -(-2\pi r\Delta r) D\left.\frac{\partial C}{\partial z}\right|_{z+\Delta z} + R_{\text{OXYGEN}} 2\pi r\Delta r\Delta z\end{aligned} \tag{4.23}$$

In addition to convective flow of the dissolved oxygen (VC terms), we also have axial and radial diffusion of the dissolved oxygen ($D\frac{\partial C}{\partial z}$ and $D\frac{\partial C}{\partial r}$ terms). Once again we divide by $2\pi r\Delta r\Delta z$, and take the limit as Δr and $\Delta z \to 0$. We then obtain the following differential equation that describes the mass balance on dissolved oxygen within the capillary.

$$\frac{\partial C}{\partial t} + V\frac{\partial C}{\partial r} = D\left[\frac{1}{r}\frac{\partial}{\partial r}\left(r\frac{\partial C}{\partial r}\right) + \frac{\partial^2 C}{\partial z^2}\right] + R_{\text{OXYGEN}} \tag{4.24}$$

The two reaction rate terms in Equations 4.22 and 4.24 are simply the negative of each other, therefore $R_{\text{HbO}} = -R_{\text{OXYGEN}}$. This is true since the rate of disappearance of dissolved oxygen must equal the rate of appearance of oxygenated hemoglobin.

We can also make use of the oxygen dissociation curve to relate C' to C as follows:

$$\begin{aligned}\frac{\partial C'}{\partial t} &= \frac{\partial C}{\partial t}\left(\frac{\partial C'}{\partial C}\right) = m\frac{\partial C}{\partial t} \\ \frac{\partial C'}{\partial z} &= \frac{\partial C}{\partial z}\left(\frac{\partial C'}{\partial C}\right) = m\frac{\partial C}{\partial z}\end{aligned} \tag{4.25}$$

where $m = \frac{dC'}{dC}$ is simply related to the slope of the oxygen dissociation curve.

Equations 4.22 and 4.24 can now be added together, using Equation 4.25 and Henry's law, Equation 4.1, to express the dissolved oxygen concentration in terms of the pO_2.

[2]Unlike the shell balances developed in Chapter 2, we will now leave in the accumulation term so that we can consider, if necessary, unsteady state problems.

We then obtain the following equation that expresses the mass balance for dissolved oxygen within the capillary.

$$(1+m)\frac{\partial pO_2}{\partial t} + V(1+m)\frac{\partial pO_2}{\partial z} = D\left[\frac{1}{r}\frac{\partial}{\partial r}\left(r\frac{\partial pO_2}{\partial r}\right) + \frac{\partial^2 pO_2}{\partial z^2}\right] \tag{4.26}$$

Before we proceed, let's take a closer look at how to evaluate the parameter m. m can be related to the blood pO_2 using Henry's law and recognizing that $Y = C'/C'_{sat}$. The following equation then provides a way to calculate m, which depends on the local value of the pO_2. Note that m is dimensionless.

$$m = \frac{dC'}{dC} H C'_{sat} \frac{dY}{dpO_2} \tag{4.27}$$

Using Hill's equation (Equation 4.6), we can evaluate $\frac{dY}{dpO_2}$ and obtain the following expression for the dependence of m on the local pO_2.

$$m = n P_{50}^n H C'_{sat} \frac{pO_2^{n-1}}{\left(P_{50}^n + pO_2^n\right)^2} \tag{4.28}$$

A shell balance on the dissolved oxygen within the tissue interstitial space of void volume (ϵ_T) surrounding the capillary may be written as follows.

$$\begin{aligned} 2\pi r \Delta r \Delta z \varepsilon_T \frac{\partial C^T}{\partial t} &= (-2\pi r \Delta z) D^T \frac{\partial C^T}{\partial r}\bigg|_r - (-2\pi r \Delta z) D^T \frac{\partial C^T}{\partial r}\bigg|_{r+\Delta r} \\ &+ (-2\pi r \Delta r) D^T \frac{\partial C^T}{\partial z}\bigg|_z - (-2\pi r \Delta r) D^T \frac{\partial C^T}{\partial r}\bigg|_{z+\Delta z} \\ &- 2\pi r \Delta r \Delta z \Gamma_{metabolic} \end{aligned} \tag{4.29}$$

Within the tissue, D^T for oxygen can once again be estimated using methods discussed in Chapter 2. After dividing by $2\pi r \Delta r \Delta z$, and taking the limit as Δr and $\Delta z \to 0$, the following differential equation is obtained for the oxygen mass balance within the tissue region. Henry's law was used to express the dissolved oxygen concentration in the tissue, C^T, in terms of the tissue region oxygen partial pressure, pO_2^T.

$$\varepsilon^T \frac{\partial pO_2^T}{\partial t} = D^T\left[\frac{1}{r}\frac{\partial}{\partial r}\left(r\frac{\partial pO_2^T}{\partial r}\right) + \frac{\partial^2 pO_2^T}{\partial z^2}\right] - \Gamma_{metabolic} H^T \tag{4.30}$$

The boundary conditions needed to solve equations 4.26 and 4.30 are as follows.

blood region: $\quad 0 \le z \le L$ and $0 \le r \le r_c$ (4.31)

$$\text{BC1: } z = 0,\ pO_2 = pO_2(t)$$

$$\text{BC2: } z = L,\ \frac{\partial pO_2}{\partial z} = 0$$

$$\text{BC3: } r = 0,\ \frac{\partial pO_2}{\partial r} = 0$$

$$\text{BC4: } r = r_c,\ pO_2 = pO_2^T \text{ and } D\frac{\partial pO_2}{\partial r} = D^T\frac{\partial pO_2^T}{\partial r}$$

Boundary condition 1 expresses the fact that the pO_2 of the blood entering the capillary is assumed to be known and may be a function of time. Boundary condition 2 simply states that oxygen cannot leave the capillary in the axial direction by axial diffusion. Boundary condition 3 assumes the oxygen profile is symmetric with respect to the capillary centerline. The last boundary condition assumes the capillary wall has negligible mass transfer resistance[3] and expresses the requirement that the dissolved oxygen concentrations and the oxygen flux are continuous at the interface between the capillary and tissue regions.

The tissue region boundary conditions may be written as follows assuming there are no anoxic regions.

$$\text{tissue region:} \quad 0 \le z \le L \text{ and } r_c \le r \le r_T \tag{4.32}$$

$$\text{BC5: } z = 0, \ \frac{\partial pO_2^T}{\partial z} = 0$$

$$\text{BC6: } z = L, \ \frac{\partial pO_2^T}{\partial z} = 0$$

$$\text{BC7: } r = r_T, \ \frac{\partial pO_2^T}{\partial r} = 0$$

Boundary conditions 5 and 6 state that oxygen cannot leave the tissue region at either end by axial diffusion. Boundary condition 7 states that the oxygen profile in the tissue region between capillaries a distance of $2r_T$ (the Krogh tissue cylinder diameter) apart is symmetric.

Under certain conditions, anoxic regions may develop within the tissue region. These regions will be defined by a *critical radius*, $r_{\text{anoxic}}(z)$, a distance beyond which there is no oxygen in the tissue. If an anoxic region exists, then boundary condition 7 becomes:

$$\text{tissue region—anoxic:} \quad \text{BC8: } r = r_{\text{anoxic}}(z), \ \frac{\partial pO_2^T}{\partial r} = 0 \text{ and } pO_2^T = 0 \tag{4.33}$$

4.10 AN APPROXIMATE SOLUTION FOR OXYGEN TRANSPORT IN THE KROGH TISSUE CYLINDER

Solution of the above equations for the oxygen concentrations within the blood and tissue regions is a formidable problem and requires a numerical solution (Lagerlund and Low 1993). However, with some simplifications, a reasonable analytical solution can be obtained that is a good starting point for exploring the key factors that govern oxygenation of the tissue surrounding a given capillary. We can also limit ourselves to a steady-state solution, thus eliminating the time derivatives. Another useful approximation is to treat m as a constant. Recall that m is related to the slope of the oxygen dissociation curve by Equation 4.28, and an average value can be used in the range of pO_2 levels of interest. Since the capillary is much longer in length than the Krogh tissue cylinder radius, we can also ignore axial diffusion within the tissue region. This greatly simplifies the equation

[3]Recall that oxygen is lipid soluble and readily permeates the entire surface of the capillary wall.

for the tissue region (Equation 4.30) to give:

$$\frac{\mathrm{d}}{\mathrm{d}r}\left(r\frac{dp\mathrm{O}_2^T}{\mathrm{d}r}\right) = \frac{r\Gamma_{\mathrm{metabolic}}H^T}{D^T} \tag{4.34}$$

Within the capillary, we can ignore axial diffusion in comparison to axial convection and from Equation 4.26 obtain the following equation.

$$(1+m)V\frac{\partial p\mathrm{O}_2}{\partial z} = D\left(\frac{1}{r}\frac{\partial}{\partial r}\left(r\frac{\partial p\mathrm{O}_2}{\partial r}\right)\right) \tag{4.35}$$

Being a partial differential equation, this equation is still difficult to solve. One approximate approach is to eliminate the radial diffusion term by lumping (i.e., integrating the equation) over the r-direction as illustrated below. Radial averaging of the capillary oxygen levels is appropriate here since the bulk of the oxygen mass transfer resistance is not within the capillary. Hence we would not expect steep gradients in the oxygen concentration in the radial direction within the blood.

$$2\pi\int_0^{r_c}(1+m)V\frac{\partial p\mathrm{O}_2}{\partial z}r\,\mathrm{d}r = 2\pi D\int_0^{r_c}\frac{1}{r}\left(\frac{\partial}{\partial r}\left(r\frac{\partial p\mathrm{O}_2}{\partial r}\right)\right)r\,\mathrm{d}r \tag{4.36}$$

This equation may then be integrated and written as follows:

$$2\pi(1+m)V\frac{\mathrm{d}}{\mathrm{d}z}\int_0^{r_c}p\mathrm{O}_2 r\,\mathrm{d}r = 2\pi Dr_c\left.\frac{\partial p\mathrm{O}_2}{\partial r}\right|_{r_c} \tag{4.37}$$

We next recognize that the radially averaged $p\mathrm{O}_2$ level in the blood, $\langle p\mathrm{O}_2\rangle$, at a given axial location, is defined by the following equation:

$$\langle p\mathrm{O}_2\rangle\pi r_c^2 = 2\pi\int_0^{r_c}p\mathrm{O}_2 r\,\mathrm{d}r \tag{4.38}$$

This allows Equation 4.37 to be rewritten in terms of the average $p\mathrm{O}_2$ level in the blood.

$$(1+m)\frac{\mathrm{d}\langle p\mathrm{O}_2\rangle}{\mathrm{d}z} = \frac{2D}{r_cV}\left.\frac{\mathrm{d}p\mathrm{O}_2}{\mathrm{d}r}\right|_{r_c} = \frac{2D^T}{r_cV}\left.\frac{\mathrm{d}p\mathrm{O}_2^T}{\mathrm{d}r}\right|_{r_c} \tag{4.39}$$

Note that the solutions for the oxygen level in the blood and tissue regions are connected through the oxygen flux terms on the right-hand side of Equation 4.39 that arises from the use of boundary condition 4 in Equation 4.31.

We can now proceed to obtain an analytical solution for the oxygen levels within the capillary and the tissue region. First we integrate Equation 4.34 two times and use boundary conditions 4 and 7 in Equations 4.31 and 4.32 to obtain the following result.

$$p\mathrm{O}_2^T(r,z) = \langle p\mathrm{O}_2(z)\rangle - \frac{r_c^2\Gamma_{\mathrm{metabolic}}H^T}{4D^T}\left[1-\left(\frac{r}{r_c}\right)^2\right] - \frac{r_T^2\Gamma_{\mathrm{metabolic}}H^T}{2D^T}\ln\left(\frac{r}{r_c}\right) \tag{4.40}$$

Although we ignored axial diffusion within the tissue region, note that the axial average capillary $p\mathrm{O}_2$ level, $\langle p\mathrm{O}_2(z)\rangle$, impresses an axial dependence on the tissue region $p\mathrm{O}_2$

level. This equation therefore depends on the local capillary oxygen pO_2 and is valid so long as $pO_2^T \geq 0$ throughout the tissue region.

In some cases, part of the tissue region may become anoxic and Equation 4.34 must be solved using the anoxic boundary condition given by Equation 4.33. In this case, the tissue pO_2 is given by:

$$pO_2^T(r,z) = \langle pO_2(z) \rangle - \frac{r_c^2 \Gamma_{\text{metabolic}} H^T}{4D^T}\left[1 - \left(\frac{r}{r_c}\right)^2\right] - \frac{r_{\text{anoxic}}(z)^2 \Gamma_{\text{metabolic}} H^T}{2D^T} \ln\left(\frac{r}{r_c}\right) \tag{4.41}$$

We can get $r_{\text{anoxic}}(z)$ from the additional condition in Equation 4.33 that requires $pO_2^T = 0$ at $r_{\text{anoxic}}(z)$. We then obtain the following nonlinear equation that can be solved for $r_{\text{anoxic}}(z)$.

$$\left(\frac{r_{\text{anoxic}}(z)}{r_c}\right)^2 \ln\left(\frac{r_{\text{anoxic}}(z)}{r_c}\right)^2 - \left(\frac{r_{\text{anoxic}}(z)}{r_c}\right)^2 + 1 = \frac{4D^T \langle pO_2(z) \rangle}{r_c^2 \Gamma_{\text{metabolic}} H^T} \tag{4.42}$$

Equations 4.40 through 4.42 provide the pO_2 level within the tissue region under both nonanoxic and anoxic conditions for a given capillary pO_2 level.

The capillary pO_2 level changes with axial position and can now be found by solving Equation 4.39. Solution of this equation requires that we know the value of $D^T \frac{dpO_2^T}{dr}|_{r_c}$ We can get this by differentiating either Equation 4.40 or 4.41 and evaluating the derivative at r_c. Upon substitution of this result for the nonanoxic case we obtain the following equation.

$$\frac{d\langle pO_2(z) \rangle}{dz} = -\frac{\Gamma_{\text{metabolic}} H^T}{(1+m)V}\left[\left(\frac{r_T}{r_c}\right)^2 - 1\right] \tag{4.43}$$

This equation may be integrated to give the following result for the axial change in the capillary oxygen partial pressure. Note that this equation predicts that the capillary pO_2 level decreases linearly with axial position in the capillary.

$$\langle pO_2(z) \rangle = \langle pO_2 \rangle_{\text{in}} - \frac{\Gamma_{\text{metabolic}} H^T}{(1+m)V}\left[\left(\frac{r_T}{r_c}\right)^2 - 1\right] z \tag{4.44}$$

For anoxic conditions we can simply replace r_T by $r_{\text{anoxic}}(z)$ in Equation 4.44.

At a given value of z, we can substitute the anoxic version of Equation 4.44 into Equation 4.42 to obtain the axial dependence of $r_{\text{anoxic}}(z)$. This would be the radial position for a given z at which the tissue pO_2 would equal zero.

$$\left(\frac{r_{\text{anoxic}}(z)}{r_c}\right)^2 \ln\left(\frac{r_{\text{anoxic}}(z)}{r_c}\right)^2 - \left(\frac{r_{\text{anoxic}}(z)}{r_c}\right)^2 + 1 = \left[\frac{4D^T \langle pO_2 \rangle_{\text{in}}}{r_c^2 \Gamma_{\text{metabolic}} H^T}\right] - \frac{4D^T}{(1+m) r_c^2 V}\left[\left(\frac{r_{\text{anoxic}}(z)}{r_c}\right)^2 - 1\right] z \tag{4.45}$$

Of particular interest would be those conditions under which anoxia first begins. Anoxia will first start at the corner of the Krogh tissue cylinder represented by the coordinates $z = L$ and $r = r_T$. This critical tissue pO_2 level, equal to zero at the *anoxic corner,* can be readily found by first solving Equation 4.44 for the pO_2 level in the blood exiting the capillary. Equation 4.40 can then be solved with $r = r_T$ to find the value of pO_2^T at the anoxic corner. One can then adjust various parameters such as V and $\Gamma_{\text{metabolic}}$ until the pO_2^T equals zero at the anoxic corner.

The Krogh tissue cylinder approach assumes that we know what the Krogh tissue cylinder radius (r_T) is. In Chapter 2, we showed that the blood perfusion rate could be related to the Krogh tissue cylinder radius, the capillary radius and length, and the average blood velocity. This relationship is given by Equation 2.58, which, when solved for r_T, gives the following equation.

$$r_T = r_c\sqrt{\frac{V}{q_b L}} \tag{4.46}$$

This completes our analysis of tissue oxygenation in the region surrounding a given capillary. Example 4.5 illustrates how these equations may be solved to obtain the pO_2 levels within the capillary and the tissue region that surrounds the capillary.

Example 4.5 Calculate the oxygen profiles within a capillary and its surrounding tissue for an islet of Langerhans. Assume blood enters the capillary at a $pO_2 =$ 95 mm Hg. Also calculate the critical blood perfusion rate. Determine the anoxic boundary for a blood perfusion rate of 0.25 ml/cm^3/min.

SOLUTION Mathcad® Chart 4.6 illustrates the solution for the islet tissue perfusion rate of 4 ml/cm^3/min. Note that the value of m is based on the average pO_2 of the blood which must be found by trial and error as shown in the example.

Chart 4.6 for **Example 4.5:** Calculating pO_2 levels within a capillary and the surrouding tissue region.

$\Gamma_{met} := 25.9$	tissue (islet) metabolic O_2 consumption rate, μM/sec
$q := 4$	tissue perfusion rate, ml/cm^3/min
$pO_{2in} := 95$	entering blood oxygen level, mm Hg
$H := .74$	Henry's constant, mm Hg/μM
$C_{sat} := 8800$	saturated hemoglobin, μM
$P_{50} := 26$	pO_2 when $Y = .5$, mm Hg
$n := 2.34$	constant in the Hill equation
$V := .05$	average blood velocity, cm/sec
$L := .1$	capillary length, cm
$r_c := .0005$	capillary radius, cm
$D_T := 6.3 \times 10^{-6}$	oxygen diffusivity in the tissue region, cm^2/sec

we will let $P = pO_2$ of the blood and $P_T = pO_2$ in the tissue region

$P := 72$ a guess of what the exiting capillary blood pO_2 will be, mm Hg

Now calculate the value of m at the average pO_2 of the blood.

$$P_{avg} := .5\left(pO_{2in} + P\right)$$

$$m := nP_{50}^n H C_{sat} \frac{P_{avg}^{(n-1)}}{\left(P_{50}^n + P_{avg}^n\right)^2}$$

$$m = 10.487$$

Now calculate the Krogh tissue cylinder radius based on the blood perfusion rate and the capillary properties.

$$r_T := r_c \sqrt{\frac{V}{\left(qL\frac{1}{60}\right)}}$$

$r_T = 0.00137$ cm, about 14 microns, small because of high perfusion rate

$z := 0, .001 \ldots L$ a range defined to be axial position in the capillary, cm

$$P(z) := pO_{2\,in} - \frac{\Gamma_{met} H}{((1+m)V)}\left[\left(\frac{r_T}{r_c}\right) - 1\right] z$$

capillary oxygen profile

The worst case for the tissue oxygen profile would be when $r = r_T$.

$$P_T(z) := P(z) - \frac{r_c^2 \Gamma_{met} H}{(4D_T)}\left[1 - \left(\frac{r_T}{r_c}\right)^2\right] - \frac{r_T^2 \Gamma_{met} H}{(2D_T)} \ln\left(\frac{r_T}{r_c}\right)$$

We can now plot the oxygen profiles in the capillary and tissue regions.

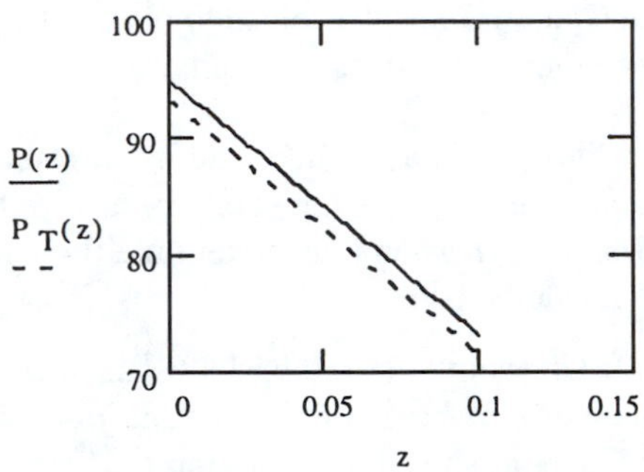

$P(L) = 73.31$

The existing pO_2 in the blood is very close to our guess value, so we know our calculation of the value of m based on the average blood pO_2 is correct.

If the tissue perfusion rate is decreased to about 0.62 ml/cm^3/min, the pO_2 level becomes zero at the anoxic corner. This is shown in Mathcad® Chart 4.7.

Chart 4.7 for **Example 4.5:** Calculating pO_2 levels within a capillary and the surrounding tissue region under the condition where the pO_2 in the anoxic corner is 0 mm Hg, find the critical blood perfusion rate.

$\Gamma_{met} := 25.9$	tissue (islet) metabolic O_2 consumption rate, μM/sec
$q := .62$	tissue perfusion rate, ml/cm^3/min
$pO_{2\,in} := 95$	entering blood oxygen level, mm Hg
$H := .74$	Henry's constant, mm Hg/μM
$C_{sat} := 8800$	saturated hemoglobin, μM
$P_{50} := 26$	pO_2 when $Y = .5$, mm Hg
$n := 2.34$	constant in the Hill equation
$V := .05$	average blood velocity, cm/sec
$L := .1$	capillary length, cm
$r_c := .0005$	capillary radius, cm
$D_T := 6.3 \times 10^{-6}$	oxygen diffusivity in the tissue region, cm^2/sec

We will let $P = pO_2$ of the blood and $P_T = pO_2$ in the tissue region

$P := 27.5$ a guess of what the exiting capillary blood pO_2 will be, mm Hg, this must be close to what we ultimately get

Now calculate the value of m at the average pO_2 of the blood.

$P_{avg} := .5(pO_{2\,in} + P)$

$$m := nP_{50}^{n}\,HC_{sat}\frac{P_{avg}^{(n-1)}}{\left(P_{50}^{n} + P_{avg}^{n}\right)^{2}}$$

$m = 26.02$

Now calculate the Krogh tissue cylinder radius based on the blood perfusion rate and the capillary properties.

$$r_T := r_c\sqrt{\frac{V}{\left(qL\frac{1}{60}\right)}}$$

$r_T = 0.00348$ cm, about 33 microns

$z := 0, .001 \ldots L$

a range defined to be axial position in the capillary, cm

$$P(z) := pO_{2\,in} - \frac{\Gamma_{met}H}{((1+m)V)}\left[\left(\frac{r_T}{r_c}\right) - 1\right]z$$

capillary oxygen profile

The worst case for the tissue oxygen profile would be when $r = r_T$.

$$P_T(z) := P(z) - \frac{r_c^2\Gamma_{met}H}{(4D_T)}\left[1 - \left(\frac{r_T}{r_c}\right)^2\right] - \frac{r_T^2\Gamma_{met}H}{(2D_T)}\ln\left(\frac{r_T}{r_c}\right)$$

We can now plot the oxygen profiles in the capillary and tissue regions.

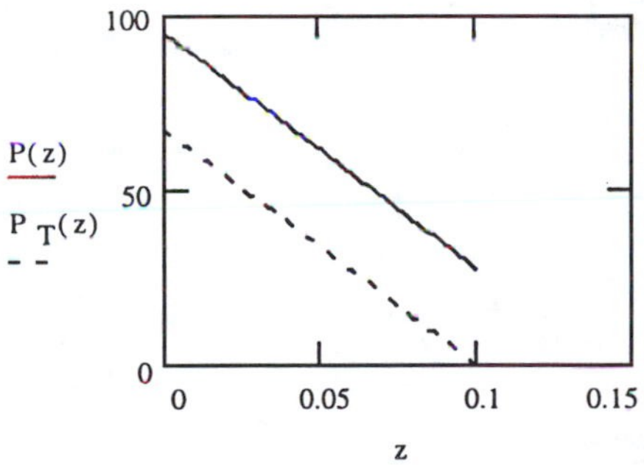

$P(L) = 27.774$ close to initial guess, so calculation of m is correct

If the tissue blood perfusion rate is decreased further, we will develop an anoxic layer of tissue whose axial boundary would be defined by $r_{anoxic}(z)$. This boundary can be found by solving Equation 4.45. Mathcad® Chart 4.8 shows the anoxic boundary for the case where the tissue perfusion rate has been decreased to 0.25 ml/cm^3/min.

Chart 4.8 for **Example 4.5:** Calculating the anoxic region boundary.

$\Gamma_{met} := 25.9$	tissue (islet) metabolic O_2 consumption rate, μM/sec
$q := .25$	tissue perfusion rate, ml/cm^3/min
$pO_{2\,in} := 95$	entering blood oxygen level, mm Hg
$H := .74$	Henry's constant, mm Hg/μM
$C_{sat} := 8800$	saturated hemoglobin, μM
$P_{50} := 26$	pO_2 when $Y = .5$, mm Hg
$n := 2.34$	constant in the Hill equation
$V := .05$	average blood velocity, cm/sec
$L := .1$	capillary length, cm

$r_c := .0005$ capillary radius, cm
$D_T := 6.3 \times 10^{-6}$ oxygen diffusivity in the tissue region, cm^2/sec
We will let $P =$ the pO_2 of the blood and $P_T = pO_2$ in the tissue region
$P := 26$ a guess of what the exiting capillary blood pO_2 will be, mm Hg, this must be close to what we ultimately get
Now calculate the value of m at the average pO_2 of the blood.

$P_{avg} := .5(pO_{2\,in} + P)$

$$m := nP_{50}^n HC_{sat} \frac{P_{avg}^{(n-1)}}{\left(P_{50}^n + P_{avg}^n\right)^2}$$

$m = 26.926$

Now calculate the Krogh tissue cylinder radius based on the blood perfusion rate and the capillary properties.

$$r_T := r_c \sqrt{\frac{V}{\left(qL\frac{1}{60}\right)}}$$

$r_T = 0.00548$ cm, about 55 microns

First we need to find at what value of z the tissue pO_2 at r_T becomes zero.

$Z := .5L$ an initial guess as to where this occurs

$$P(Z) := pO_{2\,in} - \frac{\Gamma_{met} H}{((1+m)V)}\left[\left(\frac{r_T}{r_c}\right)^2 - 1\right] Z$$

$$f(Z) := P(Z) - \frac{r_c^2 \Gamma_{met} H}{(4D_T)}\left[1 - \left(\frac{r_T}{r_c}\right)^2\right] - \frac{r_T^2 \Gamma_{met} H}{(2D_T)} \ln\left(\frac{r_T}{r_c}\right)$$

$Z := \text{root}(f(Z), Z)$

$Z = 0.00514$ cm, axial location where tissue becomes anoxic at r_T
$r_{an} := .5(r_c + r_T)$ initial guess of the anoxic radial boundary

given

$$\left(\frac{r_{an}}{r_c}\right)^2 \ln\left(\frac{r_{an}}{r_c}\right)^2 - \left(\frac{r_{an}}{r_c}\right)^2 + 1 - \left[\frac{4D_T pO_{2\,in}}{(r_c^2 \Gamma_{met} H)}\right] + \frac{4D_T}{\left[(1+m) r_c^2 V\right]}\left[\left(\frac{r_{an}}{r_c}\right) - 1\right] z = 0$$

$\text{anoxic}(z) := \text{find}(r_{an})$

$$N := \frac{L - Z}{.001}$$

$N := \text{floor}(N) + 1$

$i := 0 \cdots N$

$z_i := Z + .001i$

$r_{anoxic_i} := \text{anoxic}(z_i)$

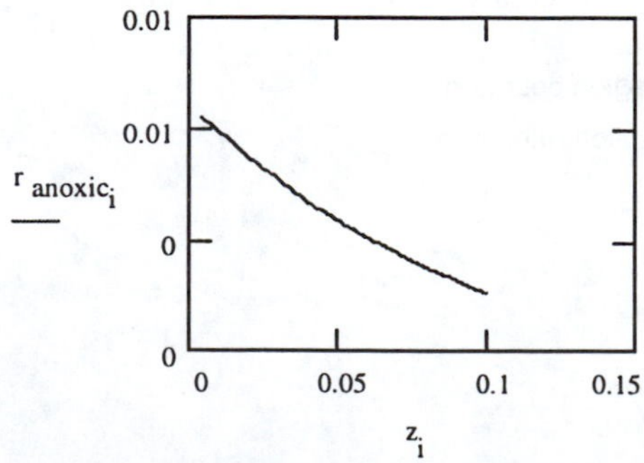

The anoxic radial boundary as a function of capillary axial position.

$$P_L := pO_{2\,\mathrm{in}} - \frac{\Gamma_{\mathrm{met}} H}{((1+m)V)} \left[\left[\frac{(r_{\mathrm{anoxic}_N})}{r_c} \right]^2 - 1 \right] z_N$$

$P_L = 27.292$ exiting blood pO_2, mm Hg, which is close to the guess, so the calculation of m is correct.

4.11 ARTIFICIAL BLOOD

Blood is crucial for survival and is frequently needed for treatment of life threatening injuries, to replace blood loss during surgery, and to treat a variety of blood disorders. The world demand for blood amounts to about 100 million units[4] of blood per year. Within the United States, a blood transfusion occurs every 3.75 seconds (Lewis 1997). Blood banks provide, in most cases, an immediate source of blood to meet these needs. However, blood banking requires an extensive infrastructure for collection, testing for disease, cross-matching, and storage. In addition to potential contamination of the blood supply with infectious agents such as human immunodeficiency virus (HIV) and hepatitis, blood itself can cause hemolytic transfusion reactions. The risks associated with blood transfusions range from 0.0004%[5] for HIV infection, 0.001% for a fatal hemolytic reaction, 0.002% to 0.03% for contracting the hepatitis virus, and 1% for a minor reaction (Intaglietta and Winslow, 1995).

Concern in recent years about the safety of the blood supply has resulted in major efforts to develop a substitute for blood (Nucci and Abuchowski 1998). A blood substitute could replace human donor blood and potentially be used for trauma, surgery, as well as for treatment of chronic blood disorders that require frequent transfusions. Blood substitutes are also of interest to the military because of the potential for a significant decrease in the special handling and logistical requirements of human blood on the modern battlefield.

An artificial blood must meet the two most important functions of blood given by transfusion. These are the replacement of lost plasma and the ability to transport sufficient amounts of oxygen. Three approaches being considered have the potential to be used as artificial blood. These are solutions containing stroma-free[6] hemoglobin, perfluorocarbon (PFC) liquids, and camouflaged RBCs. There are currently about ten different artificial blood products in development that utilize either some form of hemoglobin, PFCs, or camouflaged RBCs (Winslow 1997; Scott et al. 1997).

Several products are based on the use of stroma-free hemoglobin. The hemoglobin can be derived from a variety of sources. For example, hemoglobin can be of human or bovine origin, it may be derived from bacteria via recombinant DNA techniques, or transgenic methods applied in animals and plants. Hemoglobin, as discussed earlier, consists of four polypeptide chains, two α chains and two β chains. The hemoglobin molecule outside of the RBC is not stable. For example, if hemoglobin is diluted in

[4] A unit of blood is 500 ml.

[5] 1% is equal to 1 transfusion incident per 100 units transfused.

[6] Stroma-free in this case means without the red blood cell and its cellular framework.

comparison to its concentration in RBCs, it will spontaneously dissociate into these smaller chains. These lower-molecular-weight chains are then rapidly removed from the circulation and excreted by the kidneys. Also recall that hemoglobin by itself has a very high affinity for oxygen. Therefore, to be effective as a blood substitute, artificial bloods based on hemoglobin must chemically modify the hemoglobin in order to improve its stability and to decrease its affinity for oxygen. The P_{50} value obtained for artificial blood based on hemoglobin is comparable to that of normal human blood.

PFCs are hydrocarbons in which the bonds between C and H are replaced by the much stronger C and F bonds. These strong C–F bonds give PFCs their chemical inertness (Shah and Mehra 1996). PFCs also exhibit high oxygen solubility, they are for the most part biocompatible and low cost. However, they are immiscible with water. In order to be used as a blood substitute, they must first be emulsified into tiny droplets typically in the range of 0.1 to 0.2 microns in diameter. The droplets are also frequently coated with phospholipids derived from egg yolk to stabilize the emulsion. Oxygenated blood carries about 20 ml of oxygen (based on 37°C and 760 mm Hg) per 100 ml of blood. PFCs in equilibrium with pure oxygen at 760 mm Hg have an average oxygen solubility of about 50 ml of oxygen per 100 ml of PFC (Gabriel et al. 1996). PFCs exhibit a linear equilibrium relationship between pO_2 and oxygen solubility. The Henry constant for PFCs would be equal to about 0.04 mm Hg/μM. Unlike blood containing hemoglobin, PFCs do not saturate and can therefore transport more oxygen by simply increasing the pO_2 level in the gas breathed by the patient. For example, a 40 volume% PFC emulsion when saturated with pure oxygen at 760 mm Hg would have the same oxygen carrying capacity of saturated blood. Because of their relatively high vapor pressure, a portion of the PFCs are removed from the circulation by evaporation via respiration.

Artificial bloods based on modified hemoglobin or PFCs still face some significant development problems. Key among these problems are gastrointestinal complaints and vasoconstriction (reduction in blood vessel diameter) when using hemoglobin products and flu-like symptoms and thrombocytopenia (very low quantity of platelets) when using PFCs. These artificial bloods also exhibit short life times in the circulation. Modified hemoglobins last on the order of 12 hours to 2 days whereas PFCs are removed in about 12 hours. In comparison, normal RBCs have lifetimes of about 120 days. Until the lifetimes of these products can be significantly improved, they will only be able to be used for acute situations such as trauma and surgery (Winslow, 1997).

Recently, a promising technique was described that may overcome some of the problems of artificial blood based on modified hemoglobin or PFCs (Scott et al. 1997). This approach involves covalently binding polyethylene glycol (PEG) to the surface of intact red blood cells. The RBCs appear to be not affected by the presence of the PEG coat. The PEG molecules have the effect of masking the surface antigens on the RBCs that lead to the various blood types in the case of human blood, or that trigger rejection in the case of animal RBCs. The result is a universal blood type that could allow the use of unmatched human or animal RBCs. The camouflaged RBCs should have a lifetime that is comparable to that of a normal RBC.

PROBLEMS

1. Consider a very metabolically active tissue region supplied with arterial blood that has a $pO_2 =$ 95 mm Hg. For an exiting venous pO_2 of 40 mm Hg, what values of n and P_{50} would maximize the amount of oxygen delivered to the tissue? What if the exiting $pO_2 = 20$ mm Hg?

2. Determine the volumetric tissue oxygen consumption rate ($\Gamma_{\text{metabolic}}$, μM/sec) from the data shown in Figure 4.1 for Brockman bodies. Carefully state all assumptions.

3. Calculate the external mass transfer coefficient (k_m, cm/sec) for oxygen from the data shown in Figure 4.1. Recall that the flux of oxygen into the spherical Brockman body by diffusion must equal the oxygen transport rate from the bulk fluid as described by the following equation:

$$-D_{\text{effective}} \left.\frac{dC_{\text{tissue}}}{dr}\right|_{\text{surface}} = k_m (C_{\text{bulk}} - C_{\text{tissue}}|_{\text{surface}}) \approx \frac{1}{3} R_{\text{Brockman body}} \Gamma_{\text{metabolic}}$$

Show that, for the case without convection, the Sherwood Number (Sh) is equal to Z.

4. Derive Equation 4.18.

5. For the situation described in Example 4.4, determine the critical loading of islets, i.e., the void volume of the tissue space for which the pO_2 reaches zero at the center line of the symmetric islet layer.

6. For the results obtained in Example 4.4, estimate the blood flow rate needed to sustain a total of 750,000 islets. Each islet may be assumed to be 150 microns in diameter. What would be the diameter of such a device?

7. For the situation described in Example 4.4, determine the critical oxygen permeability of the membrane, i.e., the value for which the pO_2 reaches zero at the center line of the symmetric islet layer.

8. Write a short paper that discusses blood types (A, B, AB, O).

9. Derive Equation 4.40.

10. Derive Equation 4.44.

11. Develop a Krogh tissue cylinder model for blood oxygenation using PFCs. Using this model, determine the tissue oxygen profile for an artificial blood perfusion rate of 0.75 ml/cm^3/min.

12. Consider a long and wide planar aggregate or slab of hepatocytes (liver cells) used in a bioartificial liver. The slab is suspended in a growth medium at 37°C with no mass transfer limitations between the slab and the medium. The medium is saturated with air giving a pO_2 of 160 mm Hg or 190 μM. The cell density in the slab is 1.25×10^8 cells/cm^3. Each cell is spherical in shape and 20 microns in diameter. The oxygen consumption rate of the cells may be described by the Michaelis-Menten relationship (Equation 4.12) with $V_{\max} = 0.4$ nmol/10^6 cells/sec and a $K_m = 0.5$ mm Hg. Estimate the maximum half-thickness of the slab of hepatocytes.

13. Photodynamic therapy (PDT) involves the localized photoirradiation of dye-sensitized tissue with the goal of causing cell death in tumors (Henderson and Doughherty 1992). The tumor containing the photosensitizer is irradiated by a laser of the proper wavelength (around 630 nm) to generate excited singlet molecules of the sensitizer. In the presence of molecular oxygen, the singlet sensitizer forms a very reactive form of oxygen called singlet oxygen which destroys the tumor. Therefore sufficient levels of molecular oxygen are needed in the tissue to affect destroying of the tumor by PDT. The action of PDT on tissue oxygen levels is like having an additional sink for oxygen that is dependent on the laser fluence rate, ϕ_0. Foster et al. (1993) show that this PDT-induced oxygen consumption rate may be estimated by the following equation:

$$\Gamma_{\text{PDT}} = 0.14(\mu\text{M cm}^2/\text{sec/mW}) \times \phi_0(\text{mW/cm}^2),$$

where the fluence rate (ϕ_0) ranges between 50 and 200 mW/cm^2.

Consider a typical capillary within the tumor undergoing PDT. Assume a capillary diameter of 8 microns and a capillary length of 300 microns. The Krogh tissue cylinder radius is assumed to be 40 microns. The metabolic oxygen consumption rate is 11.5 μM/sec. The average velocity of blood in the capillary is 0.04 cm/sec. The entering blood pO_2 is 95 mm Hg. Answer the following questions:

(a) What is the exiting pO_2 of the blood from the capillary with the laser off?
(b) What is the pO_2 in the lethal corner ($z = L$ and $r = r_T$) with the laser off?
(c) At what fluence rate would the pO_2 in the lethal corner equal zero? Note that this is the point at which PDT would start to be ineffective for those cancer cells at the outer periphery of the Krogh tissue cylinder.

CHAPTER

FIVE

PHARMACOKINETIC ANALYSIS

5.1 TERMINOLOGY

In this chapter we will focus our attention on some useful techniques in the field of pharmacokinetics. *Pharmacokinetics* is the study of the processes that affect drug distribution and the rate of change of drug concentrations within various regions of the body. Although pharmacokinetics is of utmost importance in the treatment of diseases using drugs, our major focus in the remainder of this book will be on using this technique as a tool for understanding the transport processes between an artificial device and the body.

We also need to distinguish pharmacokinetics from similar terms. Therefore, *pharmaceutics* concerns the formulation and preparation of the drug to achieve a desired drug availability within the body, *pharmacodynamics* is concerned with the time course of the treatment response that results from a given drug, and the actual physiological response that results from a drug is the subject of *pharmacology*.

5.2 ENTRY ROUTES FOR DRUGS

There are two routes with which a drug can enter the body. The *enteral* route refers to drugs that are given via the gastrointestinal (GI) tract. All other routes are called *parenteral*. The enteral route includes absorption via one or more of the following components of the GI tract: the buccal cavity (mouth) and sublingually (beneath the tongue), gastrically (stomach), intestinally (small and large intestines), and rectally. Once the drug is absorbed from the GI tract, it enters the blood and is distributed throughout the body. It is important, however, to point out that only drugs absorbed from the buccal cavity and the lower rectum enter the *systemic circulation* directly. Drugs absorbed from the stomach, intestines, colon, and upper rectum enter the *splanchnic circulation*. The splanchnic

Table 5.1 Parenteral routes for drug administration

Route	Result
Intravenous (IV)	Introduced directly into the venous circulation
Intramuscular	Within the muscle
Subcutaneous	Beneath the epidermal and dermal skin layers
Intradermal	Within the dermis, usually a local effect
Percutaneous	Topical treatment applied to the skin
Inhalation	Mouth/nose, pharynx, trachea, bronchi, bronchioles, alveolar sacs, alveoli
Intraarterial	Introduced directly into an artery, regional drug delivery
Intrathecal	Cerebrospinal fluid, subarachnoid space
Intraperitoneal (IP)	Within the peritoneal cavity
Vaginal	Within the vagina

circulation then takes the drug to the liver via the portal vein and after leaving the liver, the drug enters the systemic circulation. Since the liver contains many enzymes capable of degrading the drug (metabolism), a significant portion of the drug may be removed during this first pass through the liver before the drug is available to the general circulation. Table 5.1 summarizes all the routes for drug administration that are called parenteral. For the most part, drugs given parenterally enter the systemic circulation directly.

As the drug is being absorbed into the body, its presence will be noticed within the circulation as its concentration in the plasma portion of the blood changes. Recall that plasma refers to the clear supernatant fluid that results from blood after the cellular components have been removed. If the blood sample is simply allowed to clot, then the resulting clear fluid is referred to as serum, since the clotting proteins have been removed by the clotting process. For the most part, the concentration of drug in plasma or serum is identical and no distinction is needed. Usually only the plasma drug concentration in the body is known.

Figure 5.1 illustrates typical plasma drug concentrations as a function of time after the introduction of the drug to the body. Curve I illustrates the case where the drug is first slowly absorbed, resulting in increasing plasma concentrations, followed by a plateau, and then the plasma concentration decreases as the drug is eliminated by a variety of body processes. Curve II represents a very rapid injection (bolus) of the drug, usually intravenously (IV injection), where we see the drug instantly reaches its peak

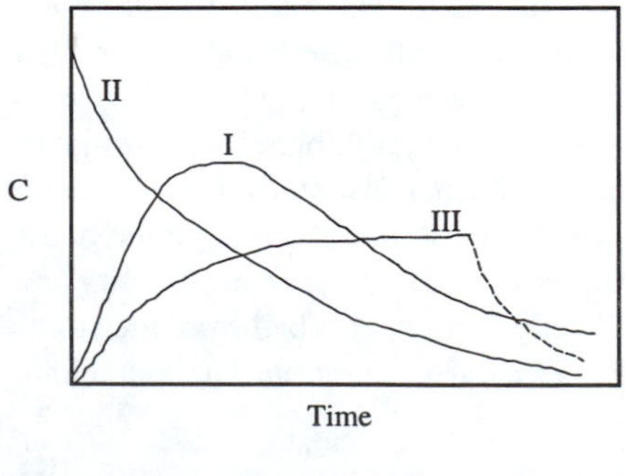

Figure 5.1 Plasma drug concentrations following I, absorption; II, rapid intravenous; and III, intravenous continuous infusion.

plasma concentration at the time of injection, and then is slowly eliminated. Curve III results from a continuous infusion, usually intravenously or through a controlled-release formulation. We see that after a short period of time, the drug reaches a steady-state plasma concentration. If the infusion is stopped, then the drug plasma concentration falls, due to its elimination as shown by the dashed curve.

5.3 MODELING APPROACHES

Our goal is then to develop mathematical models that can be used to describe profiles of plasma drug concentration such as those shown in Figure 5.1. This would allow for the description of the drug concentration in the body as a function of time. These models will incorporate several unknown parameters that we find by fitting our models to experimental drug concentration data. Development of pharmacokinetic models has generally followed three approaches. They are the *compartmental approach*, the *physiological approach*, and the *model-independent approach.*

The compartmental approach assumes the drug distributes into one or more "compartments" in the body that usually represent a particular region of the body, an organ, a group of tissues, or body fluids. The compartments are assumed to be well-mixed so that the concentration of the drug within the compartment is spatially uniform. Spatial distribution of the drug within the body is accounted for by the use of multiple compartments. The assumption of well-mixed compartments is usually valid since the cardiac output in humans is nominally 5 L/min and the blood volume is about 5 L, giving an effective residence time for one passage through the circulatory system of about 1 minute. Also recall that there is a net filtration of several milliliters per minute of the blood plasma, the interstitial fluid that bathes and circulates around all of the cells in the body. Since drug distribution occurs over periods of hours, and the body's fluids are moving over periods of minutes, the well-mixed assumption is appropriate. Movement of the drug between the compartments is usually described by simple irreversible or reversible first-order rate processes.

In physiological models, the movement of drug is based on the blood flow rate through a particular organ or tissue and includes consideraton of the rates of the mass transport processes within the region of interest (Kohn 1997). Experimental blood and tissue drug concentrations are needed to define the model parameters. Our Krogh tissue cylinder model, developed in Chapter 4, for describing the changes in oxygen levels within the capillary and the surrounding tissue region is an example of a physiological model. In this case, the "drug" is oxygen.

The model-independent approach does not try to make any physiological connection like the compartmental approach or the physiological approach. Instead one simply finds the best set of mathematical equations that describes the situation of interest (Notari 1987). These models are considered to be linear if the plasma drug concentration can be represented by a simple weighted summation of exponential decays. For example, the following equation may be used to describe the time course of the plasma drug concentration.

$$C = \sum_{i=1}^{n} C_i e^{-\lambda_i t} \tag{5.1}$$

Our focus in this chapter will be on the use of compartmental models to describe pharmacokinetic data like that shown in Figure 5.1. Compartmental models provide a convenient framework for building a pharmacokinetic model and are conceptually simple to use (Cooney 1976; Gibaldi and Perrier 1982; Welling 1986; Notari 1987).

5.4 FACTORS THAT AFFECT DRUG DISTRIBUTION

Before we can investigate some simple-compartment models, we first must discuss the following factors that influence how a particular drug is distributed throughout the body. Table 5.2 summarizes these factors and they are discussed in greater detail below.

5.4.1 Drug Distribution Volumes

Recall that Table 1.1 summarized the types of fluids found within the body. Since the drug can only distribute within these fluid volumes, we call these *true distribution volumes.* Therefore a drug could distribute within just the plasma volume of the circulatory system, i.e., a true distribution volume of about 3 liters. If the drug readily penetrates the vascular walls, then it could also distribute throughout the extracellular fluid volume for a total true distribution volume of about 15 liters. If the drug can also permeate the cell wall, then it will also be found within the intracellular fluid spaces as well, giving a total true distribution volume of about 40 liters.

The rate at which the drug is distributed throughout these fluid volumes is controlled by the rate at which drug is delivered to a region of interest by the blood, i.e., the tissue blood perfusion rate, and by the rate at which the drug diffuses from within the vascular system into the extravascular spaces. If the drug is lipid soluble, then its mass transfer across the capillary wall is usually not rate limiting and equilibrium is quickly reached between the amount of drug in the tissue region and that found in the blood. In this case, the distribution of the drug in a particular tissue region is *perfusion rate limited.* For lipid insoluble drugs, the capillary membrane permeability controls the rate at which the drug distributes between the blood and tissue regions. The distribution of the drug is then said to be *diffusion rate limited.*

Drugs are also capable of binding to proteins found in the plasma and the extravascular spaces. Albumin is perhaps the most common of the blood proteins that bind with drugs. Recall that the major difference between plasma and interstitial fluid is the amount of protein that is present. For the most part, plasma proteins have a difficult time diffusing through the capillary wall to enter the extravascular space. Therefore, a drug that binds

Table 5.2 Factors influencing drug distribution

Blood perfusion rate
Capillary permeability
Drug biological affinity
Metabolism of drug
Renal excretion

strongly with plasma proteins will for the most part be limited to just the volume of the plasma, i.e., about 3 liters. On the other hand, extravascular proteins or cell surface receptors that bind strongly to the drug will tend to accumulate the drug in the extravascular space at the expense of the plasma. Protein binding of the drug therefore can have a significant effect on the calculated or *apparent distribution volumes* as the following discussion will show.

It is important to point out that most analytical methods measure the total drug concentration (bound and unbound) that is in the plasma. Since this total plasma concentration (C_{total}) is readily measured, the *apparent distribution volume* ($V_{apparent}$) for the drug is simply given by the following equation:

$$V_{apparent} = \frac{A_{body}}{C_{total}} \tag{5.2}$$

where A_{body} is the total amount of drug within the body. The amount of drug in the body is usually in units of milligrams (mg), and the concentration of drug in micrograms per milliliter of plasma (μg/ml), which is the same as milligrams per liter of plasma (mg/L), resulting in the apparent distribution volume having units of liters (L).

It is important to note that the concentration of drug within any particular tissue region may vary significantly from the concentration of the drug in the plasma. Therefore, although the apparent distribution volume seems to have some physical significance, it really is just a factor having units of volume that, when multiplied by the plasma drug concentration, provides the total amount of drug in the body.

The apparent distribution volume calculated by Equation 5.2 can also be vastly different than the true distribution volume (V), related to the volume of the various body fluids, because of the effect of protein binding. To examine the effect of protein binding on the apparent distribution volume, consider the conceptual model shown in Figure 5.2 (Welling 1986).

Unbound drug is free to distribute between the plasma and tissue spaces, represented by a true distribution volume given by V and assumed to have a concentration represented by C. The concentration of drug bound to plasma proteins is given by C_{BP} and the concentration of drug bound to tissue proteins is C_{BT}. The concentration of total drug

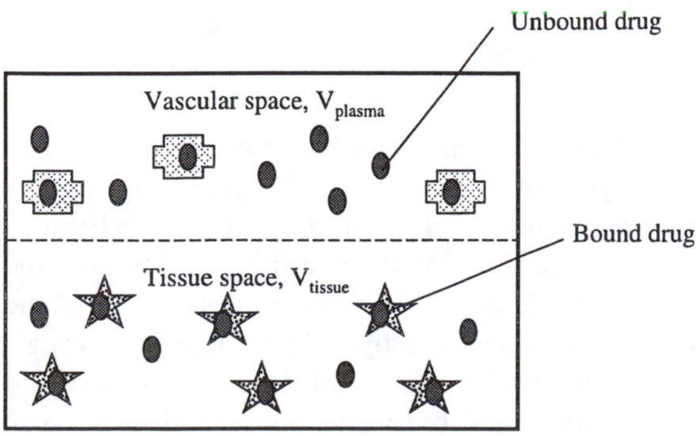

Figure 5.2 Drug distribution between plasma and tissue spaces with protein binding.

in the plasma, what is actually measured, is given by the following equation.

$$C_{\text{total}} = C + C_{BP} \tag{5.3}$$

The apparent distribution volume would be given by Equation 5.2. C_{BP} and C_{BT} are related to the total plasma concentration (C_{total}) by the binding fractions x and y, where x is defined as the fraction of the total drug in the plasma that is bound to plasma protein, and y is defined as the fraction of the total drug in the tissue region that is bound to tissue protein. A mass balance on the amount of drug in the body can be written:

$$A_{\text{body}} = VC + C_{BP}V_{\text{plasma}} + C_{BT}V_{\text{tissue}} \tag{5.4}$$

This equation may be solved for the unbound drug concentration in the plasma and tissue spaces (i.e., C in the true distribution volume, V), recognizing that $\frac{C_{BP}}{C} = \frac{x}{1-x}$ and $\frac{C_{BT}}{C} = \frac{y}{1-y}$. We also use the fact that $V_{\text{tissue}} = V - V_{\text{plasma}}$.

$$C = \frac{A_{\text{body}}}{V + \left(\frac{x}{1-x}\right)V_P + \left(\frac{y}{1-y}\right)(V - V_P)} \tag{5.5}$$

The apparent distribution volume can then be shown to be given by the following expression after combining Equations 5.2, 5.3, and 5.5.

$$V_{\text{apparent}} = (1-x)V + xV_{\text{plasma}} + y\left(\frac{1-x}{1-y}\right)(V - V_{\text{plasma}}) \tag{5.6}$$

In the absence of any protein binding, $x = y = 0$, and we see that the apparent distribution volume equals the true distribution volume, V. If the drug only binds to plasma proteins ($y = 0$), then in the limit as $x \to 1$, the apparent distribution volume approaches the plasma volume, V_{plasma}. We see in general that for any amount of plasma protein binding, the apparent distribution volume is less than the true distribution volume. On the other hand, if the drug only binds to tissue proteins ($x = 0$), then in the limit as $y \to 1$, the apparent distribution volume increases without bound, since the total concentration of the drug in the plasma is becoming vanishingly small. In this case, the apparent distribution volume will be significantly larger than the body's true fluid volumes.

5.4.2 Drug Metabolism

Once the drug is absorbed and distributed throughout the body by the circulatory system, a variety of biochemical reactions will begin to degrade the drug. This breakdown of the drug is part of the body's natural defense against foreign materials. These biochemical reactions, driven by enzymes, occur in a variety of organs and tissues. However, the major site of drug metabolism is within the liver. The enzymatic destruction of the drug reduces its pharmacological activity because the active site related to the drug's molecular structure is destroyed. Also, the *metabolites* that result tend to have increased water solubility that decreases their capillary permeability and enhances their removal from the body via the kidneys.

Enzymatic reactions that degrade the drug in the tissues tend to follow the Michaelis-Menten rate model where the reaction rate is given by the following expression:

$$r_{\text{metabolic}} = \frac{V_{\max} C}{K_m + C} \tag{5.7}$$

Usually the drug concentration is much smaller than the value of K_m. Therefore, the rate at which the drug is degraded by metabolic reactions can be represented by an irreversible first order rate law.

$$r_{\text{metabolic}} = k_{\text{metabolic}} C \tag{5.8}$$

In this equation, C represents the total plasma concentration of drug, $k_{\text{metabolic}}$ represents the first order rate constant for the metabolic reactions, and $r_{\text{metabolic}}$ represents the rate at which the drug is degraded on a per unit volume basis. We will use this expression later when we write mass balances for the drug using compartmental models.

5.4.3 Renal Excretion of the Drug

The kidneys also have a major role in the elimination of the drug from the body. Part of this elimination process is enhanced by the enzymatic degradation of the drug and the formation of more water-soluble products. The kidneys receive about 1,100 ml/min of blood or about 22% of the cardiac output. On a per mass basis, they are the highest perfused organ in the body. Their primary purpose is to remove unwanted metabolic end-products such as urea, creatinine, uric acid, and urates from the blood, and to control the concentrations of such ions as sodium, potassium, chloride, and hydrogen. The concentrated urine that is formed from the kidneys amounts to about 1.5 liters of urine per day.

Figure 5.3 illustrates the basic functional unit of the kidney called the *nephron.* Each kidney contains about one million nephrons and each nephron contributes to the formation of urine. The operation of the kidney can be explained by considering how an individual nephron functions. The nephron consists of the following major components: the *glomerulus* contained within *Bowman's capsule*, the *convoluted proximal tubule, Henle's loop*, the *convoluted distal tubule*, and finally the *collecting tubule* that branches into the *collecting duct.* The glomerulus, proximal tubule, and distal tubule have a major role in drug elimination from the kidney.

Blood enters the glomerulus via the *afferent arteriole* and leaves by the *efferent arteriole.* The glomerulus consists of a tuft of highly permeable fenestrated capillaries, several hundred times more permeable to solutes than the typical capillary. *Glomerular filtrate* is formed across these capillaries within the Bowman's capsule due to a pressure difference that exists between the glomerular capillaries and the Bowman space. The glomerular capillary membrane, for the most part, is impermeable to the plasma proteins, retaining those molecules with a molecular weight over about 69,000 daltons. Drugs that are bound to plasma proteins therefore cannot be removed from the blood stream by filtration through the glomerulus. However, unbound drugs with a molecular weight less than 69,000 daltons will be readily filtered out of the bloodstream with the glomerular filtrate. The *glomerular filtration rate* (GFR) is the sum total of all the glomerular filtrate

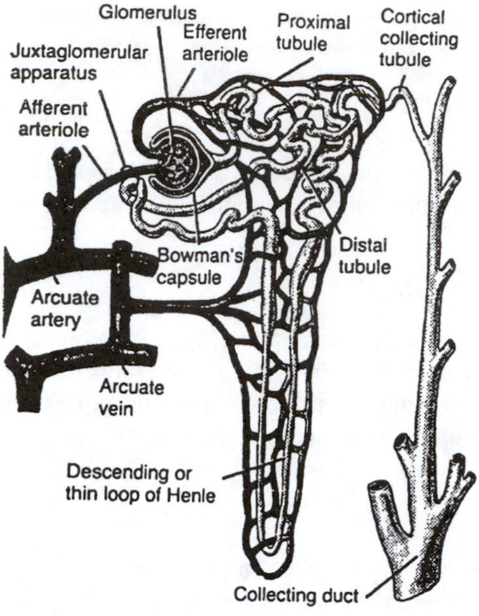

Figure 5.3 The nephron. (From Guyton, A. C. 1991. *Textbook of medical physiology*, 8th ed. Philadelphia: W. B. Saunders, with permission.)

formed per unit time by the nephrons in the two kidneys. This is about 125 ml/min or about 180 liters each day, which is more than twice the weight of the body. Clearly this volume of fluid must be recovered and we find that over 99% of the glomerular filtrate is reabsorbed in the tubules.

After the glomerular filtrate leaves Bowman's capsule, it enters the proximal tubule that is primarily concerned with the active reabsorption of sodium ions and water. Other substances such as glucose and amino acids are reabsorbed as well. The proximal tubule is important in drug elimination in that it can also actively secrete drugs from the peritubular capillaries surrounding the proximal tubule. This active transport of drugs is sufficiently fast so that even drugs that are bound to plasma proteins will dissociate, freeing up drugs for active secretion across the tubule wall. Within the descending portion of Henle's loop, continued reabsorption of water and other ions occurs primarily by passive diffusion. The ascending loop of Henle is considerably less permeable to water and urea resulting in a very dilute tubular fluid that is rich in urea. Henle's loop does not play a significant role in drug elimination. The first portion of the distal tubule is similar to that of the ascending loop of Henle in that it continues to absorb ions but is impermeable to water and urea. Within the latter portions of the distal tubule and the collecting tubule, ions and water continue to be absorbed and the urine acidified, thus controlling the acid-base balance of the body fluids.

The reabsorption rate of sodium ions is controlled by the hormone *aldosterone.* The water permeability of these distal segments is controlled by *antidiuretic hormone* and constitutes a means for controlling the final volume of urine formed. Drugs can also be reabsorbed within the distal tubule, thus affecting the rate of drug elimination from the body. The collecting duct continues to reabsorb water under the control of antidiuretic hormone and does not affect, for the most part, drug elimination. In summary, the kidney affects drug elimination through the following mechanisms: filtration through the glomerulus, secretion from the peritubular capillaries into the proximal tubule, and reabsorption within the distal tubule.

5.5 DRUG CLEARANCE

5.5.1 Renal Clearance

The elimination of drug by the kidney is described by the parameter *renal clearance.* The renal clearance is simply the volume of plasma that is totally "cleared" of the drug per unit time as a result of the drug's elimination by the kidney. Figure 5.4 illustrates the concept of renal clearance and allows for a mathematical definition for the case where the only elimination pathway for the drug is through the kidneys. As we shall see shortly, however, this figure can be generalized to represent multiple elimination pathways.

In this figure, a drug is uniformly distributed within an apparent distribution volume V_{apparent}, and has a total plasma concentration now represented by C. CL_{renal} represents the renal plasma flow rate that is totally cleared of the drug. Therefore, renal clearance is given by CL_{renal}, and the rate at which drug is removed from the body would be $CL_{\text{renal}} \times C$. We can then write an unsteady mass balance on the drug in the body's apparent distribution volume as given by the next equation.

$$V_{\text{apparent}} \frac{dC}{dt} = -CL_{\text{renal}} C \tag{5.9}$$

This equation may be integrated to provide an equation for how the drug concentration within the body changes with time starting from an initial plasma drug concentration C_0.

$$C(t) = C_0 e^{-(CL_{\text{renal}}/V_{\text{apparent}})t} = C_0 e^{-k_{\text{renal}} t} \tag{5.10}$$

We see that drug elimination by the kidneys is a first order process, exhibiting an exponential decay in plasma drug concentration with time. The renal elimination rate

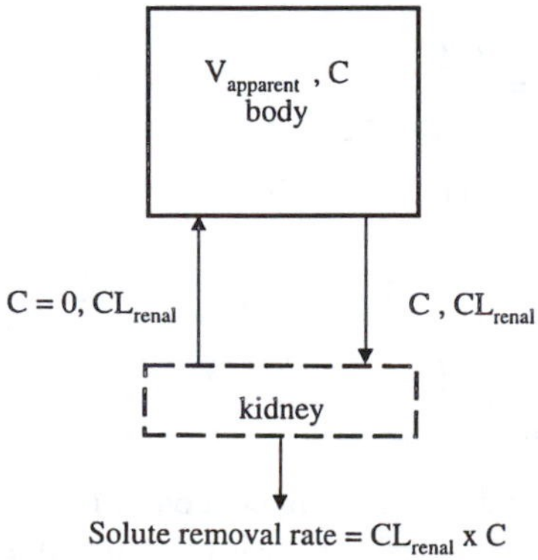

Figure 5.4 Concept of renal clearance.

constant, k_{renal} is related to the renal clearance and the apparent distribution volume by the following equation.

$$k_{\text{renal}} = \frac{CL_{\text{renal}}}{V_{\text{apparent}}} \tag{5.11}$$

If Q_{urine} represents the volumetric urine formation rate, and C_{urine} the concentration of drug in the urine, then another drug mass balance requires the following relationship.

$$CL_{\text{renal}}C = Q_{\text{urine}}C_{\text{urine}} \quad \text{or} \quad CL_{\text{renal}} = \frac{Q_{\text{urine}}C_{\text{urine}}}{C} \tag{5.12}$$

This equation then provides a connection between the renal clearance and the formation of urine containing the excreted drug.

The polysaccharide *inulin*, with a molecular weight of about 5,000 daltons, is particularly important to mention at this point because its renal clearance can be used to measure kidney function. Inulin is not metabolized by the body and readily passes through the glomerular capillaries, and is neither reabsorbed or secreted within the tubules. Therefore, the rate at which plasma is being cleared of inulin in the kidneys (renal clearance) is the same as the glomerular filtration rate (GFR).

5.5.2 Plasma Clearance

Plasma clearance (CL_{plasma}) is the term used to represent the sum of all the drug elimination processes of the body. These would include the two primary ones already discussed, i.e., metabolism and the kidneys, and other secondary processes such as sweating, bile, respiration, and the feces. Each process would have its own characteristic clearance value and a first order elimination rate constant defined as follows:

$$k_i = \frac{CL_i}{V_{\text{apparent}}} \tag{5.13}$$

i = renal, metabolic, sweat, bile, respiration, feces.

The plasma clearance would then be related to the sum of all the first-order drug elimination rate constants as given by the following relationship. The total first-order elimination rate constant is given by k_{te} and is equal to the sum of the elimination rate constants for all elimination processes.

$$k_{te} = \sum_i k_i = \frac{1}{V_{\text{apparent}}} \sum_i CL_i = \frac{CL_{\text{plasma}}}{V_{\text{apparent}}} \tag{5.14}$$

The change in drug concentration with time, taking into account all elimination processes, would be given by the following equation.

$$C(t) = C_0 e^{-k_{te}t} \tag{5.15}$$

5.5.3 Biological Half-Life

The *biological half-life* ($t_{1/2}$) of a drug is the time needed for the drug concentration in the plasma to be reduced by one-half. For first-order processes, this time is related to the first-order elimination rate constant by simply letting C/C_0 equal 0.5 in Equation 5.15. When solved for $t_{1/2}$, the following result is obtained.

$$t_{1/2} = \frac{0.693}{k_{te}} = \frac{0.693 V_{\text{apparent}}}{CL_{\text{plasma}}} \tag{5.16}$$

5.6 A MODEL FOR INTRAVENOUS INJECTION OF DRUG

Equation 5.15 now represents our first pharmacokinetic model. It is based on a single compartment with all the elimination processes lumped into a single first-order elimination rate constant. This equation predicts that the drug concentration decreases at an exponential rate with a general shape like that of Curve II in Figure 5.1. The initial drug concentration is C_0 and this would arise, for example, from a rapid bolus intravenous injection. The initial drug concentration is related to the dose (D) and the apparent distribution volume as given by the following equation.

$$C_0 = \frac{D}{V_{\text{apparent}}} \tag{5.17}$$

If the dose and the initial concentration of the drug were known, then Equation 5.17 can be used to solve for the apparent distribution volume of the drug.

The total area under a C versus t curve, such as those shown in Figure 5.1, is given the special symbol $\text{AUC}^{0\rightarrow\infty}$ and is defined mathematically as follows:

$$\text{AUC}^{0\rightarrow\infty} \equiv \int_0^{\infty} C(t)\,\mathrm{d}t \tag{5.18}$$

Equation 5.15 provides a mathematical relationship for the change of the drug concentration with time for the special case of a rapid bolus intravenous injection of a drug that distributes throughout a single compartment. Substituting this equation into Equation 5.18, we obtain the following result for the value of $\text{AUC}^{0\rightarrow\infty}$.

$$\text{AUC}^{0\rightarrow\infty} = \frac{C_0}{k_{te}} = \frac{D}{V_{\text{apparent}} k_{te}} = \frac{D}{CL_{\text{plasma}}} \tag{5.19}$$

This equation then provides a simple relationship between the total area under the curve, the drug dose, and the plasma clearance.

5.7 ACCUMULATION OF A DRUG IN THE URINE

As we discussed earlier, some of the drug will be eliminated by the kidneys and show up in the urine. In some cases it may be convenient to use additional data on the amount of drug (not its metabolites) found in the urine to supplement a pharmacokinetic analysis. The rate at which drug accumulates in the urine would be given by the following equation where M_{urine} is the mass of drug (unchanged) in the urine at any given time.

$$\frac{\mathrm{d}M_{\text{urine}}}{\mathrm{d}t} = k_{\text{renal}} V_{\text{apparent}} C \tag{5.20}$$

Continuing our discussion with the bolus intravenous injection of drug, we can substitute Equation 5.15 for the value of the drug concentration in the plasma at any time and integrate to obtain the following expression for the mass of drug in the urine.

$$M_{\text{urine}}(t) = \left(\frac{k_{\text{renal}} C_0 V_{\text{apparent}}}{k_{te}}\right)(1 - e^{-k_{te}t}) \tag{5.21}$$

After a sufficiently long period of time ($t \to \infty\cdot$), Equation 5.21 simplifies to provide the total amount of drug collected in the urine, M_{urine}.

$$M_{\text{urine}} = \frac{k_{\text{renal}} C_0 V_{\text{apparent}}}{k_{te}} = D\frac{k_{\text{renal}}}{k_{te}} = \text{AUC}^{0\to\infty} CL_{\text{renal}} \tag{5.22}$$

Thus we find a relationship exists between the total amount of drug collected in the urine, the renal clearance, and the total area under the curve. Note that if the only drug-elimination pathway is through the kidneys ($k_{\text{renal}} = k_{te}$), then the total amount of drug collected in the urine is equal to the total drug dose, as it should be.

5.8 CONSTANT INFUSION OF A DRUG

So far we have discussed the case of first-order drug elimination following a rapid bolus intravenous injection of a drug. The drug was assumed to be uniformly distributed throughout a single compartment. In this case the drug concentration versus time would be expected to follow Curve II of Figure 5.1. In many cases, a constant infusion of the drug is given. Here the concentration of the drug gradually increases and reaches a steady-state level once a balance is reached between the drug infusion rate and the drug elimination rate. The concentration-versus-time curve would follow something like Curve III of Figure 5.1. This would be similar to the so-called *controlled release* drug-delivery systems that are commonly used, for example, to treat motion sickness, for birth control, and to help people quit smoking by delivering nicotine. The constant infusion of a drug can also be used to provide a steady starting concentration of the drug.

Figure 5.5 illustrates a single-compartment model for continuous infusion of drug. I_0 represents the drug-infusion rate (mg/min), assumed to be constant here. However, one could generalize this infusion rate to be any arbitrary time-varying function. An

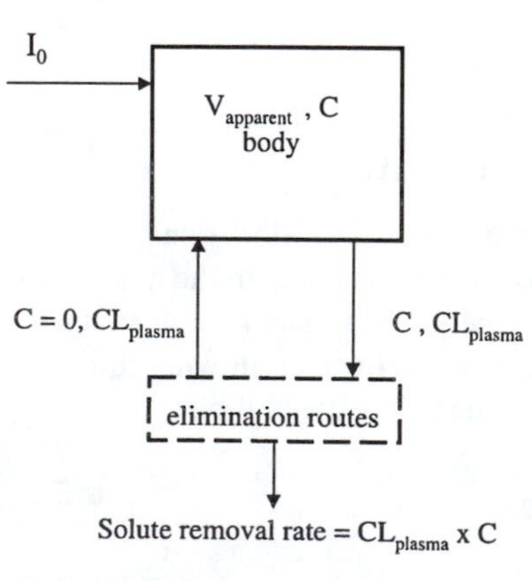

Figure 5.5 Single-compartment model for continuous infusion with first order elimination.

unsteady mass balance may be written for the drug as given by the next equation.

$$V_{\text{apparent}} \frac{dC}{dt} = I_0 - CL_{\text{plasma}}C = I_0 - V_{\text{apparent}} k_{te} C \tag{5.23}$$

This equation may be integrated under the assumption that no drug is initially present in the body to give:

$$C(t) = \left(\frac{I_0}{k_{te} V_{\text{apparent}}}\right)(1 - e^{-k_{te}t}) \tag{5.24}$$

The general shape of the curve described by this equation is that of Curve III in Figure 5.1. For long periods of time, this equation predicts a steady-state plasma drug concentration given by the next equation.

$$C_{ss} = \frac{I_0}{k_{te} V_{\text{apparent}}} = \frac{I_0}{CL_{\text{plasma}}} \tag{5.25}$$

When the infusion is stopped, then the drug is eliminated according to Equation 5.15, where the initial drug concentration, C_0 is equal to the steady-state drug concentration, (C_{ss}) that arises from the constant infusion. Now the drug concentration decreases as shown by the dotted line of Curve III in Figure 5.1.

Example 5.1 The table shown below provides some data for the elimination of radioactive inulin from a 392-g laboratory rat (Sarver, 1994). The animal was given a bolus injection at $t = 0$ equivalent to 1.01×10^5 cpm (counts per minute, radioactivity directly proportional to the inulin concentration). This was done to hasten the development of a steady-state plasma inulin level. Next, a continuous infusion of inulin was started at the rate of 2.76×10^3 cpm/min. The inulin infusion was then stopped after a total of 80 minutes. Determine from this data the glomerular filtration rate, the renal elimination rate constant, and the apparent distribution volume for inulin.

Inulin elimination from a laboratory rat (Sarver 1994)

Time (minutes)	Inulin concentration (cpm/ml)	
0	0	
30	849	
40	845	
60	903	
70	888	
75	873	
80	882	infusion stopped
90	565	
100	412	
110	271	

SOLUTION We see that after about 30 minutes, a steady-state plasma inulin concentration of about 873 cpm/ml is reached. Once the infusion is stopped, the plasma inulin concentration decreases rapidly. As discussed earlier, inulin is only removed by the kidneys and its renal clearance is the same as the glomerular filtration rate (GFR). Therefore for this experiment with inulin, Equation 5.25 can be rearranged to solve for the GFR in terms of the known values of the infusion rate and the steady-state plasma concentration.

$$k_{te} V_{\text{apparent}} = \text{GFR} = \frac{I_0}{C_{ss}}, \quad \text{for inulin only}$$

For the experimental conditions leading to the data shown in the above table, the GFR is then calculated to be 3.16 ml/min. Since the GFR is proportional to body weight (BW), this can be expressed as 0.48 ml/hr/gBW.

The renal elimination rate constant can be found by performing a regression analysis of the data obtained after the infusion is stopped, using Equation 5.15. Note that the time in this equation represents the time since the infusion was stopped. The initial concentration would be equal to 873 cpm/ml and k_{te} would be equal to k_{renal} since the only elimination process is that of the kidneys. Note that Equation 5.15 can be linearized by taking the natural logarithm of each side. The regression equation is then given by:

$$\ln C(t) = \ln C_0 - k_{te} t$$

The intercept would be the natural logarithm of the initial concentration and the slope would be the negative of the elimination rate constant. Mathcad® Chart 5.1 illustrates the regression of the data.

Chart 5.1 for **Example 5.1:** Inulin infusion study in a laboratory rat.

$i := 0 \ldots 8$ number of data points

time, minutes $\text{time}_i =$	inulin concentration, cpm/ml $C_i :=$
30	849
40	845
60	903
70	888
75	873
80	882
90	565
100	412
110	271

Calculate the steady state inulin concentration.

$$C_{ss} := \left(\frac{1}{6}\right) \sum_{i=0}^{5} C_i$$

$C_{ss} = 873.333$ cpm/ml

Perform linear regression on the data after the infusion stopped at 80 minutes, assuming the starting concentration was equal to that of C_{ss}.

$C_5 := C_{ss}$ sets the 80 minute value to the average steady state value

$j := 0 \ldots 3$ only have 4 data points after infusion stops including the C_{ss} value

$t_j := \text{time}_{(5+j)} - 80$ the time since the infusion stopped in minutes

$y_j := \ln[C_{(5+j)}]$ transform the concentration data

$k_{\text{renal}} := -\text{slope}(t, y)$

$k_{\text{renal}} = 0.038$ 1/min the renal elimination rate constant

$C_0 := \exp(\text{intercept}(t, y))$

$C_0 = 860.085$ cpm/ml — the value of the starting inulin concentration after the infusion stops compares quite well with the value of C_{ss} determined above from the data

$r := \text{corr}(t, y)$

$r = -0.998$ — Indicates that we have an excellent fit of the single compartment model that describes the elimination of inulin.

time := 0 . . . 30 — running time, minutes

$C_{pred}(\text{time}) := C_0 \exp(-k_{renal}\,\text{time})$ — the predicted inulin levels after the infusion is stopped

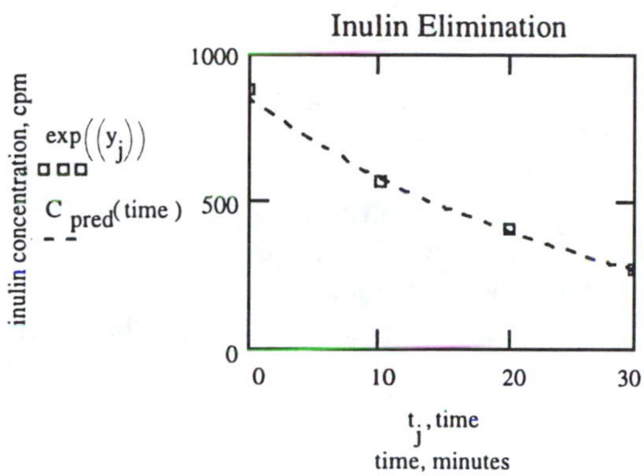

The renal elimination rate constant for inulin is found to be 0.038/min. The apparent distribution volume for inulin is given by:

$$V_{\text{apparent}} = \frac{\text{GFR}}{k_{\text{renal}}}$$

For these data, the apparent distribution volume is about 83 ml. Since the distribution volume is also proportional to body weight, the apparent distribution volume for a rat can be expressed as 0.21 ml/gBW.

5.9 FIRST-ORDER DRUG ABSORPTION AND ELIMINATION

Many drugs are given orally or by injection. Other routes may include the nasal cavity or by inhalation. In all of these non-intravenous cases, the drug is first absorbed and makes its way into the plasma by diffusion where it becomes distributed throughout the body. The plasma concentration of the drug gradually increases and finally decreases as the elimination processes begin to overcome the rate of drug absorption. Curve I of Figure 5.1 illustrates the general trend of the plasma drug concentration with time during the absorption and elimination phases. Although the drug absorption rate is strongly affected by the dosage form, i.e., solution, capsule, tablet, suspension, etc., it has been found that, for the most part, the drug absorption process can be adequately described by simple first-order kinetics.

Figure 5.6 illustrates the simplest compartmental model for the case of first-order drug absorption and elimination. Note a separate mass-balance equation is needed for the drug whose amount in the body decreases with time due to its absorption into the

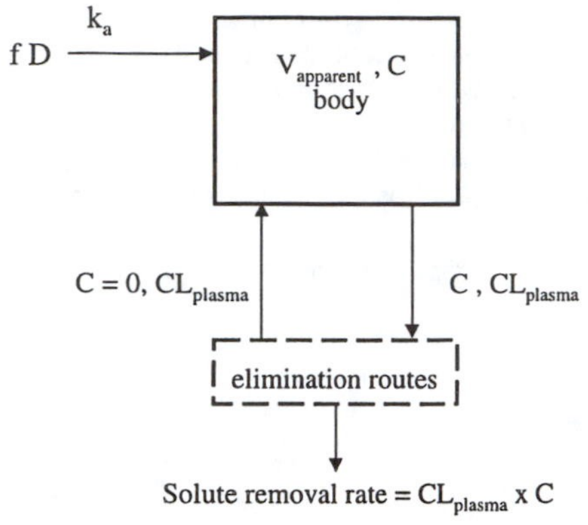

Figure 5.6 Single-compartment model with first-order absorption and elimination.

body. The factor (f) accounts for the fraction of the dose (D) that is actually absorbable. An amount of the dose equal to $(1 - f)\ D$ is assumed to be removed from the body without ever entering the apparent distribution volume.

A mass balance on the amount of drug within the apparent distribution volume can be written as follows:

$$V_{\text{apparent}} \frac{dC}{dt} = k_a A_{\text{body}} - CL_{\text{plasma}} C = k_a A_{\text{body}} - V_{\text{apparent}} k_{te} C \tag{5.26}$$

where A_{body} represents the mass of drug remaining to be absorbed, and k_a represents the drug's first order absorption rate constant.

A mass balance on the drug itself is given by the following equation.

$$\frac{dA_{\text{body}}}{dt} = -k_a A_{\text{body}} \tag{5.27}$$

When this equation is integrated, recognizing that initially $A_{\text{body}} = fD$, then the amount of drug at anytime remaining to be absorbed is given by:

$$A_{\text{body}}(t) = fDe^{-k_a t} \tag{5.28}$$

This equation may now be substituted into Equation 5.26 to obtain an expression for the total amount of drug in the plasma as a function of time, assuming the initial concentration of drug is zero.

$$C(t) = \frac{fD}{V_{\text{apparent}}} \left(\frac{k_a}{k_a - k_{te}} \right) (e^{-k_{te} t} - e^{-k_a t}) \tag{5.29}$$

We see that the concentration of drug in the plasma is dependent on the first-order rate constants for absorption and elimination. The concentration is directly proportional to the fraction of the drug dose that is absorbable and inversely proportional to the apparent distribution volume.

An expression for the time at which the concentration peaks can be obtained by finding the derivative of Equation 5.29 with respect to time, setting it equal to zero, and solving for the time. This time, called $\tau_{\max}$, is given by the following expression.

$$\tau_{\max} = \frac{1}{k_a - k_{te}} \ln \left(\frac{k_a}{k_{te}} \right) \tag{5.30}$$

Note that the time the maximum concentration is reached is independent of the dose. The corresponding concentration at the time of the peak may be found by substituting τ_{max} into Equation 5.29 for the time and simplifying to obtain:

$$C_{max} = \left(\frac{fD}{V_{apparent}}\right)\left(\frac{k_a}{k_{te}}\right)^{\frac{k_{te}}{k_{te}-k_a}} \tag{5.31}$$

We can also find the value of $AUC^{0\to\infty}$ by substituting Equation 5.29 into Equation 5.18. Upon integration, the following results.

$$AUC^{0\to\infty} = \frac{fD}{V_{apparent}k_{te}} = \frac{fD}{CL_{plasma}} \tag{5.32}$$

It is interesting to note that the value of $AUC^{0\to\infty}$ does not depend on the absorption rate constant.

To complete our mathematical description of the process shown in Figure 5.6, the rate at which the drug enters the urine would still be given by Equation 5.20. Substituting Equation 5.29 for the total plasma drug concentration into this equation gives the following result for the amount of drug in the urine at any given time.

$$M_{urine}(t) = \left(\frac{fDk_{renal}}{k_{te}}\right)\left[1 - \frac{1}{k_a - k_{te}}(k_a e^{-k_{te}t} - k_{te}e^{-k_a t})\right] \tag{5.33}$$

After all of the drug has cleared the body ($t \to \infty$), the amount of drug in the urine would be given by:

$$M_{urine}^{\infty} = fD\frac{k_{renal}}{k_{te}} \tag{5.34}$$

Note that if the only elimination pathway for the drug is through the kidneys ($k_{renal} = k_{te}$), then $M_{urine}^{\infty} = fD$.

If Equation 5.34 is subtracted from Equation 5.33, then an equation is obtained for the quantity of drug remaining to be excreted by the kidneys. This equation is of a convenient form for determining the absorption and elimination rate constants from urinary excretion data.

$$M_{urine}^{\infty} - M_{urine}(t) = \frac{M_{urine}^{\infty}}{(k_a - k_{te})}(k_a e^{-k_{te}t} - k_{te}e^{-k_a t}) \tag{5.35}$$

These parameters (k_a and k_{te}) may be determined by performing a nonlinear regression on urinary excretion data in a manner similar to that shown in Example 5.2 below. Since M_{urine}^{∞} is known from the urinary excretion data, one can then determine the renal elimination rate constant (k_{renal}) from Equation 5.34 given the dose actually absorbed, (fD) and the value of k_{te}.

Example 5.2 illustrates the use of the single-compartment model with first-order absorption and elimination to describe plasma nicotine levels produced by chewing nicotine gum (McNabb, Ebert, and McCusker 1982). This example also shows how a nonlinear regression can be used to find the three unknown parameters in Equation 5.29, i.e., $fD/V_{apparent}$, k_a, and k_{te}. To obtain the values of f and $V_{apparent}$ from the value of $fD/V_{apparent}$ requires additional data such as those obtained from urinary-excretion or intravenous-injection studies. Other techniques based on graphical methods may also be used to determine pharmacokinetic parameters as discussed in the following references (Cooney 1976; Gibaldi and Perrier 1982; Welling 1986; Notari 1987) and in the problems at the end of this chapter.

Example 5.2 The table within Mathcad® Chart 5.2 summarizes data obtained for plasma nicotine levels in a 27-year-old man chewing one piece of gum containing 4 mg of nicotine. Nicotine gum was developed to help people quit smoking by providing the same amount of plasma nicotine as obtained from cigarettes. Nicotine is absorbed from the gum over a 30-minute period of chewing and is readily absorbed within the buccal cavity. From the data provided in this table, determine the pharmacokinetic parameters that describe the absorption and elimination of nicotine.

SOLUTION Mathcad® Chart 5.2 shows the results of the nonlinear regression. The fits during the initial absorption phase and the elimination phase are excellent. However, Equation 5.29 fails to fit several of the data points in the region around the peak quite as well. This could be because a single-compartment model is not sufficient to represent the data or it could represent an error in concentration measurements or both.

Chart 5.2 for **Example 5.2:** Nicotine chewing gum example to illustrate a single compartment model with first order absorption and elimination, illustrating a nonlinear regression

$i := 0 \ldots 18$ Number of data points

min $time_i :=$	nicotine concentration, ng/ml $C_i :=$
0	0
2	2.5
3	4
6	4
8	5
10	6
12	7
13	8.5
16	10
18	11
20	12
24	13
28	12.5
32	12
35	11
41	10
48	9
52	9
65	7.5

(data obtained from McNabb, Ebert, and McCusker 1982)

a plot of the experimental data

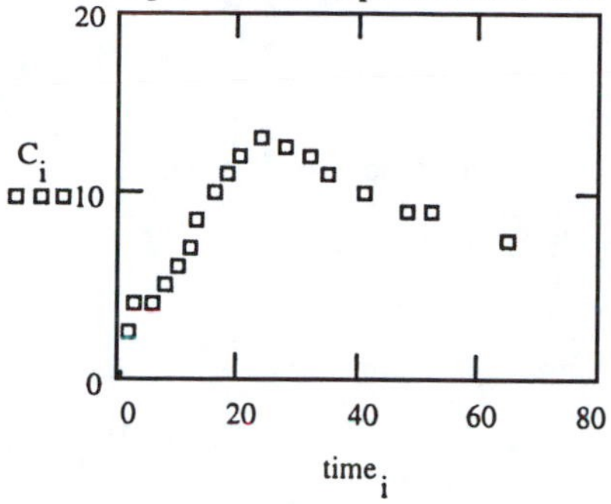

Now we perform a nonlinear regression on the 3 unknown parameters of Equation 5.29, i.e., $(f\,D/V_{\text{apparent}})$, k_a, k_{te}.

$$c(\text{time}, I_0, k_a, k_{te}) := I_0\left[\frac{k_a}{(k_a - k_{te})}\right](\exp(-k_{te}\,\text{time}) - \exp(-k_a\,\text{time}))$$

the regression equation, i.e., Equation 5.29,

$$\text{SSE}(I_0, k_a, k_{te}) := \sum_{i=0}^{18}(C_i - c(\text{time}_i, I_0, k_a, k_{te}))^2$$

Sum of the squares of the error between model and data that is to be minimized.

$I_0 := 20$
$k_{te} := .01$
$k_a := .1$

Initial guesses of the parameters
given $\text{SSE}(I_0, k_a, k_{te}) = 0$

$1 = 1 \quad 2 = 2$

dummy equations; Mathcad needs an equation for each unknown.
Here we have 3 unknowns.

$$\begin{bmatrix} I_0 \\ k_a \\ k_{te} \end{bmatrix} = \text{minerr}(I_0, k_a, k_{te})$$

$I_0 = 19.7$ ng/ml — the value of $f\,D/V_{app}$
$k_a = 0.056$ 1/min — the absorption rate constant
$k_{te} = 0.019$ 1/min — the total elimination rate constant

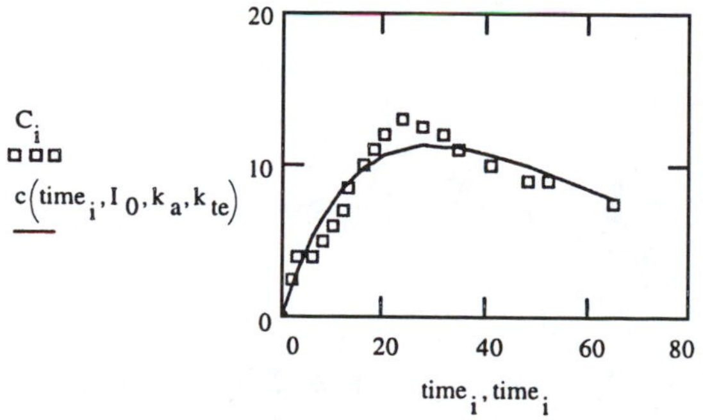

A reasonably good fit of the data is obtained.

$$\tau_{max} := \frac{1}{(k_a - k_{te})} \ln\left(\frac{k_a}{k_{te}}\right)$$

$\tau_{max} = 29.36$ min — time of peak concentration from Equation 5.30.

$$C_{max} := I_0 \left(\frac{k_a}{k_{te}}\right)^{\left[\frac{k_{te}}{(k_{te} - k_a)}\right]}$$

$C_{max} = 11.2658$ ng/ml — predicted max concentration from Equation 5.31.

5.10 TWO-COMPARTMENT MODELS

In many cases the single-compartment model is not sufficient to represent the time course of the plasma drug concentration. The single-compartment model assumes that the drug rapidly distributes throughout the homogeneous apparent distribution volume. In fact, the drug may distribute at different rates in different tissues of the body and it may require an extended period of time before the drug concentration equilibrates between the different tissue and fluid regions of the body. Clearly this is a complex process and one could envision building a model in which a central compartment representing the plasma is in direct communication with a multitude of other compartments representing other tissue and fluid regions of the body.

A reasonable simplification of this type of model is the two-compartment model shown in Figure 5.7. Drug is introduced into the central compartment that generally represents the plasma and other fluids in which the drug rapidly equilibrates. The drug then slowly distributes into the remaining tissue and fluid spaces represented by a single-tissue compartment. The constants k_{pt} and k_{tp} represent the rate constants describing

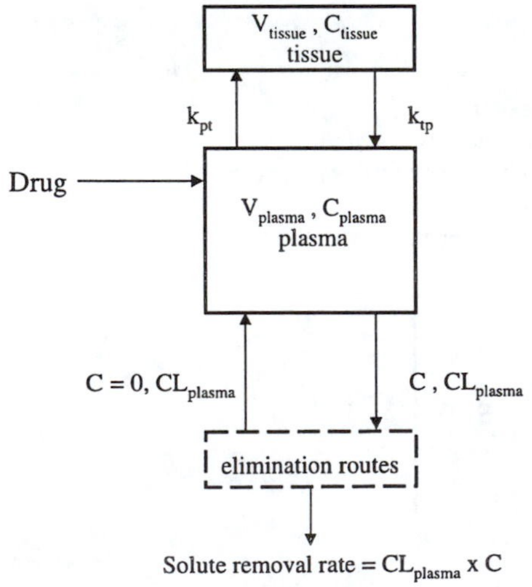

Figure 5.7 The two-compartment model.

the first-order transport between the plasma and tissue compartments. Since most or all of the drug is eliminated by the highly perfused kidneys and liver, the elimination processes (metabolic and renal) are assumed to be associated with the plasma compartment.

The two-compartment model has a characteristic biphasic temporal plasma concentration profile that distinguishes it from the single compartment model. During the early period after drug absorption, the drug is eliminated at a more rapid pace from the plasma volume due to the combined effects of the usual elimination processes as well as drug absorption by the tissues and fluids of the secondary compartment. After a period of time, the two compartments equilibrate and the drug elimination rate falls off.

5.10.1 A Two-Compartment Model for an Intravenous Injection

First let us develop a pharmacokinetic model for the case of a bolus intravenous injection. A separate unsteady mass-balance equation for the drug is then written for the plasma and the tissue compartments.

plasma compartment:

$$V_{\text{plasma}}\frac{\mathrm{d}C_{\text{plasma}}}{\mathrm{d}t} = k_{tp}V_{\text{tissue}}C_{\text{tissue}} - (k_{te} + k_{pt})V_{\text{plasma}}C_{\text{plasma}} \tag{5.36}$$

tissue compartment:

$$V_{\text{tissue}}\frac{\mathrm{d}C_{\text{tissue}}}{\mathrm{d}t} = k_{pt}V_{\text{plasma}}C_{\text{plasma}} - k_{tp}V_{\text{tissue}}C_{\text{tissue}} \tag{5.37}$$

V_{plasma} and V_{tissue} represent the apparent distribution volumes of the plasma and tissue compartments. C_{plasma} and C_{tissue} represent the concentrations within these respective compartments. These equations may be solved analytically with the following initial condition: $t = 0$, $C_{\text{plasma}} = D/V_{\text{plasma}}$, and $C_{\text{tissue}} = 0$. The equation for the concentration of drug within the plasma compartment is given by:

$$C_{\text{plasma}}(t) = \frac{D}{V_{\text{plasma}}(A - B)}[(k_{tp} - B)e^{-Bt} - (k_{tp} - A)e^{-At}] \tag{5.38}$$

and that for the concentration of drug in the tissue compartment is given by:

$$C_{\text{tissue}}(t) = \frac{Dk_{pt}}{V_{\text{tissue}}(A - B)}[e^{-Bt} - e^{-At}] \tag{5.39}$$

The constants A and B are given by the following equations in terms of the first-order rate constants.

$$\begin{aligned} A &= \frac{1}{2}\{(k_{pt} + k_{tp} + k_{te}) + [(k_{pt} + k_{tp} + k_{te})^2 - 4k_{tp}k_{te}]^{1/2}\} \\ B &= \frac{1}{2}\{(k_{pt} + k_{tp} + k_{te}) - [(k_{pt} + k_{tp} + k_{te})^2 - 4k_{tp}k_{te}]^{1/2}\} \end{aligned} \tag{5.40}$$

A nonlinear regression of plasma concentration data using Equation 5.38 can be performed to obtain the constants in this equation, i.e., V_{plasma}, A, B, and k_{tp}. The other first-order rate constants, k_{pt} and k_{te}, can then be found through the following process. We first multiply A and B in Equation 5.40 to obtain the following result.

$$k_{te} = \frac{AB}{k_{tp}} \tag{5.41}$$

Then A and B are added together to give:

$$k_{pt} = (A + B) - (k_{tp} + k_{te}) \tag{5.42}$$

Equations 5.41 and 5.42 may then be solved to obtain the values of k_{te} and k_{pt} from the known values of A, B, and k_{tp}.

The distribution volume of the tissue compartment can be found by assuming that once the two compartments are at equilibrium, then the derivative in Equation 5.37 is equal to zero and the concentrations in the two compartments are the same. Equation 5.37 then simplifies to give the following result for the volume of the tissue space.

$$V_{\text{tissue}} = V_{\text{plasma}}\left[\frac{k_{pt}}{k_{tp}}\right] \tag{5.43}$$

The overall apparent distribution volume would be equal to $V_{\text{plasma}} + V_{\text{tissue}}$ and is given by:

$$V_{\text{apparent}} = V_{\text{plasma}}\left[1 + \frac{k_{pt}}{k_{tp}}\right] \tag{5.44}$$

The data in Example 5.3 illustrate how the pharmacokinetic parameters of the two-compartment model for an intravenous dose can be determined. Other techniques based on graphical methods may also be used to determine pharmacokinetic parameters (Welling 1986; Notari 1987). Oftentimes the graphical approaches can provide reasonable starting values for the parameters in a nonlinear regression.

Example 5.3 The data shown in Mathcad® Chart 5.3 were given in Koushanpour (1976) and represents plasma inulin concentrations following a rapid intravenous injection of 4.5 grams of inulin in an 80-kg human. Note that the logarithmic graph shows that the elimination in this case is biphasic, indicating that a two-compartment model is needed to provide an adequate representation of the data. Determine the pharmacokinetic parameters of the two-compartment model.

SOLUTION Mathcad® Chart 5.3 illustrates the solution and we see that the two-compartment model fits this data extremely well. Interestingly, the single-compartment model for an IV injection given by Equation 5.15 is not capable of describing the early behavior up to about 60 minutes. Since for inulin, the only elimination pathway is through the kidneys, our earlier discussion showed that the renal clearance is the same as the GFR. Therefore, for this example, the renal clearance (CL_{renal}), or GFR, is equal to $V_{\text{plasma}} \times k_{te} = 121$ ml/min.

Chart 5.3 for **Example 5.3:** Two-compartment model for IV injection of inulin in an 80-kg human. Data from Koushanpour (1976).

$i = 0 \ldots 9$ data point index
$D = 4.5 \times 10^6$ micrograms, total dose

minutes $time_i :=$	micrograms/ml, plasma inulin concentration $C_i :=$
10	440
20	320
40	200
60	150
90	110
120	80
150	60
175	48
210	35
240	25

$y_i := \ln(C_i)$
Note the biphasic nature in the plot where the linearity of the data breaks down at small values of the time, indicating perhaps a two-compartment model is needed.

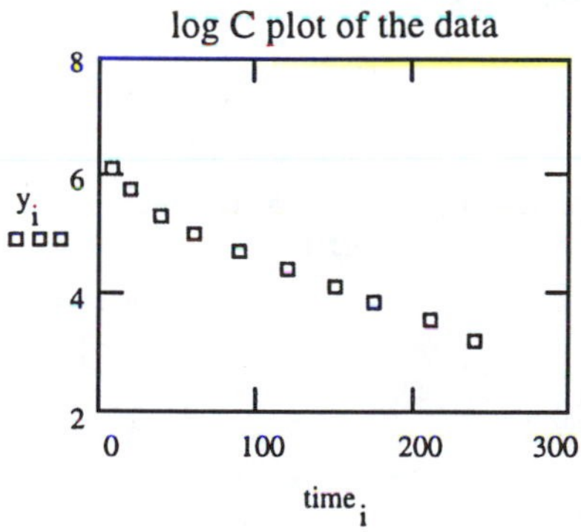

Now we perform a nonlinear regression on the 4 unknown parameters in Equation 5.38, the trailing 1/1000 in the equation below converts liters to ml.

$$c(time, V_p, A, B, k_{tp}) := \frac{D}{[V_p(A-B)]}[(k_{tp}-B)\exp(-B\,time) - (k_{tp}-A)\exp(-A\,time)]\frac{1}{1000}$$

$$SSE(V_p, A, B, k_{tp}) := \sum_{i=0}^{9}(C_i - c(time_i, V_p, A, B, k_{tp}))^2$$

sum of the squares of the error between model and data that is to be minimized

$TOL := .005$ tolerance between successive values of the unknowns
$V_p := 7.5$ liter, based on an initial concentration of about 600 μg/ml

These are the initial guesses for the other unknowns, which may require some fooling around.

with in order to get decent starting values that lead to a solution

$A := .04$ 1/min

$B := .003$ 1/min

$k_{tp} := .01$ 1/min

given $SSE(V_p, A, B, k_{tp}) = 0$

$1 = 1 \quad 2 = 2 \quad 3 = 3$ plus 3 dummy equations, need 4 equations with 4 unknowns

$$\begin{bmatrix} V_p \\ A \\ B \\ k_{tp} \end{bmatrix} := \text{minerr}(V_p, A, B, k_{tp})$$

$V_P = 7.811$ plasma distribution volume, liters

$A = 0.0362$ 1/min

$B = 0.0049$ 1/min

$k_{tp} = 0.0116$ 1/min

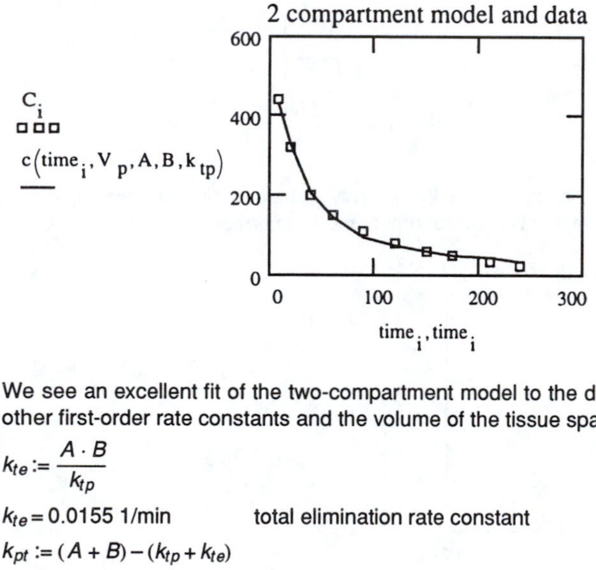

We see an excellent fit of the two-compartment model to the data. Now we can calculate the other first-order rate constants and the volume of the tissue space.

$$k_{te} := \frac{A \cdot B}{k_{tp}}$$

$k_{te} = 0.0155$ 1/min total elimination rate constant

$k_{pt} := (A + B) - (k_{tp} + k_{te})$

$k_{pt} = 0.0141$ 1/min

$$V_t := V_p \left(\frac{k_{pt}}{k_{tp}} \right)$$

$V_t = 9.502$ liters volume of the tissue compartment

$V_{total} := V_p + V_t$

$V_{total} = 17.313$ liters total distribution volume

5.10.2 A Two-Compartment Model for First-Order Absorption

Continuing our discussion of the two-compartment model, consider the case now of first-order absorption of the drug. Unsteady mass-balance equations may once again be written for each of the compartments. The equation for the tissue compartment would still be given by Equation 5.37 and the amount of drug yet to be absorbed (A_{body}) would still be given by Equations 5.27 and 5.28. The plasma compartment equation is then given by:

$$V_{\text{plasma}}\frac{dC_{\text{plasma}}}{dt} = k_a A_{\text{body}} + k_{tp} V_{\text{tissue}} C_{\text{tissue}} - (k_{te} + k_{tp}) V_{\text{plasma}} C_{\text{plasma}} \tag{5.45}$$

The initial conditions are that at time equal to zero there is no drug present in either compartment. These equations may then be integrated to obtain the following expressions for the concentration of drug within the plasma and tissue compartments. For the plasma concentration we have:

$$C_{\text{plasma}}(t) = \frac{fDk_a}{V_{\text{plasma}}}\left[\frac{(k_{tp}-A)}{(k_a-A)(B-A)}e^{-At} + \frac{(k_{tp}-B)}{(k_a-B)(A-B)}e^{-Bt} + \frac{(k_{tp}-k_a)}{(A-k_a)(B-k_a)}e^{-k_a t}\right] \tag{5.46}$$

and for the tissue concentration:

$$C_{\text{tissue}}(t) = \frac{fDk_a k_{tp}}{V_{\text{tissue}}}\left[\frac{e^{-At}}{(k_a-A)(B-A)} + \frac{e^{-Bt}}{(k_a-B)(A-B)} + \frac{e^{-k_a t}}{(A-k_a)(B-k_a)}\right] \tag{5.47}$$

The constants A and B have the same dependence on k_{pt}, k_{tp}, and k_{te} as the two-compartment IV case and are also given by Equations 5.40. Depending on the type of data one has, Equation 5.46 can be fitted to plasma concentration data to obtain the unknown parameters; fD, V_{plasma}, k_a, k_{tp}, A, and B. In some cases, it may not be possible to resolve the group fD/V_{plasma} without additional data, perhaps from an IV bolus injection and/or urinary excretion data. Once these parameters have been determined, Equations 5.41 through 5.44 may be used to find the remaining parameters. Other techniques based on graphical methods may also be used to determine pharmacokinetic parameters as discussed in the references (Cooney 1976; Gibaldi and Perrier 1982; Welling 1986; Notari 1987).

Example 5.4 reexamines the nicotine chewing gum data of Example 5.2 using the two-compartment model with first-order absorption. Example 5.5 uses the two-compartment model to examine plasma nicotine levels in a 49-year-old man after smoking one cigarette that yields 0.8 mg of nicotine.

Example 5.4 Reexamine the nicotine chewing gum data used in Example 5.2. Determine the pharmacokinetic parameters of the two-compartment model with first-order absorption.

SOLUTION Mathcad® Chart 5.4 illustrates the solution. We see that in this case, the two-compartment model offers little improvement to the fit of the data in comparison from the single-compartment model shown earlier. This illustrates that the single compartment is sufficient to represent the plamsa nicotine concentrations.

Chart 5.4 for **Example 5.4:** Reexamination of the nicotine chewing gum data using a two-compartment model with first-order absorption.

$i := 0 \ldots 18$ number of data points

minutes nicotine concentration, ng/ml

$\text{time}_i :=$ $C_i :=$

time	C
0	0
2	2.5
3	4
6	4
8	5
10	6
12	7
13	8.5
16	10
18	11
20	12
24	13
28	12.5
32	12
35	11
41	10
48	9
52	9
65	7.5

Now we perform a nonlinear regression on the 5 unknown parameters of Equation 5.46, i.e., $f\,D/V_p, k_a, k_{tp}, A, B$.

$$c(\text{time}, I_0, k_a, k_{tp}, A, B) := I_0 k_a \left[\frac{k_{tp} - A}{(k_a - A)(B - A)} \exp(-A\,\text{time}) + \frac{k_{tp} - B}{(k_a - B)(A - B)} \times \exp(-B\,\text{time}) + \frac{k_{tp} - k_a}{(A - k_a)(B - k_a)} \exp(-k_a\text{time}) \right]$$

$$\text{SSE}(I_0, k_a, k_{tp}, A, B) := \sum_{i=0}^{18} (C_i - c(\text{time}_i, I_0, k_a, k_{tp}, A, B))^2$$

Sum of the squares of the error between model and data that is to be minimized.

$I_0 := 20$
$k_a := .2$
$k_{tp} := .01$
$A := .04$
$B := .005$

Initial guess of the parameters, may need some fooling around to get reasonable starting values.
Given $\text{SSE}(I_0, k_a, k_{tp}, A, B) = 0$
$1 = 1 \quad 2 = 2 \quad 3 = 3 \quad 4 = 4$
As before we need some dummy equations

$$\begin{bmatrix} I_0 \\ k_a \\ k_{tp} \\ A \\ B \end{bmatrix} := \text{minerr}(I_0, k_a, k_{tp}, A, B)$$

$I_0 = 19.708$ ng/ml	the value of $f\,D/V_p$
$k_a = 0.0558$ 1/min	the absorption rate constant
$k_{tp} = 0.0118$ 1/min	transport constant between tissue and plasma compartments
$A = 0.0218$ 1/min	a mixed constant
$B = 0.0097$ 1/min	a mixed constant

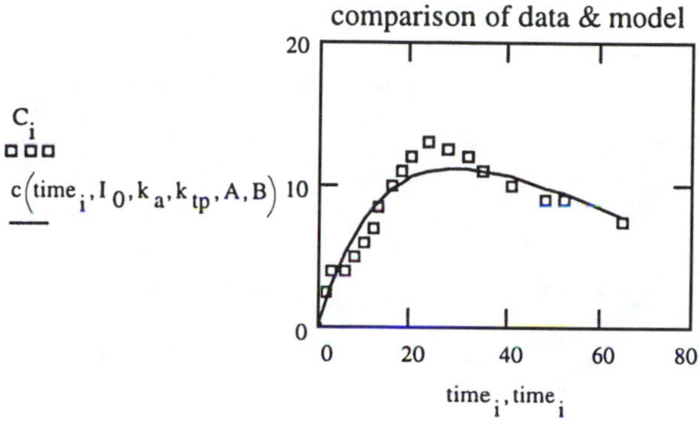

Interestingly, the two-compartment model does not fit the data much better than the single-compartment model.

Now we can evaluate the other parameters.

$$k_{te} := \frac{AB}{k_{tp}}$$

$k_{te} = 0.0178$ 1/min total elimination rate constant

$k_{pt} := (A + B) - (k_{tp} + k_{te})$

$k_{pt} = 0.0018$ 1/min transport coefficient for plasma to tissue compartment

Example 5.5 Mathcad® Chart 5.5 presents plasma nicotine levels in a 49-year-old man during and after smoking one cigarette that yields 0.8 mg of nicotine. The graph of the data in Chart 5.5 clearly shows the biphasic elimination behavior typical of a two-compartment model. Compared to the chewing gum example, there is a much faster absorption of the nicotine by smoking due to intimate contact with the blood in the lungs. Using the two-compartment model, determine the pharmacokinetic parameters and compare them to the chewing gum case of Example 5.4.

SOLUTION Mathcad® Chart 5.5 shows the solution. We see that the two-compartment model fits this data rather well.

Chart 5.5 for **Example 5.5:** Plasma nicotine levels after smoking 1 cigarette. Analysis using a two-compartment model with first-order absorption.

$i := 0 \ldots 17$ number of data points

minutes nicotine concentration, ng/ml

$time_i :=$	$C_i :=$
0	0
2	5
3	7
4	11
5	10
6	14
8	12
9	13
11	13
12	12
15	12
21	9
25	9
35	8
41	7.5
49	7
59	7
71	6

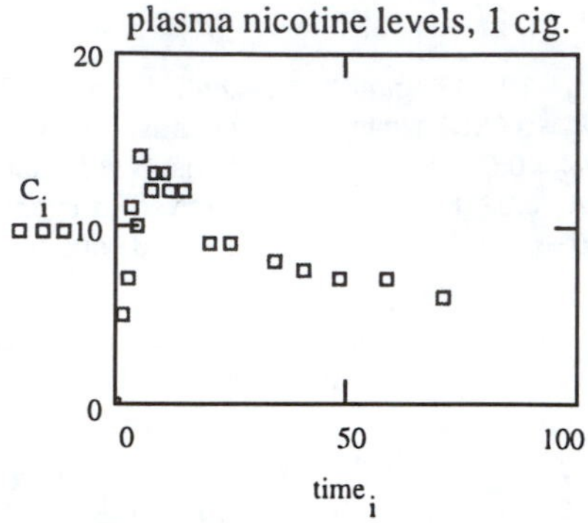

Now we perform a nonlinear regression on the 5 unknown parameters of Equation 5.46, i.e., $f\,D/V_p$, k_a, k_{tp}, A, B.

$$c(\text{time}, I_0, k_a, k_{tp}, A, B) := I_0 k_a \left[\frac{k_{tp} - A}{(k_a - A)(B - A)} \exp(-A\,\text{time}) + \frac{k_{tp} - B}{(k_a - B)(A - B)} \times \exp(-B\,\text{time}) \cdots + \frac{k_{tp} - k_a}{(A - k_a)(B - k_a)} \exp(-k_a\,\text{time}) \right]$$

$$\text{SSE}(I_0, k_a, k_{tp}, A, B) := \sum_{i=0}^{17} (C_i - c(\text{time}_i, I_0, k_a, k_{tp}, A, B))^2$$

Sum of the squares of the error between model and data that is to be minimized,

TOL $:= .005$
$I_0 := 20$
$k_a := .4$
$k_{tp} := .01$
$A := .02$
$B := .01$

Initial guesses of the parameters, may need some fooling around to get reasonable starting values.

Given $\text{SSE}(I_0, k_a, k_{tp}, A, B) = 0$.
$1 = 1 \quad 2 = 2 \quad 3 = 3 \quad 4 = 4$

As before, we need some dummy equations

$$\begin{bmatrix} I_0 \\ k_a \\ k_{tp} \\ A \\ B \end{bmatrix} := \text{minerr}(I_0, k_a, k_{tp}, A, B)$$

$I_0 = 15.3517$ ng/ml the value of $\phi D/V_p$
$k_a = 0.2851$ 1/min the absorption rate constant

$k_{tp} = 0.0125$ 1/min transport constant between tissue and plasma compartments
$A = 0.0349$ 1/min a mixed constant
$B = 0.0013$ 1/min a mixed constant

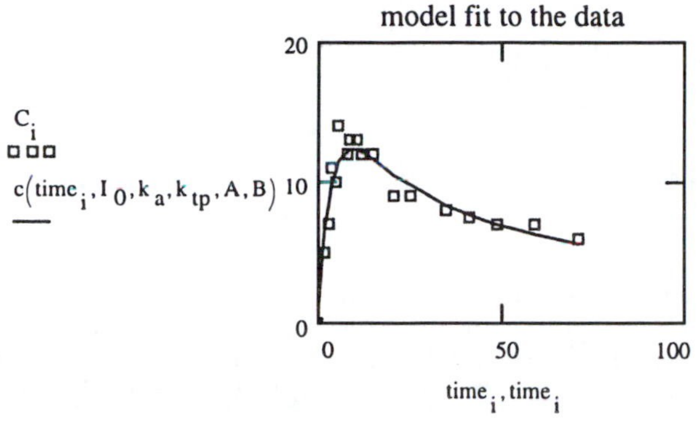

Now we can calculate the other parameters.

$$k_{te} := \frac{A \cdot B}{k_{tp}}$$

$k_{te} = 0.0037$ 1/min total elimination constant

$k_{pt} := (A + B) - (k_{tp} + k_{te})$

$k_{pt} = 0.02$ 1/min plasma to tissue

The pharmacokinetic parameters for chewing gum and one cigarette cases are compared below. One important caveat to consider in comparing the parameters for nicotine delivery by chewing gum or cigarettes in these examples is that the data were obtained from different subjects.

	Chewing gum	Cigarette
Total nicotine available	4 mg	0.8 mg
$fD/V_{apparent}$	19.73 ng/ml	15.35 ng/ml
k_a	0.056 L/min	0.285 L/min
k_{tp}	0.012 L/min	0.013 L/min
k_{pt}	0.0012 L/min	0.02 L/min
k_{te}	0.018 L/min	0.0037 L/min

First we note that the chewing gum offers nearly five times the potential dose of nicotine than from a single cigarette. Clearly, since the $fD/V_{apparent}$ (somewhat like the initial concentration in the plasma) are comparable, we conclude that only about one-fifth of the nicotine in the chewing gum is actually absorbed. However, the chewing gum at this nicotine loading provides a comparable peak plasma nicotine level as a cigarette. We next notice that the absorption rate constant for the cigarette is about five times that of chewing gum, showing that the lungs are very efficient at absorbing nicotine into the blood. Also note that delivery of nicotine via a cigarette has a marked effect on the rate at which the absorbed nicotine is transported to the tissue compartment. We see a pronounced increase with smoking in the transport rate constant (k_{pt}) of nicotine from the plasma compartment to the tissue compartment. On the other hand, the transport rate constant (k_{tp}) from the tissue compartment to the plasma compartment is about the same for the two cases. Finally, we see that the value of the total elimination rate constant (k_{te})

for the cigarette case is much smaller than that for the chewing gum case. It is difficult to discern without additional data whether this difference is due to the nicotine dosage form or due to individual differences.

PROBLEMS

1. Derive Equation 5.24.

2. Derive Equation 5.29.

3. Derive Equations 5.30 and 5.31.

4. Derive Equation 5.33.

5. The rate constants in Equation 5.29 (k_a and k_{te}) can also be found using a graphical technique called "feathering" (Notari 1987). Note that the plasma concentration in Equation 5.29 is related to the sum of two exponential terms. In most cases, k_a is larger than k_{te}, hence the second exponential in Equation 5.29 vanishes for long periods of time. Hence, for long time periods Equation 5.29 may be approximated as:

$$C(t)_{TL} \approx \frac{fD}{V_{\text{apparent}}}\left(\frac{k_a}{k_a - k_{te}}\right)e^{-k_{te}t}$$

On a semi-logarithmic plot ($\log_{10}$(C) versus t), this equation predicts that for long periods of time, the concentration data should form a straight line. This region is referred to as the terminal elimination phase. The slope of the terminal line (TL) that fits the data during this period of time would then be equal to $-k_{te}/2.303$ and the intercept would be equal to $\log_{10}[\frac{fD}{V_{\text{apparent}}}(\frac{k_a}{k_a-k_{te}})]$. We can now use the equation for $C(t)_{TL}$ to form the difference or residual, $C(t)_{TL} - C(t)$, with the following result.

$$C(t)_{TL} - C(t) = \frac{fD}{V_{\text{apparent}}}\left(\frac{k_a}{k_a - k_{te}}\right)e^{-k_a t}$$

This equation is valid for all values of time starting from time equal to zero. Note for short periods of time, $C(t)_{\text{TL}}$ is obtained either by extrapolation of the terminal line or estimating $C(t)_{\text{TL}}$ by the first equation above. For long periods of time, this difference is equal to zero. Now if a plot is made of the $\log_{10}[C(t)_{\text{TL}} - C(t)]$ versus time, the slope would equal $-k_a/2.303$ and the intercept would once again be equal to $\log_{10}[\frac{fD}{V_{\text{apparent}}}(\frac{k_a}{k_a-k_{te}})]$. With k_a and k_{te} now known, $\frac{fD}{V_{\text{apparent}}}$, may be obtained from the value of the intercept.

Use this feathering technique to obtain the parameters for the data given in Example 5.2. How do these parameters obtained using the feathering technique compare with those obtained from a nonlinear regression? Under what conditions would you expect the feathering technique to be difficult to use?

6. Rework Example 5.3 assuming a single-compartment model.

7. You are taking a final exam, it is unlike any exam you have taken before, plus you had a short night of sleep, you had car problems yesterday, you had too much to drink last night, and the phone kept ringing early this morning. The result—you have a terrible headache, unlike any headache you have ever had before. You reach for your bottle of a new headache medicine that claims maximum relief within 30 minutes and contains a well-known pain reliever. An amazing coincidence is that the exam you are taking includes a problem related to the pharmacokinetics of this pain reliever. You are given the following information:

Elimination rate constant = 0.277/hr
Apparent distribution volume = 35 liters
Fraction absorbed = 80%
Therapeutic range = 10–20μg/ml

You plan to take this pain reliever. What dose in milligrams should you take? At what time should you think about taking another dose? How much should you take for the second dose?

8. The pharmacokinetics of nicotinamide has recently been reported in humans (Petley et al. 1995). Nicotinamide, a derivative of the *B* vitamin niacin, is under investigation as a prevention for insulin-dependent diabetes mellitus. The drug was given orally to 8 male subjects with a mean body weight of 75 kg. Two formulations were used, a standard powdered form and a sustained-release form. The data below summarize the data obtained for the low-dose studies (standard = 2.5 mg/kg BW and sustained = 6.7 mg/kg BW; BW = body weight). Analyze this data using appropriate pharmacokinetic models. Explain any differences in the pharmacokinetics of the two formulations used.

Time, hours	Nicotinamide plasma	Concentration, μg/ml
	Standard dose	*Sustained release dose*
0	0	0
0.3	3.3	—
0.5	2.45	1.75
0.75	1.90	—
1.0	1.45	1.5
1.25	1.00	—
1.5	0.80	1.1
1.75	0.55	—
2.0	0.50	1.05
2.5	0.30	1.1
3.0	0.20	0.67
3.5	—	0.50
4.0	—	0.45
4.5		0.50
5.0		0.55
5.5		0.15

9. The following serum insulin levels were obtained following the intranasal administration of 150 U of insulin to 12 healthy men (Jacobs et al. 1993). Analyze this data using an appropriate pharmacokinetic model.

Time, minutes	Serum insulin concentration, μU/ml
0	10
15	150
20	220
25	230
30	260
35	250
40	210
45	170
60	140
75	105
90	80
105	50
120	45
150	35
180	35
240	35
300	35

10. The following plasma concentrations for the antibiotic imipenem were obtained from six patients suffering multiorgan failure (Hashimoto et al. 1997). All patients also were anuric because of complete renal failure and were placed on continuous venovenous hemodialysis. The antibiotic was delivered by an intravenous infusion pump for 30 minutes, resulting in an initial plasma concentration of 32.37 μg/ml. Following the infusion, blood samples were taken over the next 12 hours. The initial dose was 500 mg and 107.7 mg of imipenem was recovered in the dialysate fluid ("urine"). The table below summarizes the plasma concentrations.

Time, minutes	Plasma concentration, μg/ml
0	32.47
30	24
60	18.5
120	13
180	10
360	4.6
540	2
720	1.12

Use an appropriate compartment model to analyze the above data.

CHAPTER

SIX

EXTRACORPOREAL DEVICES

In this chapter we will discuss and analyze several examples of extracorporeal devices. These devices lie outside (*extra*) the body (*corporeal*) and are usually connected to the patient by an *arteriovenous shunt*.[1] In some respects they may be thought of as artificial organs. Their function is based on the use of physical and chemical processes to replace the function of a failed organ or to remove an unwanted constituent from the blood. The patient's blood before entering these devices is infused with *heparin* to prevent clotting. A variety of ancillary equipment may also be present to complete the system. This may include items such as pumps, flow monitors, bubble and blood detectors, as well as control systems for pressure, temperature, and concentration. It is important to note that these devices do not generally contain any living cells. Devices containing living cells are called bioartificial organs and these will be discussed in Chapter 8.

6.1 APPLICATIONS

A variety of extracorporeal devices have been developed. Perhaps the best known are *blood oxygenators* that are used in such procedures as open heart surgery and *hemodialysis* to replace the function of failed kidneys. Other examples include *hemoperfusion* wherein a bed of activated carbon particles are used for cleansing blood of toxic materials, *plasmapheresis*[2] is used to separate erythrocytes from plasma as a first step in the subsequent processing of the plasma, *immobilized enzyme reactors* to rid the body of a particular substance or to replace lost liver function, and *affinity columns* used to remove

[1] A shunt is a means of diverting flow, in this case blood, from an artery through a device and then back into the body via a vein.

[2] See Problem 11 in Chapter 2 for a discussion and analysis of a membrane plasmapheresis system.

materials such as antibodies that have been implicated as the cause of many *autoimmune diseases*.

The primary functional unit of extracorporeal devices are typically provided as a sterile disposable cartridge. However, in some cases, considering mounting healthcare costs, the functional unit can be reused many times, provided it is cleaned and sterilized between applications. Certainly there is no end to the possibilities for extracorporeal devices. However, whatever the function is of the device, there will be some commonality amongst them in terms of their construction, use of membranes, and fluid-contacting schemes. In the following discussion, we will examine in greater detail three representative devices: hemodialysis, blood pump-oxygenators, and immobilized enzyme reactors.

6.2 CONTACTING SCHEMES

Devices such as blood oxygenators and dialyzers are typically based on the use of polymeric membranes to create the surface area needed to provide mass transfer between the blood stream and another exchange fluid stream. The membrane is physically retained within the device by a support structure that also creates the flow channels. As illustrated in Figure 6.1 for hemodialyzers, the membranes are typically arranged as stacks of flat sheets, coils of membrane sheets, or bundles of hollow-fiber membranes. Hollow-fiber membranes are finding increased use because of their low cost, ease of manufacture, and consistent quality.

A variety of contacting patterns between the blood and the exchange fluid are also possible as shown in Figure 6.2. Immobilized enzyme reactors and affinity columns have the active material, i.e., an *enzyme* or other *ligand*, firmly attached to a support particle. Support particles are typically made from a wide variety of polymeric materials such as alginates, agar, carrageenin, and polyacrylamide, or inorganic materials such as glass, silicas, aluminas, and activated carbon. The particles containing the immobilized material may be arranged as a packed bed or the particles may be suspended in a well-mixed device as shown in Figure 6.3. Complete mixing of the fluid phase and suspended particles can be achieved either by a mechanical impellar or by fluid recirculation.

6.3 MEMBRANE SOLUTE TRANSPORT

In many extracorporeal devices, a solute must be transported through the blood, perhaps across a membrane, and then through an exchange fluid. These transport processes are illustrated in Figure 6.4 for the case of simple diffusion. Concentrations of the solute at the fluid-membrane interface are assumed to be at equilibrium. The partition coefficient K, discussed earlier in Chapter 2, is used to describe the solute equilibrium ($C_{mb} = KC_b^s$ and $C_{me} = KC_e^s$) at the surface of the membrane.

An overall mass-transfer coefficient (K_0) for the situation shown in Figure 6.4 may be written in terms of the individual film mass-transfer coefficients on the blood (k_b) and exchange fluid (k_e) sides along with the membrane permeability (P_m) and the partition coefficient.

$$\frac{1}{K_0} = \frac{1}{k_b} + \frac{1}{KP_m} + \frac{1}{k_e} \tag{6.1}$$

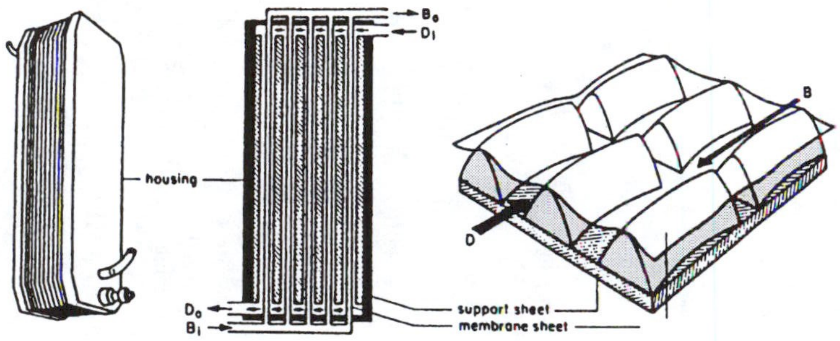

Flat-plate dialyzer.

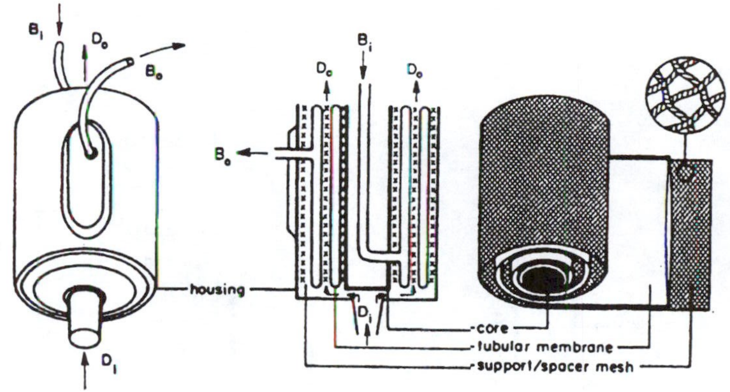

Coil dialyzer.

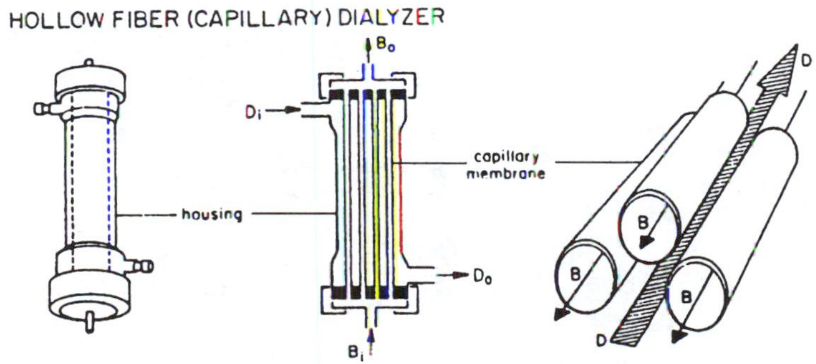

Hollow-fiber (capillary) dialyzer.

Figure 6.1 Membrane configurations. (From Shalak R. and S. Chien, 1987, *Handbook of bioengineering*, New York: McGraw-Hill. With permission.)

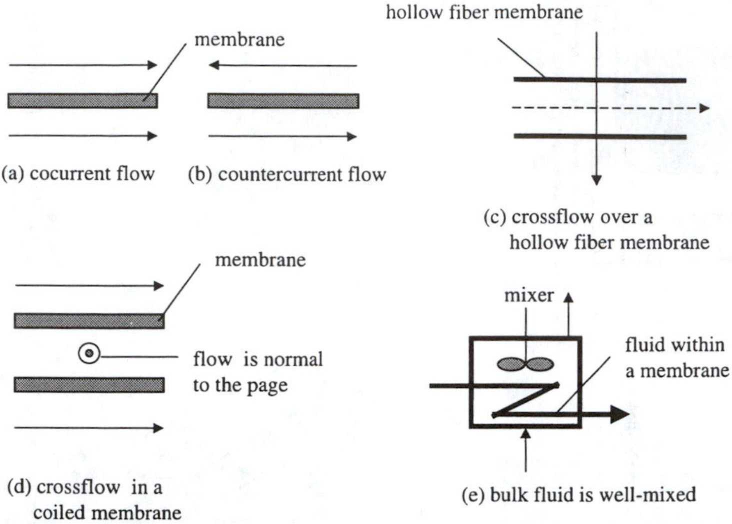

Figure 6.2 Fluid contacting patterns.

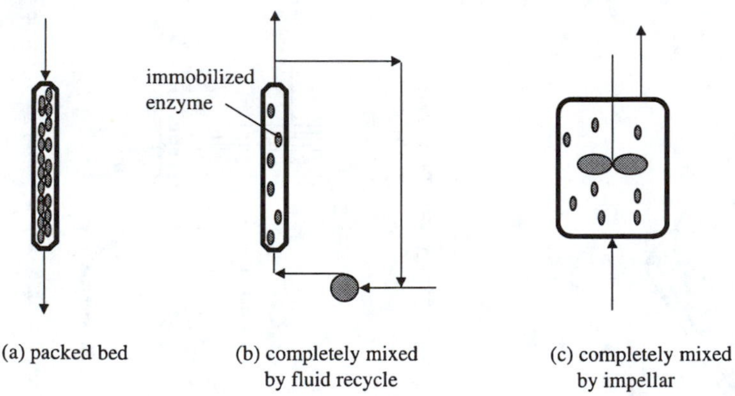

Figure 6.3 Immobilized enzyme reactor systems.

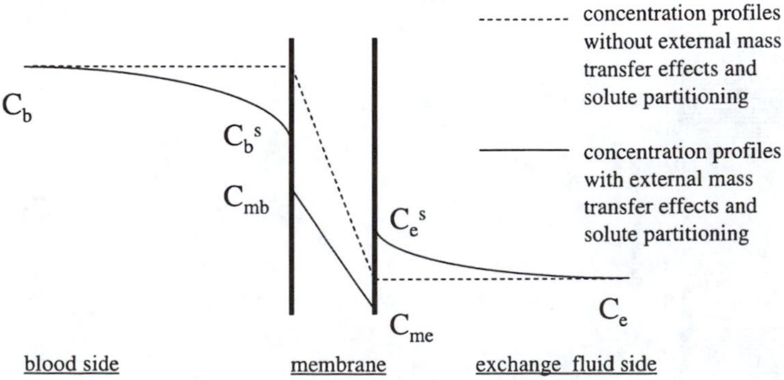

Figure 6.4 Concentrations profiles for membrane diffusion.

This equation states that the total mass-transfer resistance ($\frac{1}{K_0}$) is simply the sum of the individual mass-transfer resistances ($\frac{1}{k_b} + \frac{1}{KP_m} + \frac{1}{k_e}$). The smallest value of k_b, KP_m, or k_e for a given solute is said to be the *controlling resistance*.

Recall that we discussed in much detail the membrane permeability in Chapter 2. The membrane permeability, although best measured experimentally (Dionne et al. 1996; Baker et al. 1997), is defined by Equation 2.19. The film mass-transfer coefficients depend on the physical properties of the fluid and the nature of the flow. They also depend on position due to boundary layer growth, so we usually use length-averaged values of these coefficients in Equation 6.1.

6.4 ESTIMATING THE MASS-TRANSFER COEFFICIENTS

Techniques for estimating film mass-transfer coefficients for a variety of flow conditions may be found in Cussler (1984). For laminar flow of fluids in a tube, one may use Equation 2.4 or 2.5 to estimate the film mass-transfer coefficient. Blood flow through extracorporeal devices is typically laminar and in some cases may be fully developed. The Sherwood number (Sh) for blood flow that is fully developed ($k_b h / D_{\text{effective}}$) is equal to 3.66. In the Sherwood number, h is the flow channel thickness in a flat-plate membrane arrangement or the tube diameter in a hollow fiber. The effective diffusivity of the solute in blood is given by $D_{\text{effective}}$ and may be quite different from the solute's bulk diffusivity because of the presence of the red blood cells. This is discussed in the next section. Recall from Chapter 2 that the asymptotic limit on the Sh number for fully developed laminar flow is valid when $\frac{z}{h} > 0.05 Re Sc$ (concentration field) or $\frac{z}{h} > 0.05 Re$ (velocity field). Here, z is defined as the length of the flow path, h is the characteristic dimension of the flow channel, Re is the Reynolds number, and Sc the Schmidt number. For solutes in liquids, the Sc is on the order of 1000, so the velocity field will develop much faster than the concentration field.

The exchange fluid can be either in laminar or turbulent flow. Generally one would operate on the exchange fluid side in such a manner that its contribution to the overall mass-transfer resistance would be negligible. Therefore, $k_e \gg k_b$ and KP_m. If estimates of k_e are needed, then one could use Equation 2.4 or 2.5, if the exchange fluid flow is laminar. For turbulent flow of the exchange fluid, one can consult Cussler (1984) or use heat-transfer correlations (Bird, Stewart, Lightfoot 1960; Thomas, 1992) for the analogous mass-transfer problem. This is accomplished by replacing in the heat-transfer correlation the Nusselt number (Nu) with the Sherwood number (Sh), and the Prandtl number (Pr) with the Schmidt number ($Sc = \frac{\mu}{\rho D}$).

6.5 ESTIMATING THE SOLUTE DIFFUSIVITY IN BLOOD

Transport of a solute through blood can be complicated by the presence of the red blood cells. Estimation of the effective solute diffusivity ($D_{\text{effective}}$) needs to account for the effects of the red blood cells on the solute's bulk diffusivity. We need an estimate of solute diffusivity in order to calculate the blood side value of the Sherwood number as discussed in the previous section. Those factors influencing mass transfer within blood are summarized in Table 6.1 (Colton and Lowrie 1981).

Table 6.1 Factors that affect mass transport in blood

Suspension properties
Volume fraction of red blood cells
Volume fraction of proteins
Red blood cell solute permeability
Solute behavior
Protein binding kinetics and equilibrium
Reactions with other solutes
Flow-dependent behavior
Rheological properties
Red blood cell dynamics
Rouleaux formation
Migration from the wall
Rotation and translation

The rheological behavior of blood is not a very important factor in describing solute transport, since at the relatively high shear rates encountered in extracorporeal devices, blood behaves as a Newtonian fluid. Mass-transport theories for Newtonian fluids are well developed and we can make use of these for the special case of blood. The rotation and translation of red blood cells can influence mass transport to some degree, however these effects can usually be neglected unless sufficient data are available to warrant their inclusion.

A complete description of solute transport in blood requires that one knows the solute diffusivity and its red blood cell permeability, its equilibrium distribution between the plasma and the fluid within the red blood cell, and the kinetics of any chemical reaction it may undergo with other solutes, typically proteins, that may be present. Clearly this would be a complex problem. However, for most solutes we can treat the blood as a nonreactive fluid and assume either of the following two cases applies.

First, if the red blood cell permeability for the solute of interest is extremely large, then the solute will diffuse through the blood as if it were a pseudo-homogeneous fluid and the solute diffusivity will not be affected by the presence of the red blood cells. For this case, one only needs to know the solute diffusivity in plasma which can be obtained by estimating the aqueous diffusivity by use of Figure 2.1, followed by a correction for the difference in viscosity between an aqueous solution and plasma using Equation 2.9. Solute diffusion in plasma amounts to about a 20% reduction in comparison to the aqueous solute diffusivity.

For the second case, the red blood cell may be somewhat permeable to the solute. The solute must therefore diffuse around or through the red blood cells, increasing its diffusion path and decreasing its effective diffusivity. One could then use relationships such as those presented in Chapter 2, i.e., Equations 2.13 and 2.14. As an example,

Colton and Lowrie (1981) investigated the diffusion of urea in both stagnant and flowing blood. They presented data for the ratio of the solute permeability in blood to that in plasma alone as a function of the hematocrit. These results were then used to develop an expression for the effective solute diffusivity in nonreactive blood. The effective diffusion coefficient through blood was defined by the following equation which is similar to Fick's law (Equation 2.7).

$$N_{\text{effective}} = D_{\text{effective}} \frac{dC_{\text{effective}}}{dx} \tag{6.2}$$

The effective concentration ($C_{\text{effective}}$) represents the volume fraction–weighted sum of the solute concentration in the red blood cell phase (C_{RBC}) and the continuous plasma phase (C_{plasma}) and is given by the next equation.

$$C_{\text{effective}} = HC_{\text{RBC}} + (1 - H)C_{\text{plasma}} \tag{6.3}$$

H represents the blood hematocrit. Diffusion of solute through just the continuous or plasma phase ($H = 0$) would be given by:

$$N = D_{\text{plasma}} \frac{dC_{\text{plasma}}}{dx} \tag{6.4}$$

Substituting Equation 6.3 into Equation 6.2 for $C_{\text{effective}}$, and dividing the result by Equation 6.4, provides a relationship for the effective solute diffusivity of blood in comparison to its value in plasma alone.

$$\frac{N_{\text{effective}}}{N} = \frac{D_{\text{effective}}}{D_{\text{plasma}}}[HK_{\text{RBC}} + (1 - H)] \tag{6.5}$$

In this equation, $K_{\text{RBC}} = \frac{C_{\text{RBC}}}{C_{\text{plasma}}}$, represents the equilibrium partition coefficient for the solute between the red blood cells and the plasma. Average values of K_{RBC} for some typical solutes are summarized in Table 6.2.

Colton and Lowrie (1981) have shown that Equation 6.5 is equal to the following expression (actually Equation 2.13 with $\phi = H$) assuming the red blood cell is impermeable ($D_{\text{cell}} = 0$).

$$\frac{D_{\text{effective}}}{D_{\text{plasma}}}[HK_{\text{RBC}} + (1 - H)] = \frac{2(1 - H)}{2 + H} \tag{6.6}$$

Equation 6.6 can now be used to estimate the effective diffusivity of a solute in blood. This equation suggests that for a solute such as urea, its effective diffusivity through blood is about 50% of its value in plasma or about 40% of its aqueous diffusivity.

Table 6.2 Values of K_{RBC} for some typical solutes

Solute	K_{RBC}
Urea	0.86
Creatinine	0.73
Uric acid	0.54
Glucose	0.95

Data from Colton and Lowrie (1981).

With this background on extracorporeal devices, the following discussion will focus on a description and analysis of three representative devices: hemodialysis, blood oxygenators, and enzyme reactors.

6.6 HEMODIALYSIS

6.6.1 Background

The basic functional unit of the kidney is the nephron (shown earlier in Figure 5.3). Each kidney contains about 1 million nephrons. Only about one-third of these nephrons are needed to maintain normal levels of waste products in the blood. If about 90% of the nephrons lose their function, then the symptoms of *uremia* will develop in the patient. Uremia results when waste products normally removed from the blood by the kidneys start to accumulate in the blood and other fluid spaces. For example, water generated by normal metabolic processes is no longer removed and accumulates in the body, leading to *edema*, and in the absence of additional electrolytes, almost half of this water enters the cells rather than the extracellular fluid spaces due to osmotic effects. The normal metabolic processes also produce more acid than base, and this acid is normally removed by the kidneys. Therefore, in kidney failure, there is a decrease in the pH of the body's fluids, called *acidosis*, which can result in *uremic coma*. The end-products of protein metabolism include such nitrogenous substances as urea, uric acid, and creatinine. These materials must be removed to ensure continued protein metabolism in the body. The accumulation of urea and creatinine in the blood, although not life-threatening by themselves, is an important marker of the degree of renal failure. They are also used to measure the effectiveness of dialysis for the treatment of kidney failure. The kidneys also produce the hormone *erythropoietin* that is responsible for regulating the production of red blood cells in the bone marrow. In kidney failure, this hormone is diminished, leading to a condition called *anemia* and a lowered hematocrit. If kidney failure is left untreated, death can occur within a few days to several weeks.

Hemodialysis (HD) has been used for nearly 50 years (Kolff 1947) to treat patients with degenerative kidney failure or end-stage renal disease (ESRD). Hemodialysis has the ability to replace many functions of the failed kidneys. For example, hemodialysis removes the toxic waste products from the body, maintains the correct balance of electrolytes, and removes excess fluid from the body. Hemodialysis can keep patients alive for several years and for many patients, allows them to survive long enough to receive a kidney transplant. Nearly 500,000 patients are currently being kept alive by hemodialysis.

Figure 6.5 illustrates the basic operation of hemodialysis. Heparinized blood flows through a device containing a membrane. The dialysate or exchange fluid flows on the membrane side opposite the blood. Solutes are exchanged by diffusion between the blood and the dialysate fluid. A variety of membrane configurations are possible and some of the typical ones were shown earlier in Figure 6.1. A variety of flow patterns for the blood and dialysate have been used and these are summarized in Figure 6.2. The membranes are usually made from such materials as cellulose, cellulose acetate, polyacrylonitrile,

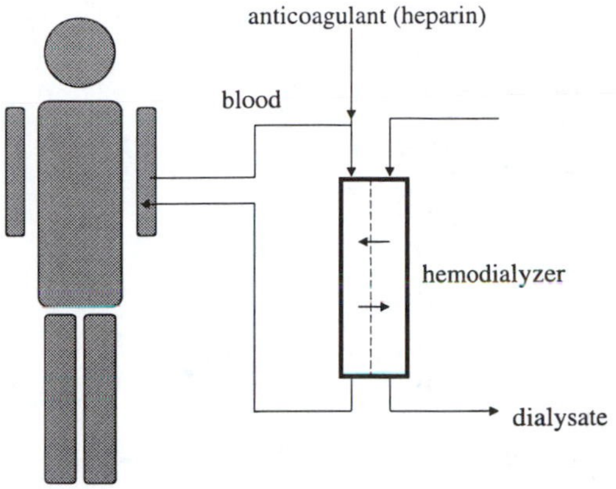

Figure 6.5 Hemodialysis.

and polycarbonate (Zelman 1987; Galletti, Colton, and Lysaght 1995). The membrane surface area is on the order of 1 m^2. Blood flow rates are in the range of several hundred ml/min and the dialysate flow rate is about twice that of the blood.

6.6.2 Dialysate Composition

Table 6.3 compares the species present in normal and uremic plasma with those of a typical dialysate fluid. The composition of the dialysate is based on the need to restore the uremic plasma to the normal state. Note in particular the high levels of urea and creatinine in the uremic plasma. For the most part, movement of the species listed in the table between the uremic plasma and the dialysate is by diffusion. The higher level of HCO_3^- in the dialysate is used to decrease the acidity of the plasma through its buffering action. Because of the reduced concentrations, the osmolarity of the dialysate is about 265 milliosmolar, in comparison to about 300 milliosmolar or so for the plasma. Hence water will tend to leave the dialysate and enter the plasma by osmosis. Since the patient needs to have excess water removed during dialysis, it is then necessary to maintain the dialysate pressure below atmospheric pressure in order to develop a transmembrane pressure gradient that is sufficient to overcome osmosis and remove water from the patient. Removal of waste products from the plasma must also be carefully controlled in order to avoid the *disequilibrium syndrome*. If waste-products are removed from the blood too fast by dialysis, then the osmolarity of the plasma becomes less than that of the cerebrospinal fluid, resulting in a flow of water from the plasma and into the spaces occupied by the brain and spinal cord. This increases the local pressure in these areas and can lead to serious side effects.

6.6.3 Role of Ultrafiltration

The rate at which water is removed (ultrafiltration) from the plasma by the pressure gradient across the dialysis membrane can be estimated using Equation 2.31. These equations can be simplified by assuming only the plasma proteins are impermeable.

Table 6.3 Composition of normal and uremic plasma and dialysate

Species	Normal plasma	Uremic plasma	Dialysate
Electrolytes (mEq[3]/liter)			
Na^{+}	142	142	133
K^{+}	5	7	1
Ca^{++}	3	2	3
Mg^{++}	1.5	1.5	1.5
Cl^{-}	107	107	105
HCO_3^{-}	27	14	35.7
$Lactate^{-}$	1.2	1.2	1.2
HPO_4^{--}	3	9	0
$Urate^{-}$	0.3	2	0
SO_4^{--}	0.5	3	0
Nonelectrolytes (mg/dl)			
Glucose	100	100	125
Urea	26	200	0
Creatinine	1	6	0

Data from Guyton, A. C. 1991. *Textbook of Medical Physiology*, 8th ed., 348. Philadelphia, W. B. Saunders.

Hence the following equation is obtained.

$$J = SL_P(\Delta P_{\text{mean}} - \pi_{\text{oncotic plasma}}) \tag{6.7}$$

$\pi_{\text{oncotic plasma}}$ represents the oncotic pressure of the plasma proteins, about 28 mm Hg, and ΔP_{mean} is the average transmembrane pressure difference which is given by the next equation.

$$\Delta P_{\text{mean}} = \left[\frac{P_{b,\text{in}} + P_{b,\text{out}}}{2} - \frac{P_{d,\text{in}} + P_{d,\text{out}}}{2}\right] \tag{6.8}$$

Note that to avoid problems with the calculations, it is best to simply use absolute pressures in Equation 6.8, recognizing that since the dialysate is subatmospheric, its reported gauge pressure will be negative. The pressure drop on the blood side is typically on the order of 20 mm Hg and that on the dialysate side is about 50 mm Hg. The mean transmembrane pressure drop is on the order of a few hundred mm Hg. The hydraulic conductance, L_P, is typically about 3 ml/hr/m^2/mm Hg. High flux membranes can have values of the hydraulic conductance as high as 20 ml/hr/m^2/mm Hg; however, protein deposition on the plasma side of the membrane can reduce this in a linear manner by about 6%/hr (Zelman 1987). Example 6.1 illustrates the calculation of the ultrafiltration rate in a typical membrane dialyzer.

[3]Equivalents are the amounts of substances that have the same combining capacity in chemical reactions.

Example 6.1 Using the membrane properties listed above, calculate the ultrafiltration rate of water, assuming a membrane surface area of 1 m^2. How much water would be removed after 6 hours of dialysis? Assume blood enters the device at 120 mm Hg (gauge) and leaves at 100 mm Hg (gauge). The dialysate fluid enters at −150 mm Hg (gauge) and leaves at −200 mm Hg (gauge).

SOLUTION

$$J = 1\,\text{m}^2 \times \frac{3\text{ ml}}{\text{hr m}^2\text{ mm Hg}} \times \left[\frac{(120+760)+(100+760)}{2} - \frac{(760-150)+(760-200)}{2} - 28\right]\text{mm Hg}$$
$$= 771\text{ ml/hr (13 ml/min)}$$

After 6 hours of dialysis, about 10 pounds of water will have been removed.

6.6.4 Clearance and Dialysance

Solute transfer in the dialyzer can be analyzed with the help of the model shown in Figure 6.6, illustrating cocurrent flow of the blood and dialysate. The Q's represent the volumetric flow rates (ml/min) of the blood (b) and the dialysate (d), and the C's represent the species concentration (usually in mg/dl). The mass-transfer rate of species i across the dialysis membrane would be given by the following equation:

$$M_i = Q_b^{\text{in}} C_{b_i}^{\text{in}} - Q_b^{\text{out}} C_{b_i}^{\text{out}} = Q_d^{\text{out}} C_{d_i}^{\text{out}} - Q_d^{\text{in}} C_{d_i}^{\text{in}} \tag{6.9}$$

The *clearance of the dialyzer* for a particular solute is defined as the volumetric flow rate of blood (ml/min) entering the dialyzer that is completely cleared of the solute. The dialyzer clearance (CL_D) for solute i is then given by the next equation.

$$CL_D = \frac{M_i}{C_{b_i}^{\text{in}}} \tag{6.10}$$

The clearance for such solutes as urea, creatinine, and uric acid is on the order of 100 ml/min. Higher-molecular-weight solutes show a reduced clearance.

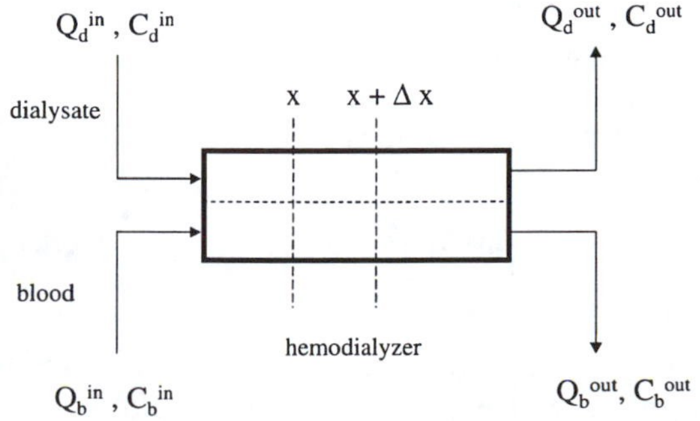

Figure 6.6 Solute mass-balance model for a cocurrent hemodialyzer.

The term *dialysance* is also used to describe the solute removal characteristics of a dialyzer. Dialysance (D_B) is defined by the next equation. We see that the mass-transfer rate of species i is now divided by the maximum solute concentration difference between the blood and the dialysate.

$$D_B = \frac{M_i}{C_{b_i}^{\text{in}} - C_{d_i}^{\text{in}}} \tag{6.11}$$

For a single pass dialyzer, $C_{d_i}^{\text{in}} = 0$, and the dialysance is then the same as the clearance.

Because of water removal, the inlet and outlet flow rates of the blood and dialysate streams are not equal. Rather, the following equation holds between the ultrafiltration rate and the flow rates of these streams.

$$J = Q_b^{\text{in}} - Q_b^{\text{out}} = Q_d^{\text{out}} - Q_d^{\text{in}} \tag{6.12}$$

If ultrafiltration is important, then we can use this relationship to derive an alternative expression for the clearance that allows for examination of the importance of ultrafiltration on the clearance of a given solute. Explicitly including ultrafiltration, the solute clearance is given by the following equation.

$$CL_D = Q_b^{\text{in}}\left(\frac{C_{b_i}^{\text{in}} - C_{b_i}^{\text{out}}}{C_{b_i}^{\text{in}}}\right) + J\left(\frac{C_{b_i}^{\text{out}}}{C_{b_i}^{\text{in}}}\right) \tag{6.13}$$

Since the value of $C_{b_i}^{\text{out}} \ll C_{b_i}^{\text{in}}$, the contribution of the ultrafiltration flow to the solute clearance is less than the value of J. As shown in Example 6.1, J is on the order of 10 ml/min, and earlier it was stated that the clearance of such solutes as urea is on the order of a 100 ml/min. Therefore, one concludes that the effect of ultrafiltration on the clearance of small-molecular-weight solutes is negligible. However, for higher-molecular-weight solutes, the effect of ultrafiltration may be significant.

6.6.5 Solute Transfer

We can now develop relationships between solute clearance and the mass-transfer characteristics of the membrane dialyzer. In order to perform this analysis, we first must choose one of the contacting patterns shown in Figure 6.2 to describe the flow of blood and dialysate in the dialyzer. Here we will use the cocurrent pattern already shown in Figure 6.6. We will also assume that the overall mass-transfer coefficient given by Equation 6.1 is constant. The flow rates of the blood and dialysate are also assumed to be constant, that is, ultrafiltration is ignored.

A shell mass balance for the blood and dialysate can then be written for a given solute as shown next.

$$\begin{aligned} Q_b C_b|_x - Q_b C_b|_{x+\Delta x} &= K_0 W \Delta x (C_b - C_d) \\ Q_d C_d|_x - Q_d C_d|_{x+\Delta x} &= -K_0 W \Delta x (C_b - C_d) \end{aligned} \tag{6.14}$$

The quantity $W\Delta x$ represents the available mass-transfer area within the shell volume ΔV. If L is the length of the dialyzer, then the quantity $A = L \times W$ would be the total membrane area within the dialyzer. Dividing by Δx, and taking the limit as Δx

approaches zero, results in the following two differential equations that provide the position-dependence of the solute concentration in the dialyzer blood and dialysate.

$$\begin{aligned} Q_b \frac{\mathrm{d}C_b}{\mathrm{d}x} &= -K_0 W (C_b - C_d) \\ Q_d \frac{\mathrm{d}C_d}{\mathrm{d}x} &= K_0 W (C_b - C_d) \end{aligned} \tag{6.15}$$

These equations can be integrated analytically, provided we first can relate C_d and C_b. This can be obtained by adding the above two equations, multiplying through by $\mathrm{d}x$, and then integrating between the entrance and any arbitrary value of x. The following result is then obtained for the value of C_d in terms of C_b. Note that $z \equiv \frac{Q_b}{Q_d}$.

$$C_d(x) = C_d^{\text{in}} - z\left(C_b(x) - C_b^{\text{in}}\right) \tag{6.16}$$

Substituting this equation for C_d in Equation 6.15, the differential equation describing the solute concentration on the blood side can then be written as follows. We see that the only dependent variable is the blood concentration.

$$Q_b \frac{\mathrm{d}C_b}{\mathrm{d}x} = -K_0 W \left[(1+z)C_b - \left(C_d^{\text{in}} + zC_b^{\text{in}}\right)\right] \tag{6.17}$$

This equation can now be integrated analytically and rearranged to give the following result. This equation is known as the performance equation for a cocurrent dialyzer.

$$E = \frac{D_B}{Q_b} = \frac{C_b^{\text{in}} - C_b^{\text{out}}}{C_b^{\text{in}} - C_d^{\text{in}}} = \frac{1 - \exp[-N_T(1+z)]}{(1+z)} \tag{6.18}$$

Two additional dimensionless parameters have been defined in the derivation of this equation, the extraction ratio, $0 \le E \le 1$, and the number of transfer units, $N_T = \frac{K_0 A}{Q_b}$. The number of transfer units provides a measure of the mass-transfer effectiveness. The concentration of solute i in the blood exiting the dialyzer would be given by: $C_{b_i}^{\text{out}} = C_{b_i}^{\text{in}}(1 - E) + EC_{d_i}^{\text{in}}$.

If the membrane area is infinite for a given value of z (i.e., for given value of the dialysate and blood flow rates), then we achieve the maximum possible extraction ratio, $E_{\text{maximum}} = \frac{1}{1+z}$. This corresponds to the best possible performance in a dialyzer for a given set of blood and dialysate flow rates, and is independent of the solute concentrations. This also implies that the solute concentration in the exiting blood and dialysate are in equilibrium, i.e., $C_{b_i}^{\text{out}} = C_{d_i}^{\text{out}} = C_{\text{equilibrium}}$. For these conditions, the maximum achievable dialysance for any combination of the blood and dialysate flows is also given by $D_{B_{\text{maximum}}} = \frac{z}{z+1} = \frac{Q_b Q_d}{Q_b + Q_d}$.

If $z = 0$, i.e., the dialysate flow rate is significantly higher than that of the blood, and if the membrane area were also infinite, then we would obtain $E = 1$. For these conditions, the solute concentration in the blood exiting the dialyzer would achieve its lowest possible value, i.e., the value of the solute concentration in the entering dialysate fluid, $C_{d_i}^{\text{in}}$.

Performance equations and curves for other contacting patterns such as countercurrent, a well-mixed dialysate, and cross-flow are summarized in Table 6.4. The following example illustrates the use of these performance equations.

Table 6.4 Performance equations for kidney dialyzers

Contacting pattern	Performance equation
Cocurrent	$E = \frac{1-\exp[-N_T(1+z)]}{1+z}$
Countercurrent	$E = \frac{\exp[N_T(1-z)]-1}{\exp[N_T(1-z)]-z}$
Well-mixed dialysate	$E = \frac{1-\exp(-N_T)}{1+z[1-\exp(-N_T)]}$
Crossflow	$E = \frac{1}{N_T}\sum_{n=0}^{\infty}[S_n(N_T)S_n(N_T z)]$ where $S_n(x) = 1-\exp(-x)\sum_{m=0}^{n}\left[\frac{x^m}{m!}\right]$

From Colton and Lowrie (1981).

Example 6.2 Estimate the clearance of urea in a 1 m^2 cocurrent and countercurrent hollow-fiber dialyzer. The dialyzer hollow fibers are 25 cm in length and they have an inside diameter of 250 microns. The wall thickness of the fibers is about 40 microns. The void volume of the fibers (ε) in the dialyzer is 50%. Blood flows within the hollow fibers and the dialysate flows in a single- pass on the shell side of the device parallel to the fibers. The blood flow rate is 200 ml/min and the dialysate flowrate is 500 ml/min. The membrane permeability, KP_m, is estimated for urea to be about 10^{-3} cm/sec.

SOLUTION For the blood flowing through the fibers, the characteristic dimension in the Sh number is the internal diameter of the fiber. For flow outside and parallel to the fibers, the characteristic dimension in the dialysate Sh number is the hydraulic diameter, D_H, defined earlier by Equation 2.6 shown below (Yang and Cussler 1986).

$$D_H = 4 \times \frac{\text{cross-sectional area}}{\text{wetted perimeter}}$$

Assuming the fibers are arranged on a square pitch, the following relationship may be developed to express the relationship between the diameter of the hollow fiber (d), the dialyzer void fraction (ε), and the hydraulic diameter (D_H).

$$D_H = d\left(\frac{\varepsilon}{1-\varepsilon}\right) \quad \text{square pitch}$$

The apparent velocity of the dialysate through the void space between the fibers would be equal to the dialysate volumetric flowrate (Q_d) divided by the cross-sectional area of the void space. The cross-sectional area of the void space is equal to the void fraction (ε) times the cross-sectional area of the dialyzer itself. A little algebra provides the following result for the apparent dialysate velocity. N_{fiber} is the number of hollow fibers in the dialyzer.

$$V_d = \frac{4(1-\varepsilon)Q_d}{N_{\text{fiber}}\varepsilon\pi d^2}$$

The diffusivity of urea in aqueous solution at 37°C from Figure 2.1 is about 1.8×10^{-5} cm^2/sec. This would be a reasonable value for the dialysate urea diffusivity and as mentioned before, the diffusivity of urea in blood would be about 40% of this value.

Mathcad® Chart 6.1 shows the rest of the calculation leading to a predicted urea dialyzer clearance of about 120 ml/min for the cocurrent dialyzer and about 133 ml/min for the countercurrent case. These urea-clearance values are typical of what is observed in actual practice. In comparison, the urea clearance of the two normal kidneys is about 70 ml/min. In general, the countercurrent dialyzer will provide the higher clearance, everything else being equal. This is because the countercurrent flow pattern maintains a relatively constant mass-transfer driving force along the ength of the dialyzer. The cocurrent pattern provides a larger driving force at the entrance of the dialyzer, but this difference rapidly decreases along the length of the dialyzer, resulting in less mass transfer of the solute.

Chart 6.1 for **Example 6.2:** Analysis of a hollow fiber dialyzer.

$A := 1$ — m², surface area of dialyzer based on internal diameter of the fibers

$L := 25$ — cm, length of the hollow fibers $\quad A := A \cdot 100^2$

$d := .025$ — cm, internal fiber diameter

$D_{aq} := 1.8 \times 10^{-5}$ — cm²/sec, urea diffusivity in aqueous media at 37°C

$D_b := .4 \cdot D_{aq}$ — calculates urea diffusivity in blood

$P := .001$ — cm/sec, membrane permeability to urea

$\varepsilon := .5$ — dialyzer void fraction

$Q_b := 200$ — ml/min, blood flow rate $\quad Q_b := \frac{Q_b}{60}$

$Q_d := 500$ — ml/min, dialysate flow rate $\quad Q_d := \frac{Q_d}{60}$

$\rho_{plasma} := 1.02$ — g/cm³, plasma density

$\mu_{plasma} := 1.2$ — cP, plasma viscosity

$\rho_{dialysate} := 1$ — g/cm³, dialysate density

$\mu_{dialysate} := 1$ — cP, dialysate viscosity

$$N_f := \frac{A}{(\pi d L)}$$

$N_f = 5093$ — number of hollow fibers in the dialyzer

$$V_b := Q_b \frac{1}{N_f} \frac{4}{(\pi d^2)}$$

$V_b = 1.333$ — cm/sec, the velocity of blood in a given hollow fiber

$$Re_b := \frac{\rho_{plasma} d V_b}{\mu_{plasma}} 100 \qquad Re_b = 2.833$$

$$Sc_b := \frac{\mu_{plasma} \cdot .01}{\rho_{plasma} D_b} \qquad Sc_b = 1.634 \times 10^3$$

$z_{critical} := .05 d\, Re_b Sc_b \qquad z_{critical} = 5.787$ cm,

Since $L > z_{critical}$, we conclude fully-developed laminar flow

$Sh_b := 3.66$

$$k_b := \frac{Sh_b D_b}{d}$$

$k_b = 1.054 \times 10^{-3}$ — cm/sec, blood side mass-transfer coefficient

$$d_e := d\left[\frac{\varepsilon}{(1-\varepsilon)}\right]$$

$d_e = 0.025$ — cm, the equivalent diameter of the space between the fibers through which the dialysate flows

$$D_{dialyzer} := d\left[\frac{N_f}{(1-\varepsilon)}\right]^{.5}$$

$D_{\text{dialyzer}} = 2.523$ cm, the dialyzer gross diameter

$$V_d := \frac{4Q_d(1-\varepsilon)}{(\pi \varepsilon N_f d^2)}$$

$V_d = 3.333$ cm/sec, dialysate velocity in the space between the fibers

$$Re_d := \frac{\rho_{\text{dialysate}} d_e V_d}{\mu_{\text{dialysate}}} 100 \qquad Re_d = 8.333$$

$$Sc_d := \frac{\mu_{\text{dialysate}} \cdot .01}{\rho_{\text{dialysate}} D_{aq}} \qquad Sc_d = 555.556$$

$z_{\text{critical}} := .05 d_e Re_d Sc_d \qquad z_{\text{critical}} = 5.787$ cm

Since $L > z_{\text{critical}}$, we conclude fully-developed laminar flow

$Sh_d := 3.66$

$$k_d := \frac{4 D_{aq}}{d_e}$$

$k_d = 2.88 \times 10^{-3}$ cm/sec, dialysate mass-transfer coefficient

$$R_b := \frac{1}{k_b} \qquad R_m = \frac{1}{P} \qquad R_d = \frac{1}{k_d}$$

$R_b = 948.695 \quad R_m = 1000 \quad R_d = 347.222$ sec/cm, resistances of the blood, membrane, and dialysate.

We see the greatest resistance to mass-transfer resides in the blood and the membrane.

$$K_0 := \frac{1}{(R_b + R_m + R_d)}$$

$K_0 = 4.356 \times 10^{-4}$ cm/sec, the overall mass-transfer coefficient

$$N_T := \frac{K_0 A}{Q_b}$$

$N_T = 1.307$ number of mass-transfer units

$$Z := \frac{Q_b}{Q_d}$$

$Z = 0.4$ ratio of blood flow to dialysate flow

$$E := \frac{1 - \exp[-N_T(1+Z)]}{(1+Z)}$$

$E = 0.6$ the extraction ratio assuming the flow pattern is cocurrent

$CL_D := E Q_b 60$

$CL_D = 119.926$ ml/min, the dialyzer urea clearance for cocurrent flow

$$E := \frac{\exp[N_T(1-Z)] - 1}{[\exp[N_T(1-Z)] - Z]}$$

$E = 0.665$ the extraction ratio assuming the flow pattern is countercurrent

$CL_D := E Q_b 60$

$CL_D = 132.969$ ml/min, the dialyzer urea clearance for countercurrent flow

6.6.6 A Single Compartment Model of Urea Dialysis

Urea distributes throughout the total body water for a total distribution volume of about 40 liters. The distribution volume (in liters) can also be estimated to be 58% of body weight (in kg) (Galletti, Colton, and Lysaght 1995). We can treat the body as a single well-mixed compartment and use the dialyzer performance equations to predict how the amount of urea within the body decreases with time during dialysis. Neglecting the effects of ultrafiltration, the clearance of urea from the body during dialysis would be similar to the IV bolus injection examined in Chapter 5, that is, Equation 5.15, where the only elimination pathway for urea would be the dialyzer. Therefore k_{te} would equal the dialyzer clearance divided by the distribution volume for urea.

$$C_{\text{urea}} = C^0_{\text{urea}} e^{-\left(\frac{CL_D}{V_{\text{apparent}}}\right)t} \tag{6.19}$$

Using the urea dialyzer clearance value for a countercurrent unit obtained from Example 6.2, we can show that after 4 hours of dialysis that the urea concentration in the body has been reduced by about 55%. It should be pointed out that during this dialysis period, the amount of urea generated is assumed negligible in comparison to the amount removed by dialysis. However, after dialysis, urea continues to accumulate at the rate of about 10 grams each day (G_{urea}). If we assume the kidney has no residual clearance for urea, then the urea concentration will increase linearly with time after dialysis according to the following equation.

$$C_{\text{urea}}(t) = C_{\text{urea}} \begin{pmatrix} \text{end of} \\ \text{dialysis} \end{pmatrix} + \left(\frac{G_{\text{urea}}}{V_{\text{apparent}}}\right)t \tag{6.20}$$

This equation can be used to show that dialysis will be required about 3 times per week. For some solutes, the removal from the dialyzer is so fast that the rate of solute transfer from the extravascular spaces into the plasma will be the rate-limiting step. For these cases, a two-compartment model may be needed to describe the kinetics of solute removal.

The following references provide additional information on the artificial kidney: (Ramachandran and Mashelkar, 1980; Colton and Lowrie 1981; Armer and Hanley 1986; Abbas and Tyagi 1987; Zelman 1987; Lee and Chang 1988; Lee, Hsu, and Hu 1989; Shaoting et al. 1990; Capello et al. 1994; Galletti, Colton, and Lysaght 1995).

6.6.7 Peritoneal Dialysis

An alternative approach for treating kidney failure is called continuous ambulatory peritoneal dialysis (CAPD) (Lysaght and Farrell 1989; Lysaght and Moran 1995). Unlike the hemodialysis technique that was just described, CAPD has the advantage of the patient not being severely restricted by the regimen of three-times-per-week treatment at a dialysis center. Rather, the patient is responsible for performing a relatively simple treatment process that can take place at home or even at work. Use of CAPD is growing rapidly. CAPD was used in about 4% of the total dialysis population in 1979 and in 1992 commanded 15% of the dialysis population. Overall, about 90,000 patients used CAPD in 1995 (Lysaght and Moran 1995).

CAPD is based on the addition to the peritoneal cavity of a sterile hypertonic solution of glucose and electrolytes. The peritoneal cavity is a closed space formed by the peritoneum, a membrane-like tissue that lines the abdominal cavity and covers the internal organs such as the liver and the intestines. The peritoneum has excellent mass-transfer characteristics for a process like CAPD. The peritoneum has the appearance of a fairly transparent sheath that is smooth and quite strong. The surface area (S) of the peritoneum is about 1.75 m^2. Its thickness ranges from 200 to over 1000 microns. The surface of the peritoneum presented to the CAPD dialysate solution consists of a single layer of mesothelial cells that is densely covered with microvilli or tiny hair-like projections. Beneath this layer of cells is the interstitium that has the characteristics of a gel-like material. Within the interstitium there is a rich capillary network providing a total blood flow on the order of 50 ml/min.

The CAPD solution is added and later removed from the peritoneal cavity through an in-dwelling catheter. This process can also be automated such that all fluid exchanges are carried out by a pumping unit, even while the patient is asleep. This automated process is called automated peritoneal dialysis (APD).

The intraperitoneal CAPD fluid partially equilibrates across the peritoneal membrane with waste-products in the plasma. Excess water from the patient's body is also removed by ultrafiltration across the peritoneal membrane as a result of osmotic gradients.

Typically, four 2-liter exchanges are made each day with an additional 2 liters of water removed as a result of ultrafiltration. The average mass-removal rate of a particular solute is given by the following equation.

$$\dot{m}_{\text{CAPD}} = \frac{V_{\text{CAPD}} C_{\text{CAPD}}}{t_{\text{CAPD}}} \tag{6.21}$$

In this equation, V_{CAPD} represents the volume of the CAPD solution and C_{CAPD} is the final concentration of the solute in this solution. The time the solution was in the peritoneal cavity is given by t_{CAPD}. Recall that the clearance of a solute would equal its mass removal rate divided by the solute concentration in the blood. Therefore, solute clearance during CAPD would be given by the following expression.

$$CL_{\text{CAPD}} = \frac{\dot{m}_{\text{CAPD}}}{C_b} = \frac{V_{\text{CAPD}} C_{\text{CAPD}}}{t_{\text{CAPD}} C_b} \tag{6.22}$$

Since the CAPD solution is usually completely equilibrated with urea, i.e. $C_b = C_{\text{CAPD}}$ the urea clearance is then simply $\frac{V_{\text{CAPD}}}{t_{\text{CAPD}}}$. About 10 liters of solution are used each day which equates to a continuous urea clearance of about 7 ml/min.

The mass-transfer characteristics of CAPD can be described using relationships we have already developed. For example, the ultrafiltration rate would be given by Equation 2.31a:

$$J = SL_P \left[\Delta P - RT \sum_i \sigma_i (C_{\text{blood}} - C_{\text{CAPD}})_i \right] \tag{6.23}$$

CAPD usually begins with a solution containing about 4% dextrose. The initial ultrafiltration rate for this solution is on the order of 20 ml/min and decreases rapidly with time as the glucose initially in the peritoneal cavity is removed by the blood. Since ΔP is small during CAPD, and the osmotic pressure of the initial dextrose solution is about 4300 mm Hg, then the hydraulic conductance is on the order of 0.15 ml/hrm^2 mm Hg.

The solute transport rate for species i would be given by Equation 2.31b, recognizing that initially significant quantities of solute may be transported by ultrafiltration.

$$\dot{m}_{\text{CAPD}} = C_i \left(1 - \sum_i \sigma_i \right) J + P_{m_i} S (C_{\text{blood}} - C_{\text{CAPD}})_i \tag{6.24}$$

Table 6.5 Transport properties of the peritoneal membrane

Solute	Molecular weight	Reflection coefficient	$P_m S$, cm^3/min
Urea	60	0.26	21
Creatinine	113	0.35	10
Uric acid	158	0.37	10
Vitamin B_{12}	1355		5
Inulin	5200	0.5	4
β_2-Microglobulin	12000		0.8
Albumin	69000	0.99	

From Lysaght and Moran (1995).

Typical values of the reflection coefficient and permeability surface area product for several solutes are summarized in Table 6.5.

Mathematical models describing CAPD can also be developed to describe the time-course of the treatment process. The patient is considered to be represented by a single well-mixed compartment of apparent distribution volume (V_{apparent}) and solute concentration (C_{body}) which is assumed to be the same as the solute concentration in the blood (C_{blood}). The CAPD solution in the peritoneal cavity is in a much smaller variable volume (about 2 liters at the start) represented by V_{CAPD} and solute concentration C_{CAPD}. The peritoneal membrane separates these two compartments. The peritoneal lymphatic system also removes fluid from the peritoneal cavity (Q_{lymph}) at a rate of about 0.1 to as high as 10 ml/min. During the initial phases of CAPD, the volume of fluid in the peritoneal cavity will increase due to ultrafiltration into the cavity; however, as time progresses and glucose is removed from the peritoneal cavity, the volume of fluid in the peritoneal cavity will decrease as a result of the lymphatic flow.

Models for describing CAPD are reviewed by Lysaght and Farrell (1989). These models range from relatively simple analytical solutions to much more complex models that include selective transport of solutes, ultrafiltration, and the effects of lymphatic

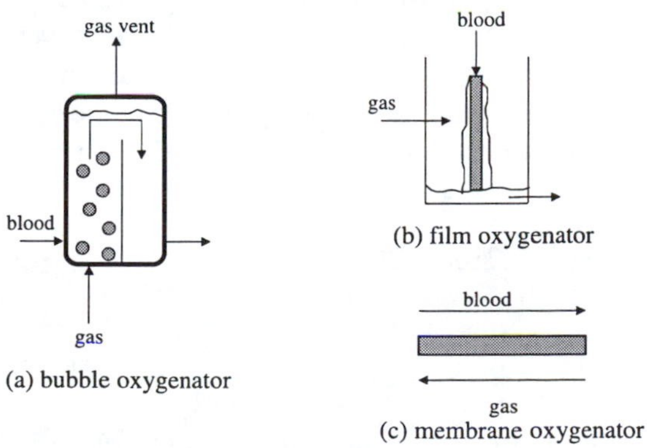

Figure 6.7 Types of blood oxygenators.

flow. These models all require numerical solutions. We will discuss two relatively simple analytical models for CAPD below.

The simplest analytical model neglects ultrafiltration and lymphatic flow and assumes that only the dialysate concentration changes with time. This diffusion-only model, with its assumption that V_{CAPD} is constant, limits use of this approach to the constant volume phase (isovolemic) that occurs about an hour or so after the exchange begins. For this diffusion-only CAPD model, the solute balance equation only needs to be written for the dialysate fluid in the peritoneal cavity.

$$\frac{\mathrm{d}(V_{\mathrm{CAPD}}C_{\mathrm{CAPD}})}{\mathrm{d}t} = V_{\mathrm{CAPD}}\frac{\mathrm{d}C_{\mathrm{CAPD}}}{\mathrm{d}t} = P_m S\left(C_{\mathrm{body}} - C_{\mathrm{CAPD}}\right) \tag{6.25}$$

This equation may be easily integrated to obtain the following equations. In these equations, the initial concentration of the solute in the CAPD solution is given by C^0_{CAPD}.

$$P_m S = \frac{V_{\mathrm{CAPD}}}{t}\ln\left(\frac{C_{\mathrm{body}} - C^0_{\mathrm{CAPD}}}{C_{\mathrm{body}} - C_{\mathrm{CAPD}}(t)}\right) \tag{6.26}$$

$$C_{\mathrm{CAPD}}(t) = C_{\mathrm{body}} - \left(C_{\mathrm{body}} - C^0_{\mathrm{CAPD}}\right)\exp\left(-\frac{P_m S t}{V_{\mathrm{CAPD}}}\right) \tag{6.27}$$

The first equation above allows for determination of the value of $P_m S$, given information on the solute concentrations at a particular time. The second equation allows for prediction of the solute concentration in the CAPD dialysate solution as a function of time assuming the transport properties of the peritoneal membrane are known.

A slightly more complex model than the one shown above can be developed that includes the effect of ultrafiltration (Garred, Canaud, and Farrell 1983). However, the solute concentration in the body is still assumed to be constant, the peritoneal membrane is not selective, i.e., the reflection coefficients are zero, and lymphatic flow is ignored. The solute balance for the CAPD solution for these conditions may be written as follows.

$$\frac{\mathrm{d}\left(V_{\mathrm{CAPD}}C_{\mathrm{CAPD}}\right)}{\mathrm{d}t} = P_m S\left(C_{\mathrm{body}} - C_{\mathrm{CAPD}}\right) + C_{\mathrm{body}}\frac{\mathrm{d}V_{\mathrm{CAPD}}}{\mathrm{d}t} \tag{6.28}$$

The second term on the right hand side $C_{\mathrm{body}}\frac{\mathrm{d}V_{\mathrm{CAPD}}}{\mathrm{d}t}$ represents the ultrafiltration flow that carries with it the solute in the body. Integration of this equation still requires knowledge of how V_{CAPD} changes with time. This could be obtained from Equation 6.23; however, an analytical solution would then be difficult to obtain. If one assumes that the initial and final CAPD volumes are known, then this equation can be integrated to give the next two equations. $\bar{V}_{\mathrm{CAPD}}$ is the average volume defined as the mean of the initial and final CAPD volumes.

$$P_m S = \frac{\bar{V}_{\mathrm{CAPD}}}{t}\ln\left[\frac{V^0_{\mathrm{CAPD}}\left(C_{\mathrm{body}} - C^0_{\mathrm{CAPD}}\right)}{V_{\mathrm{CAPD}}(t)(C_{\mathrm{body}} - C_{\mathrm{CAPD}}(t))}\right] \tag{6.29}$$

$$C_{\mathrm{CAPD}}(t) = C_{\mathrm{body}} - \frac{V^0_{\mathrm{CAPD}}}{V_{\mathrm{CAPD}}(t)}\left(C_{\mathrm{body}} - C^0_{\mathrm{CAPD}}\right)\exp\left(-\frac{P_m S t}{\bar{V}_{\mathrm{CAPD}}}\right) \tag{6.30}$$

Once again the first equation may be used to estimate the mass-transport properties of the peritoneal membrane from given values of the solute concentration and volumes at a particular time. The second equation allows for determination of the time course of the solute concentration in the dialysate solution.

6.7 BLOOD OXYGENATORS

6.7.1 Background

Surgery on the heart to repair internal defects or to improve blood flow to the heart muscle by coronary artery bypass grafting requires that the heart be stopped or arrested. Accordingly, blood is no longer pumped throughout the body and is no longer oxygenated by the lungs. Special devices called heart-lung machines or blood pump-oxygenators (Richardson 1987; Galletti and Colton 1995) have been developed and used for over forty years (Gibbon 1954) to replace the gas-exchange function of the lungs and the pumping action of the heart during these cardiac surgical procedures. Blood flow to the blood pump-oxygenator is usually collected by a pump from the large systemic veins such as the vena cava. This blood is then passed through the oxygenator and returned by a second pump to the aorta. Blood flow to the chambers of the heart itself is stopped, providing a dry and bloodless field for surgery. Nearly a million of these cardiopulmonary bypass procedures are performed worldwide each year.

6.7.2 Operating Characteristics

There are several important differences between the operational characteristics of the lungs and blood pump-oxygenators. These are summarized in Table 6.6. The blood flow to the blood pump-oxygenator is the same as to the lungs since it must provide the same level of blood oxygenation; that is, increasing the pO_2 of the venous blood from about 40 mm Hg to the arterial pO_2 of 95 mm Hg. The capillaries in the lungs present the blood to the alveolar gas as a very thin film of about 5–10 microns in thickness. This provides

Table 6.6 Operational characteristics of the lungs and blood pump-oxygenators

Characteristic	Lungs	Blood oxygenator
Blood flow rate	5 L/min	5 L/min
Pressure head	12 mm Hg	0–200 mm Hg
Blood volume	1 liter	1–4 liters
Blood film thickness	5–10 microns	100–300 microns
Length of blood flow channel	100 microns	2–30 cm
Blood contact time	0.7 sec	3–30 sec
Surface area for mass transfer	70 m^2	2–10 m^2
Gas flow rates	7 L/min	2–10 L/min
pO_2 and pCO_2 blood in	40 and 45 mm Hg	40 and 45 mm Hg
pO_2 and pCO_2 blood out	95 and 40 mm Hg	100–300 and 30–40 mm Hg
Gas pO_2 and pCO_2 in	149 and 0.3 mm Hg	250–713 and 0–20 mm Hg
Gas pO_2 and pCO_2 out	120 and 27 mm Hg	150–675 and 10–30 mm Hg
pO_2 gradient	40–50 mm Hg	650 mm Hg
pCO_2 gradient	3–5 mm Hg	30–50 mm Hg

From Cooney 1976; Galletti and Colton 1995.

rapid gas transport in a very short time, typically a few tenths of a second. In a blood pump–oxygenator, the blood film thickness is governed by the number and thickness of the blood flow channels, typically providing film thicknesses on the order of a few hundred microns and contact times on the order of tens of seconds. Effective gas transport then occurs over greater oxygenator lengths and requires greater pressures to force the blood through the oxygenator.

In the lungs, the pO_2 of the alveolar air is about 104 mm Hg. The average oxygen partial pressure driving force in the lungs is therefore on the order of 40–50 mm Hg. Carbon dioxide has a pCO_2 of about 40 mm Hg in the alveolar space and its level in the blood varies from 40 mm Hg in arterial blood to 45 mm Hg in venous blood. Therefore, the driving force for carbon dioxide transport is only a few mm Hg and considerably smaller than that for oxygen. However, this smaller driving force is offset by the much higher permeability of carbon dioxide through the respiratory membranes which is about 20 times that for oxygen. One significant advantage of the blood pump–oxygenator, as compared to the lungs, is the driving force for oxygen and carbon dioxide transport. In a blood pump–oxygenator, humidifed oxygen at atmospheric pressure can be used, resulting in a pO_2 of about 713 mm Hg [760 mm Hg − 47 mm Hg (water saturation pressure at 37°C) = 713 mm Hg]. Also, little or no carbon dioxide is present in this gas, so the driving force for carbon dioxide transport is considerably higher. This results in a considerable reduction in the mass-transfer surface area of the blood oxygenator.

6.7.3 Types of Oxygenators

The design goal of an oxygenator is to present as large a surface area as possible between the blood and the oxygen-carrying gas stream. Several approaches have been developed to accomplish this and they are summarized in Figure 6.7. *Bubble oxygenators* were the earliest systems developed for cardiopulmonary bypass surgery. Bubble oxygenators develop the required surface area for gas exchange by the production of numerous small bubbles of gas. These gas bubbles are in direct contact with the blood and are typically several millimeters in diameter. Of particular concern in the operation of bubble oxygenators is the removal of foam and gas bubbles prior to the return of the blood to the patient. This is achieved through the use of filters and defoaming sponges as the blood leaves the oxygenator. Design of bubble oxygenators is complicated by the complex nature of bubble motion. For example, bubbles exist in many different sizes and they tend to breakup and coalesce during their passage in the oxygenator.

To avoid the problem with antifoaming and bubble removal, *film oxygenators* have also been developed. In these systems, a film of blood is created on a surface that may be stationary or rotating. The blood film is then exposed directly to the gas. To increase the mass-transfer rates, stationary films often employ irregular surfaces to enhance internal mixing of the blood film. Rotating discs whose bottom surface is in contact with a pool of blood create very thin films that are continuously renewed, resulting in efficient exposure of the blood to the gas stream.

Membrane oxygenators have for the most part now replaced the earlier bubble and film oxygenators. In these systems, the membrane physically separates the blood and gas streams. Membrane systems offer the advantages of minimizing the trauma to the

Table 6.7 Oxygen and carbon dioxide permeability of selected membrane materials

Material	O_2 permeability, ml (STP) micron/min/m^2/atm	CO_2 permeability, ml (STP) micron/min/m^2/atm
Air	1.27×10^9	1.02×10^9
Polydimethysiloxane (silicone)	27,900	140,000
Water	3,810	68,600
Polystyrene	1,397	6,985
Polyisoprene (natural rubber)	1,270	7,620
Polybutadiene	1,016	7,112
Cellulose (cellophane)	635	11,430
Polyethylene	305	1,524
Polytetrafluoroethylene (teflon)	203	610
Polyamide (nylon)	2.54	10.2
Polyvinylidene chloride (saran)	0.25	1.52

From Galletti and Colton (1995).

blood that is created in direct-contact systems such as bubble and film oxygenators. The membrane and the associated fluid boundary layers, however, do present an additional mass-transfer resistance. However, recent advances in membrane technology has resulted in membranes with significant oxygen and carbon dioxide permeabilities. The gas permeablites for various membrane materials are shown in the Table 6.7.

Typical units of the thickness normalized membrane gas permeability ($\widehat{P}_m$) are ml (STP) micron/min/m^2/atm, where the STP (standard temperature and pressure) conditions are for the pure gas at 0°C and 1 atmosphere. The volumetric transport rate of gas is then given by the following equation:

$$N_V = \frac{\widehat{P}_m \times (\text{membrane surface area}) \times \Delta(\text{gas partial pressure})}{\text{membrane thickness}} \tag{6.31}$$

The most common material used for early versions of membrane devices is silicone; these devices have wall thicknesses on the order of 50–200 microns. The oxygen permeability for a 130-micron thick silicone membrane is about 215 ml (STP)/min m^2 atm and that for carbon dioxide is about 1100 ml (STP)/min m^2 atm (Cooney 1976; Gray 1981, 1984).

Recent hollow-fiber systems employ hydrophobic microporous polypropylene membranes. These membranes allow for free passage of gas by diffusion through the membrane pores, but because of the small size of the pores, there is sufficient surface tension to prevent plasma filtration. The pores in these membranes are typically on the order of 0.1 microns in diameter. Note from Table 6.7 that a column of air within the pores of these membranes has a very high gas permeability in comparison to diffusion through the polymeric material itself. Therefore, these membranes have very high gas permeabilities.

These membrane-based systems come in a variety of configurations, for example flat plate, coil, and hollow-fiber arrangements as shown earlier in Figure 6.1. Blood and gas stream contact can include cocurrent, countercurrent, and cross flow. Cross-flow systems typically have the gas flowing through the lumen of the hollow fiber, with blood flowing at right angles across the outer surface of the fibers. Cross flow of the blood

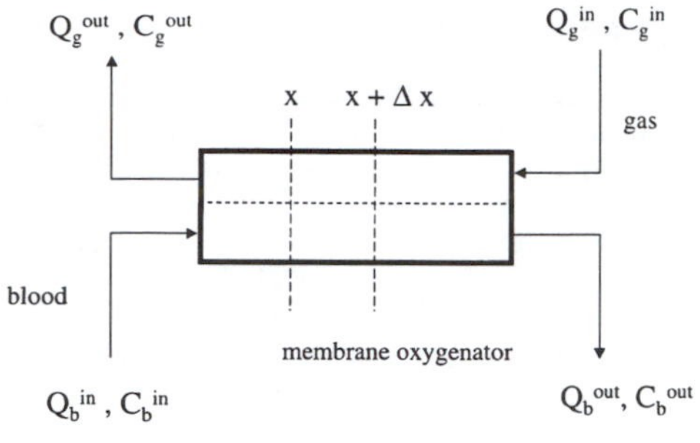

Figure 6.8 Solute mass-balance model for a countercurrent membrane blood oxygenator.

results in significant improvement in performance as measured by gas-exchange rates (Catapano, Wodetzki, and Baurmeister 1992; Vaslef et al. 1994).

6.7.4 Analysis of a Membrane Oxygenator, Oxygen Transfer

A mathematical model for a membrane blood gas oxygenator can be developed as follows. The following model is useful for preliminary design calculations and for exploring the effects of operating conditions on device performance. The model is developed in a manner similar to that used earlier for hemodialysis; however, now we must make use of techniques developed in Chapter 4 to account for the binding of oxygen with hemoglobin.

Figure 6.8 shows the mass-balance model where now we assume countercurrent flow of the blood and gas streams. We also assume the total flow-rates of these streams are unchanged. C_b and C_g are the bulk concentrations on the blood (b) and gas (g) sides of the membrane. C_b' represents the amount of oxygen in the blood that is bound to hemoglobin. C_{bm} and C_{gm} are the corresponding concentration values at the membrane surface on the blood and gas sides. A shell balance on oxygen may be written as follows for the blood and gas sides of the membrane of length Δx.

$$\begin{aligned} Q_b\left(C_b + C_b'\right)\big|_x - Q_b\left(C_b + C_b'\right)\big|_{x+\Delta x} &= k_b W_b \Delta x (C_b - C_{bm}) \\ Q_g C_g\big|_{x+\Delta x} - Q_g C_g\big|_x &= k_g W_g \Delta x (C_g - C_{gm}) \end{aligned} \tag{6.32}$$

In these equations, k_b and k_g are the blood and gas-side mass-transfer coefficients. W is the membrane area per unit length of membrane, with the subscript denoting either the blood (b) or gas (g) sides. This allows application of this discussion to cylindrical fibers where the area normal to the direction of transport changes with radial position. For planar membranes, we simply have that $W_b = W_g = W$.

After dividing by Δx, and taking the limit as $\Delta x \to 0$, we obtain the following differential equations.

$$\begin{aligned} Q_b \frac{\mathrm{d}\left(C_b + C_b'\right)}{\mathrm{d}x} &= -k_b W_b (C_b - C_{bm}) \\ Q_g \frac{\mathrm{d}C_g}{\mathrm{d}x} &= k_g W_g (C_g - C_{gm}) \end{aligned} \tag{6.33}$$

Once again we can make use of the fact that $m = \frac{dC'_b}{dC_b}$. Recall that m is related to the slope of the oxygen-dissociation curve. A value for m may be estimated by Equation 4.28. Assuming m is constant and evaluated at some suitable combination of the venous and arterial pO_2 values, the blood-side equation simplifies to the following result.

$$Q_b(1+m)\frac{dC_b}{dx} = -k_b W_b (C_b - C_{bm}) \tag{6.34}$$

It is now convenient to convert from concentration to partial pressure in our description of the mass-transfer process. Recall for blood that the partial pressure of oxygen is related to the dissolved oxygen concentration by Henry's law, $pO_{2b} = HC_b$. The value of H is 0.74 mm Hg/μM. On the gas side, $pO_{2g} = RTC_g$, using the ideal gas law. Our equations for the blood and gas side now may be written as:

$$\begin{aligned} Q_b(1+m)\frac{dpO_{2b}}{dx} &= -k_b W_b (pO_{2b} - pO_{2bm}) \\ Q_g\frac{dpO_{2g}}{dx} &= k_g W_g (pO_{2g} - pO_{2gm}) \end{aligned} \tag{6.35}$$

These equations may be rewritten in terms of the overall partial pressure driving force, $pO_{2g} - pO_{2b}$, through definition of the overall mass-transfer coefficient, K_0, given by the next equation:

$$\frac{1}{K_0} = \frac{H}{k_b} + \frac{W_b}{\rho^{STP} P_m \bar{W}_L} + \frac{RT W_b}{k_g W_g} \tag{6.36}$$

ρ^{STP} represents the density of the gas at STP conditions. The units on K_0 would typically be mol/cm^2 mm Hg sec. K_0 is also now based on the membrane area on the blood side. The log mean area of the membrane is given by $\bar{W}_L \equiv \dfrac{W_g - W_b}{\ln(W_g/W_b)}$.

In terms of K_0, the blood and gas-side equations become:

$$\begin{aligned} \frac{Q_b(1+m)}{H}\frac{dpO_{2b}}{dx} &= K_0 W_b (pO_{2g} - pO_{2b}) \\ \frac{Q_g}{RT}\frac{dpO_{2g}}{dx} &= K_0 W_b (pO_{2g} - pO_{2b}) \end{aligned} \tag{6.37}$$

We can now solve these equations to provide an analytical solution to describe the performance of the oxygenator. First, we subtract the gas-side equation from the blood-side equation. This allows us to obtain a relationship between pO_{2g} and pO_{2b}.

$$\frac{Q_b(1+m)}{H}\frac{dpO_{2b}}{dx} - \frac{Q_g}{RT}\frac{dpO_{2g}}{dx} = 0 \tag{6.38}$$

This equation may now be integrated from the entrance of the blood stream, $x = 0$, to any arbitrary value of x to give:

$$pO_{2g}(x) = \left(\frac{Q_b}{Q_g}\right)\left(\frac{RT}{H}\right)(1+m)\left[pO_{2b}(x) - pO_{2b}^{in}\right] + pO_{2g}^{out} \tag{6.39}$$

Now we can substitute this equation for pO_{2g} in the blood-side Equation (6.37) to give:

$$\frac{dpO_{2b}}{dx} = \left(\frac{K_0 W_b H}{Q_b(1+m)}\right)\left\{\left[\left(\frac{Q_b}{Q_g}\right)\left(\frac{RT}{H}\right)(1+m) - 1\right]pO_{2b} + \left[pO_{2g}^{out} - \left(\frac{Q_b}{Q_g}\right)\left(\frac{RT}{H}\right)(1+m)pO_{2b}^{in}\right]\right\} \tag{6.40}$$

This equation may then be integrated to give the following result for the blood-side membrane area, A_{oxygen} required for a given change in blood oxygenation.

$$A_{oxygen} = \frac{\alpha}{\beta}\ln\left(\frac{\beta pO_{2b}^{out} + \gamma}{\beta pO_{2b}^{in} + \gamma}\right) \tag{6.41}$$

α, β, and γ are given by:

$$\alpha = \frac{Q_b(1+m)}{K_0 H}$$
$$\beta = \left(\frac{Q_b}{Q_g}\right)\left(\frac{RT}{H}\right)(1+m) - 1 \tag{6.42}$$
$$\gamma = pO_{2g}^{out} - \left(\frac{Q_b}{Q_g}\right)\left(\frac{RT}{H}\right)(1+m)pO_{2b}^{in}$$

The value of α would typically have units of cm^2, β is dimensionless, and γ would be in mm Hg.

6.7.5 Analysis of a Membrane Oxygenator, Carbon Dioxide Transfer

We also need to develop a similar set of equations for the transport of carbon dioxide in the oxygenator. Carbon dioxide can exist in blood in a variety of forms, such as a dissolved gas, and in combinations with water, hemoglobin, and other proteins. Like oxygen, the total amount of carbon dioxide in the blood depends on its partial pressure, i.e., pCO_2. The carbon dioxide dissociation curve shown in Figure 6.9 provides a relationship between the total amount of carbon dioxide in the blood and its pCO_2. The ordinate expresses the volume of carbon dioxide gas as a percentage of the blood volume. The volume percents are based on body conditions (BTP) of 37°C and 1 atm of pure gas. For example, there is about 50 ml of carbon dioxide at BTP in each 100 ml of blood, i.e., 50 volume per cent, at a pCO_2 of about 42 mm Hg.

The change in the carbon dioxide concentration for normal blood is very narrow. For example, venous blood has a pCO_2 of about 45 mm Hg and a corresponding concentration of 52 volume percent. Arterial blood has a pCO_2 of about 40 mm Hg and a corresponding concentration of 48 volume percent. These values can be used to show that the Henry's constant for carbon dioxide in blood for this range of pCO_2 is about 0.0022 mm Hg/μM.

Figure 6.8 can also be used to write shell balances on the blood and gas sides for carbon dioxide. A set of equations similar to those given by Equations 6.37 are obtained where m is now equal to zero. Henry's law is also assumed to describe the relationship between the carbon dioxide concentration in the blood and its pCO_2. The

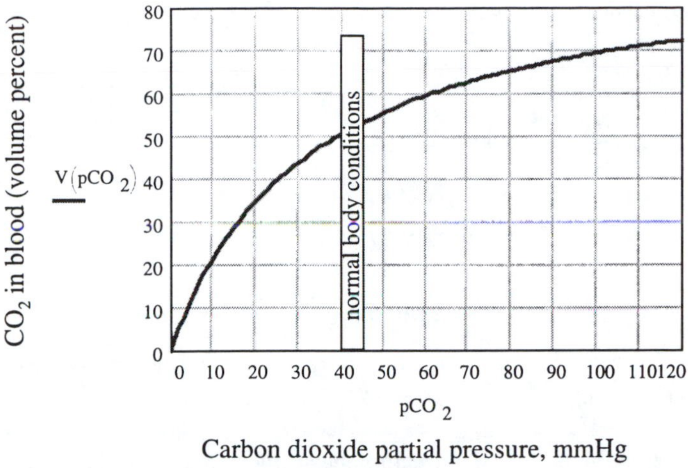

Figure 6.9 Carbon dioxide blood solubility curve.

overall mass-transfer coefficient, K_0, is still given by Equation 6.36, recognizing that the properties of carbon dioxide are to be used.

$$\frac{Q_b}{H}\frac{\mathrm{dpCO}_{2b}}{\mathrm{d}x} = K_0 W_b(\mathrm{pCO}_{2g} - \mathrm{pCO}_{2b})$$
$$\frac{Q_g}{RT}\frac{\mathrm{dpCO}_{2g}}{\mathrm{d}x} = K_0 W_g(\mathrm{pCO}_{2g} - \mathrm{pCO}_{2b}) \tag{6.43}$$

As carried out before for oxygen, the gas-side equation can be subtracted from the blood-side equation and the result integrated to provide the following expression relating pCO_{2g} to pCO_{2b}.

$$\mathrm{pCO}_{2g}(x) = \left(\frac{Q_b}{Q_g}\right)\left(\frac{RT}{H}\right)\left[\mathrm{pCO}_{2b}(x) - \mathrm{pCO}_{2b}^{\mathrm{in}}\right] + \mathrm{pCO}_{2g}^{\mathrm{out}} \tag{6.44}$$

This equation may be used in the blood equation (6.43) to obtain the following differential equation for pCO_{2b}.

$$\frac{\mathrm{dpCO}_{2b}}{\mathrm{d}x} = \left(\frac{K_0 W_b H}{Q_b}\right)\left\{\left[\left(\frac{Q_b}{Q_g}\left(\frac{RT}{H}\right) - 1\right)\right]\mathrm{pCO}_{2b} + \left[\mathrm{pCO}_{2g}^{\mathrm{out}} - \left(\frac{Q_b}{Q_g}\right)\left(\frac{RT}{H}\right)\mathrm{pCO}_{2b}^{\mathrm{in}}\right]\right\} \tag{6.45}$$

When this equation is integrated over the length of the oxygenator, the following result is obtained for the blood-side membrane surface area needed for the required carbon dioxide removal.

$$A_{\mathrm{carbon\,dioxide}} = \frac{\alpha'}{\beta'}\ln\left[\frac{\beta'\mathrm{pCO}_{2b}^{\mathrm{out}} + \gamma'}{\beta'\mathrm{pCO}_{2b}^{\mathrm{in}} + \gamma'}\right] \tag{6.46}$$

The constants α', β', and γ' are given by the next set of equations and have the same typical units as described earlier for oxygen.

$$\alpha' = \frac{Q_b}{K_0 H}$$
$$\beta' = \frac{Q_b}{Q_g}\left(\frac{RT}{H}\right) - 1 \qquad (6.47)$$
$$\gamma' = \mathrm{pCO}_{2g}^{\mathrm{out}} - \left(\frac{Q_b}{Q_g}\right)\left(\frac{RT}{H}\right)\mathrm{pCO}_{2b}^{\mathrm{in}}$$

6.7.6 Example Calculations for Membrane Oxygenators

The following examples illustrate the calculation of the area of a membrane blood oxygenator using the above equations.

Example 6.3 Determine the hollow-fiber membrane surface area for a blood oxygenator operating under the following conditions. Assume the fibers are made from microporous polypropylene having a membrane porosity of 40%. The gas permeabilities are then based on diffusion through a stagnant layer of gas trapped within the pores of the membrane. The length of each fiber is 50 cm with a wall thickness of 50 microns. The inside diameter of a fiber is 400 microns. The blood flow rate through the lumen of the fibers in the device is 5,000 ml/min and the gas flow rate is 5,000 ml/min, both at 37°C and 1 atm. The pO_2 of the entering blood is 40 mm Hg and the exiting blood must be at a pO_2 of 95 mm Hg. The pCO_2 of the entering blood is 45 mm Hg and the exiting blood has a pCO_2 of 40 mm Hg. The entering gas contains no carbon dioxide and is saturated with water. The pO_2 of the entering gas is therefore 713 mm Hg. Estimate the surface area required to meet these specifications for both oxygen and carbon dioxide. The gas-side mass-transfer resistance may be considered negligible because of the negligible gas solubility.

SOLUTION Mathcad® Chart 6.2 illustrates the solution. The calculation starts off by determining the exiting partial pressures of oxygen and carbon dioxide in the gas stream using the mass balances given by Equations 6.39 and 6.44. Note that the membrane area required for oxygen uptake is about 12.3 m^2 whereas that for carbon dioxide removal is 2.5 m^2.

Chart 6.2 for **Example 6.3:** Blood gas oxygenator design.

$L := 50$	cm, length of the hollow fibers
$T := 310$	K, operating temperature
$d := .04$	cm, internal fiber diameter
$d_{external} := .05$	cm, external fiber diameter
$D_{O_2} := 1.7 \times 10^{-5}$	cm^2/sec, oxygen diffusivity in blood
$D_{CO_2} := 1.7 \times 10^{-5}$	cm^2/sec, carbon dioxide diffusivity in blood
$H_{O_2} := .74$	mm Hg/μM, oxygen Henry's constant for blood
$H_{CO_2} := .0022$	mm Hg/μM, carbon dioxide Henry's constant for blood
$P_{O_2} := 10^7$	ml/min m^2 atm, membrane gas permeability for oxygen
$P_{CO_2} := 8 \times 10^6$	ml/min m^2 atm, membrane gas permeability for carbon dioxide
$\varepsilon := .5$	oxygenator void fraction

$Q_b := 5000$ ml/min, blood flow rate

$Q_b = \frac{Q_b}{60}$

$Q_g := 5000$ ml/min, gas flow rate

$Q_g = \frac{Q_g}{60}$

$R := 82.06$ cm^3 atm/mol K, gas constant

$pO_{2b\,in} := 40$ mm Hg, oxygen entering blood partial pressure

$pO_{2b\,out} := 95$ mm Hg, oxygen leaving blood partial pressure

$pO_{2g\,in} := 713$ mm Hg, oxygen entering gas partial pressure

$pCO_{2b\,in} := 45$ mm Hg, carbon dioxide entering blood partial pressure

$pCO_{2b\,out} := 40$ mm Hg, carbon dioxide leaving blood partial pressure

$pCO_{2g\,in} := 0$ mm Hg, carbon dioxide entering gas partial pressure

$n := 2.34$ constant in Hill equation

$P_{50} := 26$ mm Hg, constant in Hill equation

$C_{sat} := 8800$ μM, saturated hemoglobin oxygen

$P_{avg} := 62$ guess of average pO_2 of blood

$$m := n \cdot P_{50}^{n} \cdot H_{O_2} \cdot C_{sat} \cdot \frac{P_{avg}^{(n-1)}}{(P_{50}^{n} + P_{avg}^{n})^2}$$

$m = 25.151$

$$pO_{2g\,out} := pO_{2g\,in} - \left(\frac{Q_b}{Q_g}\right) \cdot \frac{R \cdot T}{H_{O_2}} \cdot (1+m) \cdot (pO_{2b\,out} - pO_{2b\,in}) \cdot \left(\frac{1}{10^6}\right) \cdot \left(\frac{1}{1000}\right) \cdot 760$$

$pO_{2g\,out} = 675.423$ mm Hg, gas stream exiting oxygen partial pressure

$$pCO_{2g\,out} := pCO_{2g\,in} - \left(\frac{Q_b}{Q_g}\right) \cdot \frac{R \cdot T}{H_{CO_2}} \cdot (pCO_{2b\,out} - pCO_{2b\,in}) \cdot \left(\frac{1}{10^6}\right) \cdot \left(\frac{1}{1000}\right) \cdot 760$$

$pCO_{2g\,out} = 43.939$ mm Hg, gas stream exiting carbon dioxide partial pressure

$Sh_b := 3.66$ assume laminar fully developed blood flow

$$k_b := \frac{Sh_b \cdot D_{O_2}}{d}$$

$k_b = 0.0016$ cm/sec, blood side mass-transfer coefficient for oxygen

$$\rho := \frac{1}{(R \cdot 310)} \qquad \rho_{stp} := \frac{1}{(R \cdot 273)}$$

$\rho = 3.931 \times 10^{-5}$ mol/cm^3, gas density at 37°C and 1 atm

$\rho_{stp} = 4.464 \times 10^{-5}$ mol/cm^3, gas density at 0°C and 1 atm

$$K_0 := \frac{1}{\left[\frac{H_{O_2}}{k_b} \times 10^6 \times 1000 + \frac{d}{(\rho_{stp} \cdot P_{O_2}) \cdot \frac{d_{external} - d}{\ln\left(\frac{d_{external}}{d}\right)}} \times 760 \times 100^2 \times 60\right]}$$

$K_0 = 2.102 \times 10^{-12}$ mol/cm^2 sec mm Hg, overall oxygen mass transfer coefficient

$$\alpha := \frac{Q_b \cdot (1+m)}{K_0 \cdot H_{O_2}} \cdot \left(\frac{1}{10^6}\right) \cdot \left(\frac{1}{1000}\right)$$

$\alpha = 1.401 \times 10^6$ cm^2

$$\beta := \left(\frac{Q_b}{Q_g}\right) \cdot \frac{R \cdot T}{H_{O_2}} \cdot (1+m) \cdot 760 \cdot \left(\frac{1}{1000}\right) \cdot \left(\frac{1}{10^6}\right) - 1$$

$\beta = -0.317$

$$\gamma := pO_{2g\,out} - \left(\frac{Q_b}{Q_g}\right) \cdot \frac{R \cdot T}{H_{O_2}} \cdot (1+m) \cdot pO_{2b\,in} \cdot 760 \cdot \left(\frac{1}{1000}\right) \cdot \left(\frac{1}{10^6}\right)$$

$\gamma = 648.094$ mm Hg

$$A_{O_2} := \frac{\alpha}{\beta} \cdot \ln\left(\left(\frac{\beta \cdot pO_{2b\,out} + \gamma}{\beta \cdot pO_{2b\,in} + \gamma}\right)\right) \cdot \left(\frac{1}{100^2}\right)$$

$A_{O_2} = 12.296$ m^2, the membrane area required for oxygen transport

$$k_b := \frac{Sh_b \cdot D_{CO_2}}{d}$$

$k_b = 0.0016$ cm/sec, carbon dioxide blood side mass transfer coefficient

$$K_0 := \frac{1}{\left[\frac{H_{CO_2}}{k_b} \times 10^6 \times 1000 + \frac{d}{(\rho_{stp} \cdot P_{CO_2}) \cdot \frac{d_{external} - d}{\ln\left(\frac{d_{external}}{d}\right)}} \times 760 \times 100^2 \times 60\right]}$$

$K_0 = 7.065 \times 10^{-10}$ mol/cm^2 sec mm Hg, overall carbon dioxide mass-transfer coefficient

$$\alpha := \frac{Q_b}{(K_0 \cdot H_{CO_2})} \cdot \left(\frac{1}{10^6}\right) \cdot \left(\frac{1}{1000}\right)$$

$\alpha = 5.362 \times 10^4$ cm^2

$$\beta := \left(\frac{Q_b}{Q_g}\right) \cdot \frac{R \cdot T}{H_{CO_2}} \cdot 760 \cdot \left(\frac{1}{1000}\right) \cdot \left(\frac{1}{10^6}\right) - 1$$

$\beta = 7.788$

$$\gamma := pCO_{2g\,out} - \left(\frac{Q_b}{Q_g}\right) \cdot \frac{R \cdot T}{H_{CO_2}} \cdot pCO_{2b\,in} \cdot 760 \cdot \left(\frac{1}{1000}\right) \cdot \left(\frac{1}{10^6}\right)$$

$\gamma = -351.515$ mm Hg

$$A_{CO_2} := \frac{\alpha}{\beta} \cdot \ln\left[\frac{\beta \cdot pCO_{2b\,out} + \gamma}{(\beta \cdot pCO_{2b\,in}) + \gamma}\right] \cdot \left(\frac{1}{100^2}\right)$$

$A_{CO_2} = 2.499$ m^2, membrane area for carbon dioxide transport

We can now calculate the amount of gas transferred based on change in gas compositions.

$$Q_{O_2} := Q_g \cdot \frac{pO_{2g\,in} - pO_{2g\,out}}{760} \cdot 60$$

$Q_{O_2} = 247$ ml oxygen (37°C and 1 atm)/min

$$Q_{CO_2} := Q_g \cdot \frac{pCO_{2g\,out} - pCO_{2g\,in}}{760} \cdot 60$$

$Q_{CO_2} = 289$ ml carbon dioxide (37°C and 1 atm)/min

Based on the required change in blood compositions, the gas flow rates are as follows:

$$Q_{O_2} := C_{sat}\left[\frac{pO^n_{2b\,out}}{(P^n_{50} + pO^n_{2b\,out})} - \frac{pO^n_{2b\,in}}{(P^n_{50} + pO^n_{2b\,in})}\right] \cdot Q_b \cdot 60 \cdot \left(\frac{1}{1000}\right) \cdot \left(\frac{1}{10^6}\right) \cdot \left(\frac{1}{\rho}\right)$$

Q_{O_2} = 248 ml (37°C and 1 atm)/min of oxygen delivered to the blood. This value checks with the one calculated above based on gas compositions. Since the value based on the blood compositions is fixed, and the gas one based on mass transfer, we must adjust the value of *m*, or P_{avg}, to make them agree since we have approximated the hemoglobin oxygen-dissociation curve through the value of *m*

$$Q_{CO_2} := \frac{Q_b \cdot 60}{H_{CO_2}} \cdot (pCO_{2b\,in} - pCO_{2b\,out}) \cdot \left(\frac{1}{10^6}\right) \cdot \left(\frac{1}{1000}\right) \cdot \frac{1}{\rho}$$

Q_{CO_2} = 289 ml (37°C and 1 atm)/min, checks with the gas value above mass balances, therefore checkout.

The respiratory exchange ratio is defined as the ratio of carbon dioxide output to oxygen uptake, i.e., Q_{CO2}/Q_{O2}, and should be about 0.8.

$$R := \frac{Q_{CO_2}}{Q_{O_2}}$$

$R = 1.167$ We are removing too much carbon dioxide!

Calculate number of fibers needed

$$N_f := \frac{A_{O_2} \cdot 100^2}{(\pi \cdot d \cdot L)}$$

$N_f = 19570$ number of hollow fibers required in the oxygenator to provide the surface area for oxygen mass transfer

$$V_b := Q_b \cdot \frac{1}{N_f} \cdot \frac{4}{(\pi \cdot d^2)}$$

$V_b = 3.389$ cm/sec, the velocity of blood in a given fiber

$\rho_{blood} := 1.06$ g/cm^3, blood density

$\mu_{blood} := 3.0$ cP, blood viscosity

$$V_b := Q_b \cdot \frac{1}{N_f} \cdot \frac{4}{(\pi \cdot d^2)}$$

$V_b = 3.389$ cm/sec, the velocity of blood in a given hollow fiber

$$Re_b := \frac{\rho_{blood} \cdot d \cdot V_b}{\mu_{blood}} \cdot 100$$

$Re_b = 4.789$

$$Sc_b := \frac{\mu_{blood} \times .01}{\rho_{blood} \cdot D_{O_2}}$$

$Sc_b = 1.665 \times 10^3$

$z_{critical} := .05 \cdot d \cdot Re_b \cdot Sc_b$

$z_{critical} = 15.947$ cm

Since $L > z_{critical}$ we conclude fully-developed flow is reasonable assumption for conservative estimation of the blood-side mass-transfer coefficient.

As shown on the above chart, we need to check our mass balances by calculating the amount of oxygen and carbon dioxide actually transported. In this example, we must change the blood pO_2 from 40 to 95 mm Hg in order to sustain life for the adult cardiac output of 5,000 ml/min. We see that the total amount of oxygen transported, Q_{oxygen}, is about 250 ml/min based on the required change in blood pO_2 levels. This is the value typically reported for an adult.

A proper calculation of Q_{oxygen} based on the gas side changes in pO_2 requires a value of m which is an approximation of the hemoglobin-oxygen dissociation curve, (see Equation 4.28). Therefore, we must adjust the value of m (or $P_{average}$) until the amount of oxygen transferred is the same based on the changes in the blood pO_2 and the gas pO_2 as shown in the chart. This results in a $P_{average}$ of 62 mm Hg which is close to the average of the arterial and venous pO_2 levels of 67.5 mm Hg.

The amount of carbon dioxide exchanged is calculated to be 289 ml/min and is the same whether based on the blood or gas sides. This value for an adult should be about 200 ml/min. Therefore, we are removing too much carbon dioxide which will tend to make the blood basic and lead to alkalosis.

The respiratory exchange ratio (R) is defined as the ratio of carbon dioxide output to oxygen uptake and should be about 0.8. Here we see that R is about 1.17. For a given membrane area, we can control the amount of carbon dioxide removed by adjusting the pCO_2 of the incoming gas. In this example, its value was zero. The next example illustrates the effect of the entering gas pCO_2 on the carbon dioxide removal rate for a given membrane area.

Example 6.4 Using the membrane area calculated for oxygen from Example 6.3, find the entering gas pCO_2 needed to provide an oxygen uptake of 250 ml/min and a carbon dioxide removal of 200 ml/min.

SOLUTION Mathcad® Chart 6.3 shows the solution. Note the solution requires the use of the Mathcad root finder (root) to obtain the exiting gas partial pressures. Equations 6.41 and 6.46 are rearranged to solve for the exiting blood partial pressures. By adjusting the entering gas pCO_2 from 0 up to 15 mm Hg, we can decrease the carbon dioxide output to the required value of 200 ml/min, while keeping the oxygen uptake at 250 ml/min, thus achieving the normal respiratory exchange ratio of 0.8. We also had to increase the membrane area by 0.3 m^2 to maintain the delivery of the same amount of oxygen. This is due to the decreased driving force for oxygen transport as a result of the carbon dioxide that was added to the feed gas.

Chart 6.3 for **Example 6.4:** Blood gas oxygenator design based on membrane area calculated from Chart 6.2 for oxygen.

$L := 50$	cm, length of the hollow fibers
$T := 310$	K, operating temperature
$d := .04$	cm, internal fiber diameter
$d_{external} := .05$	cm, external fiber diameter
$D_{O_2} := 1.7 \times 10^{-5}$	cm^2/sec, oxygen diffusivity in blood
$D_{CO_2} := 1.7 \times 10^{-5}$	cm^2/sec, carbon dioxide diffusivity in blood
$H_{O_2} := .74$	mm Hg/μM, oxygen Henry's constant for blood
$H_{CO_2} := .0022$	mm Hg/μM, carbon dioxide Henry's constant for blood
$P_{O_2} := 10^7$	ml/min m^2 atm, membrane gas permeability for oxygen
$P_{CO_2} := 8 \times 10^6$	ml/min m^2 atm, membrane gas permeability for carbon dioxide
$A := 12.6$	m^2, membrane area based on oxygen
$A := A \times 100^2$	
$\varepsilon := .5$	oxygenator void fraction
$Q_b := 5000$	ml/min, blood flow rate
$Q_b := \frac{Q_b}{60}$	
$Q_g := 5000$	ml/min, gas flow rate
$Q_g := \frac{Q_g}{60}$	
$R := 82.06$	cm^3 atm/mol K, gas constant
$pO_{2b\,in} := 40$	mm Hg, oxygen entering blood partial pressure
$pO_{2b\,out} := 95$	mm Hg, estimate of oxygen leaving blood partial pressure
$pCO_{2g\,in} := 15$	mm Hg, carbon dioxide entering gas partial pressure
$pO_{2g\,in} := 713 - pCO_{2g\,in}$	mm Hg, oxygen entering gas partial pressure
$pO_{2g\,out} := 685$	mm Hg, estimate of oxygen exiting the gas stream
$pCO_{2b\,in} := 45$	mm Hg, carbon dioxide entering blood partial pressure
$pCO_{2b\,out} := 40$	mm Hg, carbon dioxide leaving blood partial pressure
$pCO_{2g\,out} := 35$	mm Hg, estimated exiting carbon dioxide gas partial pressure
$n := 2.34$	constant in Hill equation
$P_{50} := 26$	mm Hg, constant in Hill equation
$C_{sat} := 8800$	μM, saturated hemoglobin oxygen
$P_{avg} := 62$	estimate of average blood pO_2

$$m := n \cdot P_{50}^{n} \cdot H_{O_2} \cdot C_{sat} \cdot \frac{P_{avg}^{(n-1)}}{\left(P_{50}^{n} + P_{avg}^{n}\right)^2}$$

$m = 25.151$

$Sh_b := 3.66$ assume laminar fully developed blood flow

$k_b := \frac{sh_b \cdot D_{O_2}}{d}$

$k_b = 0.0016$ cm/sec, blood side mass-transfer coefficient for oxygen

$\rho := \frac{1}{(R \cdot 310)} \qquad \rho_{stp} := \frac{1}{(R \cdot 273)}$

$\rho = 3.931 \times 10^{-5}$ mol/cm³, gas density at 37° C and 1 atm

$\rho_{stp} = 4.464 \times 10^{-5}$ mol/cm³, gas density at 0° C and 1 atm

$$K_0 := \frac{1}{\left[\frac{H_{O_2}}{k_b} \times 10^6 \times 1000 + \frac{d_{external} - d}{(\rho_{stp} \cdot P_{O_2}) \cdot \ln\left(\frac{d_{external}}{d}\right)} \cdot 760 \times 100^2 \times 60\right]}$$

$K_0 = 2.102 \times 10^{-12}$ mol/cm² sec mm Hg, overall oxygen mass-transfer coefficient

$$\alpha := \frac{Q_b \cdot (1 + m)}{K_0 \cdot H_{O_2}} \cdot \left(\frac{1}{10^6}\right) \cdot \left(\frac{1}{1000}\right)$$

$\alpha = 1.401 \times 10^6$ cm²

$$\beta := \left(\frac{Q_b}{Q_g}\right) \cdot \frac{R \cdot T}{H_{O_2}} \cdot (1 + m) \cdot 760 \cdot \left(\frac{1}{1000}\right) \cdot \left(\frac{1}{10^6}\right) - 1$$

$\beta = -0.317$

Solve for pO_2 in the exiting blood and gas streams

$$\gamma(pO_{2g\,out}) := pO_{2g\,out} - \left(\frac{Q_b}{Q_g}\right) \cdot \frac{R \cdot T}{H_{O_2}} \cdot (1 + m) \cdot pO_{2b\,in} \cdot 760 \cdot \left(\frac{1}{1000}\right) \cdot \left(\frac{1}{10^6}\right)$$

$$pO_{2b\,out}(pO_{2g\,out}) := \frac{\left[(\beta \cdot pO_{2b\,in} + \gamma(pO_{2g\,out})) \cdot \exp\left(\frac{\beta \cdot A}{\alpha}\right)\right] - \gamma(pO_{2g\,out})}{\beta}$$

$$f(pO_{2g\,out}) = pO_{2g\,in} - \left(\frac{Q_b}{Q_g}\right) \cdot \frac{R \cdot T}{H_{O_2}} \cdot (1 + m) \cdot ((pO_{2b\,out})(pO_{2g\,out}) - pO_{2b\,in}) \cdot \left(\frac{1}{10^6}\right) \cdot \frac{760}{1000} - pO_{2g\,out}$$

$pO_{2g\,out} := \mathrm{root}(f(pO_{2g\,out}), pO_{2g\,out})$

$pO_{2g\,out} = 660.416$ mm Hg, exiting oxygen partial pressure in the gas stream

$pO_{2b\,out} := pO_{2b\,out}(pO_{2g\,out})$

$pO_{2b\,out} = 95.01$ mm Hg, exiting oxygen partial pressure in the blood stream

$k_b := \frac{Sh_b \cdot D_{CO_2}}{d}$

$k_b = 0.0016$ cm/sec, carbon dioxide blood side mass-transfer coefficient

$$K_0 := \frac{1}{\left[\frac{H_{CO_2}}{k_b} \times 10^6 \times 1000 + \frac{d_{external} - d}{(\rho_{stp} \cdot P_{CO_2}) \cdot \ln\left(\frac{d_{external}}{d}\right)} \times 760 \times 100^2 \times 60\right]}$$

$K_0 = 7.07 \times 10^{-10}$ mol/cm² sec mm Hg, overall carbon dioxide mass-transfer coefficient

$$\alpha := \frac{Q_b}{(K_0 \cdot H_{CO_2})} \cdot \left(\frac{1}{10^6}\right) \cdot \left(\frac{1}{1000}\right)$$

$\alpha = 5.358 \times 10^4$ cm²

$$\beta := \left(\frac{Q_b}{Q_g}\right) \cdot \frac{R \cdot T}{H_{CO_2}} \cdot 760 \cdot \left(\frac{1}{1000}\right) \cdot \left(\frac{1}{10^6}\right) - 1$$

$\beta = 7.788$

$$\gamma(\mathrm{pCO}_{2g\,\mathrm{out}}) := \mathrm{pCO}_{2g\,\mathrm{out}} - \left(\frac{Q_b}{Q_g}\right)\cdot\frac{R\cdot T}{H_{\mathrm{CO}_2}}\cdot \mathrm{pCO}_{2b\,\mathrm{in}}\cdot 760\cdot\left(\frac{1}{1000}\right)\cdot\left(\frac{1}{10^6}\right)$$

$$\mathrm{pCO}_{2b\,\mathrm{out}}(\mathrm{pO}_{2g\,\mathrm{out}}) := \frac{\left[(\beta\cdot \mathrm{pCO}_{2b\,\mathrm{in}} + \gamma(\mathrm{pCO}_{2g\,\mathrm{out}}))\cdot \exp\left(\frac{\beta\cdot A}{\alpha}\right)\right] - \gamma(\mathrm{pCO}_{2g\,\mathrm{out}})}{\beta}$$

$$f(\mathrm{pCO}_{2g\,\mathrm{out}}) := \mathrm{pCO}_{2g\,\mathrm{out}} - \mathrm{pCO}_{2g\,in}\left(\frac{Q_b}{Q_g}\right)\cdot\frac{R\cdot T}{H_{\mathrm{CO}_2}}\cdot(\mathrm{pCO}_{2b\,\mathrm{out}}(\mathrm{pCO}_{2g\,\mathrm{out}}) - \mathrm{pCO}_{2b\,\mathrm{in}})\cdot\left(\frac{1}{10^6}\right)\cdot\frac{760}{1000}$$

$\mathrm{pCO}_{2g\,\mathrm{out}} := \mathrm{root}\,(f(\mathrm{pCO}_{2g\,\mathrm{out}}), \mathrm{pCO}_{2g\,\mathrm{out}})$

$\mathrm{pCO}_{2g\,\mathrm{out}} = 45$ mm Hg, exiting carbon dioxide partial pressure in the gas stream

$\mathrm{pCO}_{2b\,\mathrm{out}} := \mathrm{pCO}_{2b\,\mathrm{out}}(\mathrm{pCO}_{2g\,\mathrm{out}})$

$\mathrm{pCO}_{2b\,\mathrm{out}} = 41.586$ mm Hg, exiting carbon dioxide partial pressure in the blood stream

We can now calculate the amount of gas transferred based on change in gas composition.

$\mathrm{pO}_{2g\,\mathrm{in}} = 698$

$$Q_{\mathrm{O}_2} := Q_g\cdot\frac{\mathrm{pO}_{2g\,\mathrm{in}} - \mathrm{pO}_{2g\,\mathrm{out}}}{760}\cdot 60$$

$\mathrm{pO}_{2g\,\mathrm{out}} = 660.416$

$Q_{\mathrm{O2}} = 247.266$ ml oxygen (37°C and 1 atm)/min

$$Q_{\mathrm{CO}_2} := Q_g\cdot\frac{\mathrm{pCO}_{2g\,\mathrm{out}} - \mathrm{pCO}_{2g\,\mathrm{in}}}{760}\cdot 60$$

$Q_{\mathrm{CO}_2} = 197.368$ ml carbon dioxide (37°C and 1 atm)/min

based on the required change in blood composition

$$Q_{\mathrm{O}_2} := C_{\mathrm{sat}}\cdot\left[\frac{\mathrm{pO}^n_{2b\,\mathrm{out}}}{(P^n_{50} + \mathrm{pO}^n_{2b\,\mathrm{out}})} - \frac{\mathrm{pO}^n_{2b\,\mathrm{in}}}{(P^n_{50} + \mathrm{pO}^n_{2b\,\mathrm{in}})}\right]\cdot Q_b\cdot 60\cdot\left(\frac{1}{1000}\right)\cdot\left(\frac{1}{10^6}\right)\cdot\left(\frac{1}{\rho}\right)$$

$Q_{\mathrm{O}_2} := 247.791$ ml (37°C and 1 atm)/min of oxygen delivered to the blood, this value checks with the one calculated above based on gas compositions. Since this value for the blood is fixed, and the gas one based on mass transfer, we must adjust the value of m, or P_{avg}, to make them agree, since we have approximated the hemoglobin oxygen-dissociation curve through the value of m

$$Q_{\mathrm{CO}_2} := \frac{Q_b\cdot 60}{H_{\mathrm{CO}_2}}\cdot(\mathrm{pCO}_{2b\,\mathrm{in}} - \mathrm{pCO}_{2b\,\mathrm{out}})\cdot\left(\frac{1}{10^6}\right)\cdot\left(\frac{1}{1000}\right)\cdot\frac{1}{\rho}$$

$Q_{\mathrm{CO}_2} := 197.374$ ml (stp)/min, checks with the gas value above, this value should only be about 200 ml/min

Mass balances therefore checkout.

The respiratory exchange ratio is defined as the ratio of carbon dioxide output to oxygen uptake, i.e., $Q_{\mathrm{CO}_2}/Q_{\mathrm{O}_2}$, and should be about 0.8,

$$R := \frac{Q_{\mathrm{CO}_2}}{Q_{\mathrm{O}_2}}$$

$R := 0.8$

which is the desired value, therefore we are removing the appropriate amount of carbon dioxide and providing the appropriate amount of oxygen for the given blood flow rate of 5000 ml/min

Hollow-fiber membrane blood oxygenators typically have surface areas in the range from about 1.5–5 m^2. In the above examples, it was assumed that blood flowed through the lumen of the hollow fibers in fully developed laminar flow. This assumption provides a conservative or worst-case estimate of oxygenator performance. Hence, the surface areas calculated in the above examples are several times larger than what can be achieved in actual practice. Since the major resistance for oxygen transport resides in the blood, increasing the blood side oxygen mass-transfer coefficient can have a significant effect on the size of the oxygenator. This can be accomplished by such methods as pulsating flow, small obstructions in the fiber to cause radial mixing, or a helical arrangement of the hollow fibers to induce secondary flows (swirls). Having the blood flow across the outside of the hollow fibers can also significantly increase the mass transfer rate of oxygen into the blood (Catapano, Wodetzki, and Baurmeister 1992; Vaslef et al. 1994). This is a result of repetitive mixing of the blood as it crosses successive hollow fibers, effectively decreasing the boundary layer resistance to oxygen transport.

Generally, it is recommended that experiments be performed to measure the oxygen-transport rates for a given oxygenator design. Mathematical models, such as the one developed above, can then be used to correlate the results, and provide accurate measurements of the blood-side mass-transfer coefficient. A calibrated model can then be used to explore operation under a wide range of operating conditions and assist in the optimal development of the device. For example, Vaslef et al. (1994) were able to assess the performance of cross flow hollow-fiber oxygenators using water. Subsequent tests using bovine blood provided excellent comparisons between their measurements and model predictions of the oxygen transport rates.

6.8 IMMOBILIZED ENZYME REACTORS

6.8.1 Background

Our final example of extracorporeal devices is the use of immobilized enzyme reactors to chemically change a species found in the blood. An enzyme is a protein that acts as a biochemical catalyst and offers great specificity in terms of the types of chemical species or substrates it acts on. Enzymes are usually named after the substrate whose reaction is catalyzed. For example, the enzyme that hydrolyzes urea is called *urease*. Note that the suffix *-ase* is usually added to a portion of the name for the substrate the enzyme acts on.

Figure 6.3 shows several reactor arrangements that are possible when an enzyme is attached to or immobilized within a solid support. Enzyme immobilization offers several advantages. First, immobilization keeps the enzyme out of the bulk solution, which in the case of blood returning to the body, could result in an allergic reaction to the foreign protein of the enzyme. Secondly, immobilization offers the potential to reuse the enzyme, which may in fact be quite expensive. Finally, in many cases, the enzyme is stabilized (less labile) when immobilized, retaining its activity for longer periods of time.

6.8.2 Examples of Medical Applications of Immobilized Enzymes

One example where immobilized enzyme reactors have been proposed is in the treatment of neonatal jaundice (Lavin et al. 1985; Sung et al. 1986). Newborns tend to have higher levels of the greenish-yellow pigment *bilirubin* than those found in adults. Bilirubin is a natural product derived from red blood cells after these cells have lived out their lifespan of about 120 days. It is formed from the heme portion (the four pyrole rings) of the hemoglobin molecule after removal of the iron. The bilirubin binds to plasma albumin for transport to the liver where it is finally excreted from the body in the bile fluids.

The fetus's bilirubin readily crosses the placenta and is removed by the mother's liver. However, in the period after birth, the infant's liver is not fully functional for the first week, resulting in increased levels of plasma bilirubin. In some cases, the infant's bilirubin levels are sufficiently high resulting in a *jaundiced* (yellow) appearance to the skin. High levels of plasma bilirubin can be toxic to a variety of tissues and in these cases jaundiced infants are commonly treated by *phototherapy* or blood transfusions. In phototherapy, the infant is placed under a blue light that converts the bilirubin to a less toxic byproduct. Phototherapy through the skin is not capable of controlling cases of severe jaundice. However, blood transfusions can replace the infant's blood with adult blood effectively diluting the infant's plasma bilirubin levels. However, blood transfusions pose their own risk, particularly infectious diseases such as hepatitis and acquired immunodeficiency syndrome (AIDS).

An alternative approach to the treatment of neonatal jaundice would be the use of a bilirubin-specific enzyme for removal of bilirubin from the infant's blood (Lavin et al. 1985; Sung et al. 1986). The enzyme bilirubin oxidase catalyzes the oxidation of bilirubin according to the following reaction stoichiometry.

$$\text{bilirubin} + \frac{1}{2}O_2 \rightarrow \text{biliverdin} + H_2O \qquad (6.48)$$

Calculations indicate that the amount of oxygen needed to convert all of the bilirubin found in the blood is about 100 times less than the actual oxygen content of blood. Therefore, no external supply of oxygen is needed within the enzyme reactor to carry out this reaction. Biliverdin itself is much less toxic than bilirubin and in fact is further oxidized by bilirubin oxidase to other less toxic substances. Experiments using a water-jacketed reactor (much like that in Figure 6.3a) containing bilirubin oxidase covalently attached to agarose beads showed that plasma bilirubin levels in rats decreased by 50% after 30 minutes of treatment. The rat's blood was recirculated through the 6-ml reactor volume at a flow rate of 1 ml/min. Clearly these results indicate that an immobilized bilirubin oxidase reactor could be a new approach for the treatment of neonatal jaundice. It also shows the feasibility of using immobilized enzyme reactors for the specific removal of species present in the blood.

Cells found in the liver and other organs carry out a wide variety of life-sustaining enzymatic reactions. There is considerable interest in using these cells or their enzymes to treat liver failure and other enzyme-deficiency diseases. The discussion here focuses on the use of just the key enzymes. The use of immobilized cells as bioartificial organs, perhaps using liver cells (hepatocytes), is discussed in greater detail in Chapter 8. However, it is important to recognize that some of the techniques used in this chapter to design immobilized enzyme reactors are directly applicable to systems that employ immobilized cells.

Another example of an immobilized enzyme reactor that we will look at in considerably more detail is that for the removal of *heparin* (Bernstein, Yang, and Langer

1987a,b); recall that heparin is used as an anticoagulant in extracorporeal treatments such as hemodialysis and blood oxygenators. Heparin is a large negatively-charged conjugated polysaccharide molecule that is produced by many types of cells in the body. By itself, heparin has little anticoagulant activity at the typical concentrations found in blood. However, in some regions of the body such as the liver and lungs, it is produced in greater amounts. It therefore has an important role in preventing blood clots in the slow moving venous blood flow entering the capillaries of the lungs and liver. By combining with *antithrombin III*, it increases by several orders of magnitude the ability of antithrombin to remove *thrombin*. Thrombin is an enzyme that converts the plasma protein fibrinogen into fibrin leading to the fibrous mesh-like structure characteristic of a blood clot. Therefore, this synergistic combination of heparin with antithrombin III results in a powerful anticoagulant.

Over 20 million extracorporeal procedures using heparin are performed each year and in about 15% of these, complications due to heparin arise. Certainly removal of heparin from the blood before it is returned to the body could significantly improve the safety of these procedures. The enzyme heparinase has the ability to degrade heparin into less harmful byproducts and one could envision an immobilized heparinase reactor for the removal of heparin from the blood returning to the patient's body.

6.8.3 Enzyme Reaction Kinetics

Our goal now is to develop relationships that can be used to describe the rates or kinetics of chemical reactions that are catalyzed by enzymes. Description of the enzyme reaction kinetics will allow us to develop mathematical models for immobilized enzyme reactors that can be used to analyze experimental data, provide information for scaleup, and explore the effects of operating conditions on reactor performance. The heparinase reactor will be used as a model reaction system for these discussions.

The successful development of an immobilized enzyme reactor requires knowledge of the enzyme kinetics, understanding of the effects of diffusion, and a design equation for the specific reactor that is used (Fogler, 1992). Each of these is discussed in greater detail in the following discussion.

First we need to define the kinetics or the rate of the enzyme reaction and how it depends on the concentration of the substrate (reactant). This usually entails defining the free enzyme kinetics (enzyme in solution and not on a solid support) and the kinetics after the enzyme has been immobilized on the support material. In some cases, immobilization has no effect on the activity of the enzyme, whereas sometimes immobilization can significantly alter the kinetics of the conversion process. Generally, the kinetics of enzyme reactions are described by the Michaelis-Menten model given by the following equation.

$$r_S = \frac{k_{cat} E_t S}{K_m + S} = \frac{V_{max} S}{K_m + S} \tag{6.49}$$

The reaction rate, r_S, has units of moles/reaction volume/time. S represents the substrate or reactant concentration in units of moles/volume. V_{max} represents the maximum reaction rate for a given total enzyme concentration E_t (Units/reaction volume), where

$V_{max} = k_{cat}E_t$, and k_{cat} is the reaction rate constant (moles/Units/time). Enzyme activity is commonly expressed in terms of "Units" (U), and for heparinase, a unit of activity is defined as the amount of enzyme required to degrade 1 mg of heparin per hour. The amount or mass of enzyme needed is dependent on the enzyme purity and is usually reported as so many units of activity per mg of enzyme. K_m is the Michaelis constant and may be thought of as that substrate concentration at which the reaction rate is equal to one-half the maximum rate (V_{max}). It is important to note that at high substrate concentrations, the reaction rate saturates at V_{max} because in total, there are not enough active sites on the enzyme molecules for substrate reactions. The reaction rate is then independent of the substrate concentration and is therefore said to be zero order. At low substrate concentrations ($S \ll K_m$), the reaction rate is linearly proportional to the substrate concentration and is therefore said to be first order.

Example 6.5 The following table of data was obtained (Bernstein, Yang, and Langer 1987b) for the kinetics of heparin degradation by heparinase in solution at 37°C and a pH of 7.4. Heparinase activity was also found not to be affected by immobilization on a cross-linked agarose, therefore these soluble or free enzyme kinetics can be used as well when the enzyme is immobilized. Note that it is convenient for the subsequent analysis to divide the enzyme reaction rate by the corresponding enzyme concentration, thus expressing the rate on a per unit amount of enzyme that is present in the reactor. One unit of heparinase (1U) is defined as the amount of enzyme required to degrade 1 mg heparin per hour. From this data determine the value of K_m and k_{cat}.

Heparin degradation kinetics

Enzyme loading (Units/ml)	Heparin concentration (mg/ml)	Reaction rate (mg/ml/hr)	Normalized reaction rate (mg/U/hr)
24	1	20.83	0.868
	0.5	18.18	0.758
	0.25	13.70	0.571
	0.1	11.11	0.463
	0.05	7.58	0.316
43	1	38.46	0.894
	0.5	32.26	0.750
	0.25	30.30	0.705
	0.1	23.81	0.554
	0.05	16.67	0.388
57	1	50	0.877
	0.5	40	0.702
	0.25	38.46	0.675
	0.1	26.32	0.462
67	1	55.56	0.829
	0.5	52.63	0.786
	0.25	50	0.746
	0.1	35.71	0.533

From Bernstein et al. 1987b.

SOLUTION Mathcad® Chart 6.4 shows the solution. For heparin degradation by heparinase, we find that $K_m = 0.078$ mg/ml and $k_{cat} = 0.891$ mg/U/hr.

Chart 6.4 for **Example 6.5:** Calculation of Michaelis-Menten rate parameters using the data for heparin degradation.

$i := 0 \ldots 17$ index of data points

mg/Uhr, reaction rate and mg/ml, substrate concentration

$R_i :=$	$S_i :=$
.868	1
.758	.5
.571	.25
.463	.1
.316	.05
.894	1
.750	.5
.705	.25
.554	.1
.388	.05
.877	1
.702	.5
.675	.25
.462	.1
.829	1
.786	.5
.746	.25
.533	.1

Use the Lineweaver-Burke method to find k_{cat} and K_m. This requires that we make a plot of $1/R$ versus $1/S$. This plot of the data should be linear. A linear regression can also be performed to find the values of k_{cat} and K_m as shown below.

$x_i := \frac{1}{S_i}$ $y_i := \frac{1}{R_i}$ dummy variables for linear regression

$m := \text{slope}(x, y)$ $m = 0.088$

$b := \text{intercept}(x, y)$ $b = 1.123$

$r := \text{corr}(x, y)$ $r = 0.961$ indicates a very good fit of the data

$k_{cat} := \frac{1}{b}$ $k_{cat} = 0.891$ mg/U/hr, the rate constant

$K_m := m \cdot k_{cat}$ $K_m = 0.078$ mg/ml, the Michaelis-Menten constant

Now let's compare the data to the Michaelis-Menten rate expression

$R_{prediction_i} := \frac{k_{cat} \cdot S_i}{K_m + S_i}$ rate based on regression values of k_{cat} and K_m

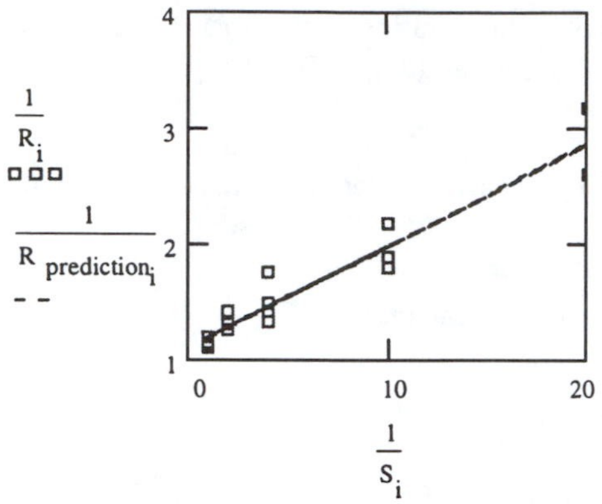

We observe in the plot that the Michaelis-Menten model provides excellent representation of the heparin degradation kinetics.

6.8.4 Reaction and Diffusion in Immobilized Enzyme Systems

Next a reaction-diffusion model is needed to describe how the substrate diffuses through the porous support particle and reacts through the action of the immobilized enzyme. We also need to approximate the support geometry and here we will assume it is spherical. Extensions to other support geometries is straightforward once a given geometry has been examined in detail. In some cases, we may even need to know how the enzyme is distributed throughout the support particle (Bernstein, Yang, and Langeret 1987a). In the absence of such information, we simply assume the enzyme is uniformly distributed.

Diffusion through the support material also requires that we know the substrate diffusivity (D) in the bulk or free solution (see Figure 2.1). We also need to know how the solute diffusivity is affected by the properties of the support itself, specifically the porosity, pore size, and tortuosity. For substrates that have molecular dimensions that are comparable to the pore size, we will also need to include the effects of steric exclusion and hindered diffusion as discussed earlier in Section 2.6. These factors combine to define the effective diffusivity (D_e) of the substrate in the support particle as given by the next equation.

$$D_e = \frac{\varepsilon D}{\tau}\omega_r \tag{6.50}$$

In this equation, ε represents the porosity of the support, τ is the tortuosity of the pores, and ω_r represents the reduction of the substrate diffusivity due to the proximity of the pore wall (Equation 2.11). In this discussion, the substrate is already in the pore so there is no steric exclusion. We will see that steric exclusion shows up in the surface boundary equation of the differential equation that describes the reaction-diffusion phenomena occurring in the porous support.

Figure 6.10 illustrates a spherical shell of thickness Δr lying within the support particle containing a uniformily distributed and immobilized enzyme. The particle itself

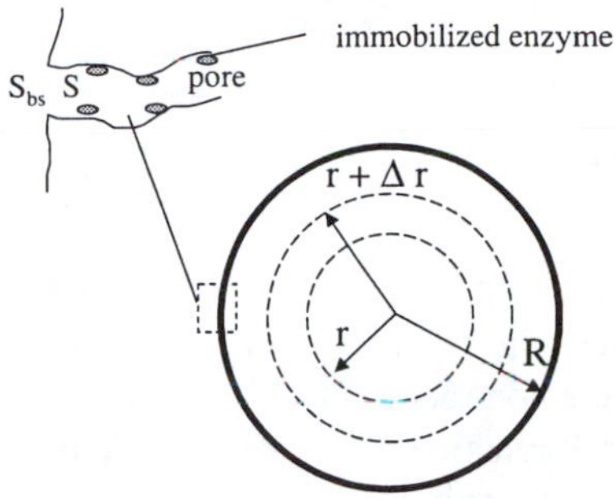

Figure 6.10 Shell balance on a spherical immobilized enzyme particle.

is porous and we can write the following steady-state shell balance on the substrate.

$$4\pi r^2 D_e \left.\frac{dS}{dr}\right|_{r+\Delta r} - 4\pi r^2 D_e \left.\frac{dS}{dr}\right|_{r} = 4\pi r^2 \Delta r \left(\frac{V_{\max} S}{K_m + S}\right) \tag{6.51}$$

This equation simply states that the amount of substrate reacted within the shell volume is equal to the difference in the amount of substrate entering and leaving the shell volume by diffusion. After dividing by $4\pi \Delta r$, and taking the limit as Δr approaches zero, we obtain the reaction-diffusion equation for the substrate in the particle.

$$D_e \frac{d^2S}{dr^2} + \frac{2D_e}{r}\frac{dS}{dr} = \frac{V_{\max} S}{K_m + S} \tag{6.52}$$

The boundary conditions are:

$$\begin{aligned} &\text{BC1:} \quad r = 0, \frac{dS}{dr} = 0 \\ &\text{BC2:} \quad r = R, S = KS_{bs} \end{aligned} \tag{6.53}$$

Once again K represents the partition coefficient which may include effects other than simple steric exclusion, for example electrostatic attraction or repulsion due to the presence of surface charges on the substrate and the particle. It is defined as the ratio of the substrate concentration on the pore side of the support surface (S) to that on the surface on the bulk side (S_{bs}) as shown in Figure 6.10. The substrate concentration at the surface of the support (S_{bs}) may differ from the bulk substrate concentration (S_b) because of external mass-transfer effects. This is discussed later.

Solution of Equations 6.52 and 6.53 is facilitated through the definition of the following dimensionless variables: $s' = S/KS_{bs}$ and $r' = r/R$. Upon substitution of these dimensionless variables the following result is obtained:

$$\frac{d^2s'}{dr'^2} + \frac{2}{r'}\frac{ds'}{dr'} = 9\phi^2 \frac{s'}{1 + \beta s'} \tag{6.54}$$

The dimensionless boundary conditions are given by:

$$\begin{aligned} &\text{BC1:} \quad r' = 0, \frac{ds'}{dr} = 0 \\ &\text{BC2:} \quad r' = 1, s' = 1 \end{aligned} \tag{6.55}$$

Two additional parameters, ϕ and β also result.

$$\phi = \frac{R}{3}\left[\frac{V_{\max}}{K_m D_e}\right]^{1/2} \qquad \beta = \frac{S_{bs} K}{K_m} \tag{6.56}$$

The quantitity ϕ is especially important and is known as the *Thiele modulus*. The square of this quantity represents the ratio of the substrate reaction rate to its diffusion rate. Its magnitude allows one to determine whether the overall reaction rate is *reaction limited* (small ϕ, minimal intraparticle substrate concentration gradient) or *diffusion limited* (large ϕ, significant intraparticle substrate concentration gradient). The factor $R/3$ is the ratio of the support volume (V) to its external surface area (A). For non-spherical geometries, one can replace $R/3$ with this ratio, i.e. V/A. β is a dimensionless Michaelis constant. Large values of β indicate a zero order reaction, whereas $\beta \to 0$ indicates a first order reaction.

6.8.5 Solving the Immobilized Enzyme Reaction-Diffusion Model

The solution to Equation 6.54 for the dimensionless substrate concentration profile in the immobilized enzyme support is usually defined in terms of the effectiveness factor represented by the symbol η. Because of substrate reaction and diffusional effects, the substrate concentration will generally decrease in the radial direction as one enters the support. This radial decrease in substrate concentration in the support will result in a decrease in the volume-averaged reaction rate for the support in comparison to the reaction rate that would be possible if the substrate concentration was uniform throughout the support and equal to its surface concentration. The effectiveness factor is used to describe this decrease in observed reaction rate and is defined by the following expression:

$$\eta = \frac{\text{observed particle rate}}{\text{rate obtained with no concentration gradient within the particle}} = \frac{4\pi R^2 D_e \left.\frac{dS}{dr}\right|_{r=R}}{\frac{4}{3}\pi R^3 \left(\frac{V_{\max} S|_R}{K_m + S|_R}\right)} = \frac{\left.\frac{ds'}{dr'}\right|_{r'=1}}{\left(\frac{3\phi^2}{1+\beta}\right)} \tag{6.57}$$

The effectiveness factor (η) varies from 0 (diffusion limited) to 1 (reaction limited). Determination of the effectiveness factor using the nonlinear Michaelis-Menten rate equation requires a numerical solution of Equations 6.54 and 6.55 for various values of β. It is also important to point out that η, which depends on S_{bs}, will depend on position within the reactor. So the solution for η will be dependent on the local value of S_{bs} as found by the solution to the reactor design equation to be discussed below.

Example 6.6 Determine the effectiveness factor for the case where the Thiele modulus and the dimensionless Michaelis constant are both unity. Assume the particle is spherical and the kinetics are described by the Michaelis-Menten model.

SOLUTION Mathcad® Chart 6.5 illustrates the solution. The value of η for these conditions is found to be 0.857. The solution of Equations 6.54 and 6.55 represents a two-point boundary value problem and we must adjust the dimensionless concentration at the particle center to achieve the required surface boundary condition, that is, the dimensionless surface concentration is unity.

Chart 6.5 for **Example 6.6:** Calculation of the effectiveness in a spherical particle for Michaelis-Menten kinetics.

$\phi := 1$ the Thiele modulus

$\beta := 1$ dimensionless Michaelis constant

$s := \begin{pmatrix} .445 \\ 0 \end{pmatrix}$ defines initial conditions at the particle center, the first value is an estimate of the dimensionless concentration and will be adjusted manually to satisfy the surface boundary condition that requires the dimensionless concentration to equal unity, the second value is the requirement that the first derivative be equal to zero at the particle center

$\varepsilon := 10^{-6}$ a dummy variable to prevent divide by zero at the particle center

$N := 20$ number of integration intervals

The following function $D(r, s)$ defines the first derivative and the second derivative of the dimensionless concentration.

$$D(r, s) := \begin{bmatrix} s_1 \\ -\dfrac{2}{(r+\varepsilon)} \cdot s_1 + 9 \cdot \phi^2 \cdot \dfrac{s_0}{(1+\beta \cdot s_0)} \end{bmatrix}$$

$Z := RKFIXED(s, 0, 1, N, D)$ evaluate the solution at N points between 0 and 1

Z is a matrix containing the solution, the first column of data is the dimensionless radial position, the second column is the dimensionless concentration, and the third column is the first derivative.

Now plot the dimensionless concentration profile as a function of the dimensionless radial distance.

$i := 0 \ldots N$

Manually adjust the value of s_0 until the dimensionless concentration at the particle surface is equal to 1.

$(Z^{\langle 1 \rangle})_N = 1.001$ this value should equal 1, which is close enough

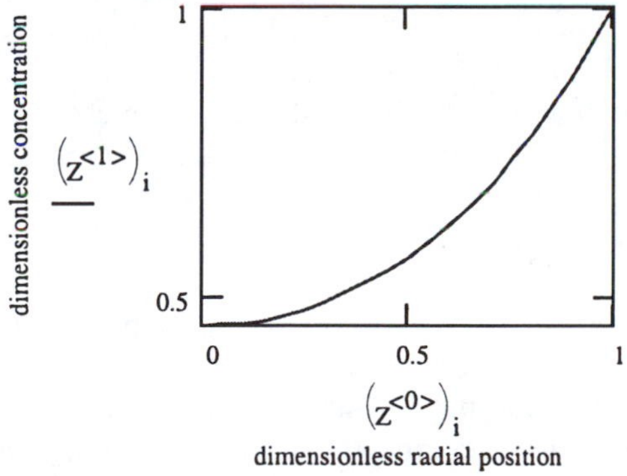

With the dimensionless concentration profile known we can determine the effectiveness factor through the use of Equation 6.57.

$$\eta = \frac{(Z^{\langle 2 \rangle})_N}{\left[\dfrac{3 \cdot \phi^2}{(1+\beta)}\right]}$$

$\eta = 0.857$ The value of the effectiveness factor for the given values of β and ϕ.

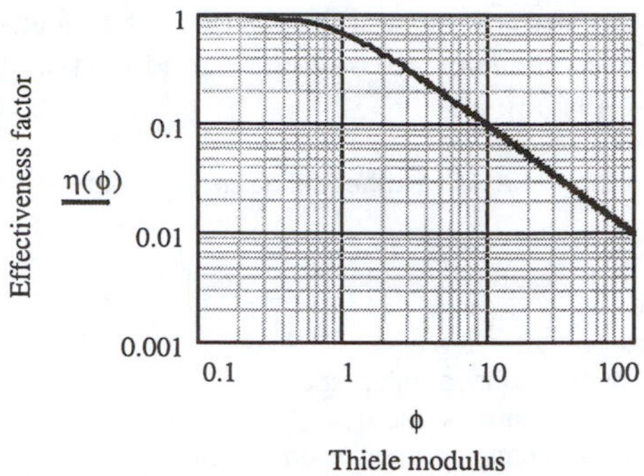

Figure 6.11 Effectiveness factor for a first-order reaction.

6.8.6 Special Case of a First-Order Reaction

If $K_m > KS_{bs}$, then the Michaelis-Menten kinetic model becomes that of a first-order reaction. For a first-order reaction, Equations 6.54 and 6.55 may be solved analytically resulting in the following relationship between η and ϕ. This relationship between the effectiveness factor and the Thiele modulus for a first-order reaction is also shown in Figure 6.11. Note that this solution is independent of the particle surface concentration and hence position in the reactor.

$$\eta = \frac{1}{\phi}\left(\frac{1}{\tanh 3\phi} - \frac{1}{3\phi}\right) \tag{6.58}$$

For large values of ϕ, the value of η approaches $1/\phi$.

6.8.7 Observed Reaction Rate

With the effectiveness factor known, we can write the observed reaction rate for the immobilized enzyme as follows:

$$r_S = \eta \frac{V_{\max} KS_{bs}}{K_m + KS_{bs}} \tag{6.59}$$

where S_{bs} is the substrate concentration at the support surface and η accounts for the reduction in the reaction rate due to the internal substrate concentration gradient.

6.8.8 External Mass-Transfer Resistance

Flow of blood or other fluids around the immobilized enzyme support results in the formation of a thin stagnant layer of fluid along the surface of the enzyme particle. This stagnant fluid layer provides an additional resistance for mass transfer that must be accounted for. An external mass-transfer coefficient (k_{ext}) is defined to account for this flow-induced external mass-transfer resistance. Conservation of mass requires for a given particle that the following steady state relationship holds between the external mass transfer rate and the reaction rate within the support.

$$4\pi R^2 k_{\text{ext}}(S_b - S_{bs}) = \frac{4}{3}\pi R^3 \eta \left(\frac{V_{\max} KS_{bs}}{K_m + KS_{bs}}\right) \tag{6.60}$$

This equation may be solved for the substrate concentration on the surface of the particle (S_{bs}) in terms of the bulk substrate concentration (S_b). Recall it is this surface concentration that is used in the calculation of the effectiveness factor as discussed above. Correlations for the external mass-transfer coefficient may be found in the literature (Cussler 1984). The following equation is given by Cussler for the external mass-transfer coefficient in packed beds.

$$\frac{k_{\text{ext}}}{V_0} = 1.17\left(\frac{dV_0}{\nu}\right)^{-0.42}\left(\frac{\nu}{D}\right)^{-0.67} \tag{6.61}$$

In this equation, V_0 is the superficial velocity of the fluid, i.e., the volumetric flow rate of the fluid divided by the cross-sectional area of the vessel that contains the support particles of diameter d. The kinematic viscosity (υ) is defined as the ratio of the fluid viscosity and the fluid density, i.e., μ/ρ. Generally, one would want to operate the reactor in such a manner that the external mass-transfer resistance is negligible, hence $S_{bs} = S_b$. This can be accomplished by high bulk flow rates in the case of a plug flow reactor or intense mixing in a well-mixed reactor.

6.8.9 Reactor Design Equations

Finally we need a reactor design equation that, in concert with the above information on the enzyme kinetics and diffusional effects, provides a relationship as to how the substrate concentration varies within the reaction volume. One usually assumes that the reaction mixture either flows through the reactor in plug flow, leading to a change in substrate concentration in the direction of flow, or that the reaction volume is well-mixed, resulting in a uniform substrate concentration throughout the reaction volume.

6.8.9.1 Packed-bed reactor Figure 6.12 shows a packed-bed immobilized enzyme reactor. Q_b is the entering flow rate of the fluid, which could for example be blood or plasma. The substrate concentration entering the reactor is represented by S_b^{in}. A steady-state shell balance may also be performed on a section of the reactor volume ($A_{xs}\Delta x$) where A_{xs} is the cross-sectional area of the reactor. It is assumed that the fluid flows through the reactor in plug flow, hence the velocity does not change with radial position.

$$Q_b S_b|_x - Q_b S_b|_{x+\Delta x} = (1 - \varepsilon_R)\Delta x A_{xs}\eta\left(\frac{V_{\max} K S_{bs}}{K_m + K S_{bs}}\right) \tag{6.62}$$

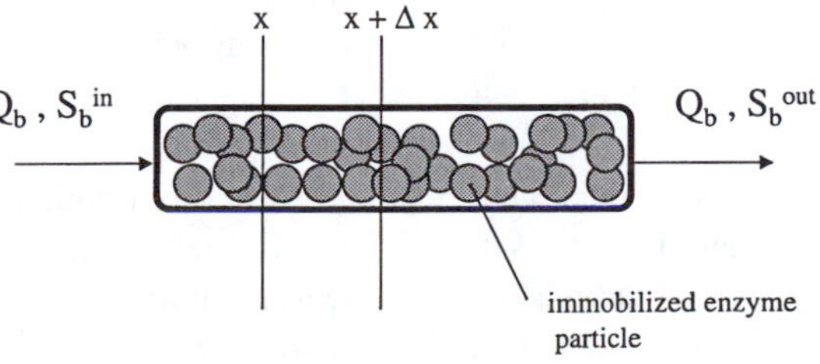

Figure 6.12 Shell balance for a packed-bed immboilized enzyme reactor.

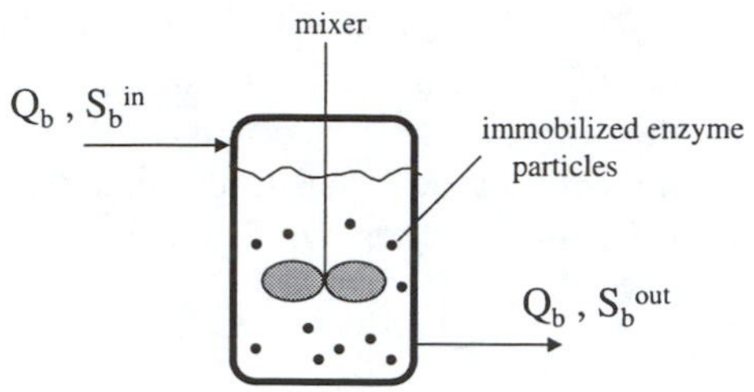

Figure 6.13 A well-mixed immobilized enzyme reactor.

In this equation, ε_R represents the void volume in the reactor. Dividing by Δx and taking the limit as $\Delta x \to 0$, results in the following differential equation:

$$Q_b \frac{\mathrm{d}S_b}{\mathrm{d}x} = -(1-\varepsilon_R) A_{xs} \eta \left(\frac{V_{\max} K S_{bs}}{K_m + K S_{bs}} \right) \tag{6.63}$$

$$\text{BC1:} \quad x = 0,\ S_b = S_b^{\text{in}}$$

This equation may be solved numerically along with Equation 6.60 which provides the relationship between S_b and S_{bs} due to the effects of external mass transfer resistance. Since β also depends on S_{bs} (see Equation 6.56) which depends on x, we also need to evaluate the value of η as a function of x, as well.

6.8.9.2 Well-mixed reactor The other type of commonly used immobilized enzyme reactor is the well-mixed reactor shown in Figure 6.13. Perfect mixing results in a uniform substrate concentration throughout the reactor volume. The exiting concentration of the substrate is also the same as that within the reactor. A steady state substrate mass balance provides the following equation that can be solved for the exiting substrate concentration.

$$Q_b S_b^{\text{in}} = (1-\varepsilon_R) V_{\text{reactor}} \eta \left(\frac{V_{\max} K S_{bs}^{\text{out}}}{K_m + K S_{bs}^{\text{out}}} \right) + Q_b S_b^{\text{out}} \tag{6.64}$$

This equation may be solved, along with Equation 6.60, and a value of the effectiveness factor evaluated at S_{bs}^{out}, to determine the exiting substrate concentration S_b^{out}.

The substrate conversion for both types of reactors (X_S) is defined by the next equation.

$$X_S \equiv 1 - \frac{S_b^{\text{out}}}{S_b^{\text{in}}} \tag{6.65}$$

Substrate conversion is a convenient measure of reactor performance since it lies between 0 (no conversion) and 1 (complete conversion). The following example illustrates the solution to the above equations for a well-mixed reactor.

Example 6.7 The following experimental data were obtained for an immobilized heparinase reactor (Bernstein, Yang, and Langer 1987b) for two different inlet heparin concentrations and different enzyme loadings. Heparinase was immobilized on a cross-linked agarose support. The flow pattern in the reactor was well-mixed.

Experimental data for a heparin reactor

Total support volume (ml)	Enzyme loading (U/ml) in the support	Substrate conversion (X_S) $S_b^{in} = 0.2$ or 0.5 mg/ml
80	100	0.58 or 0.54
85	150	0.66 or 0.62
96	120	0.69 or 0.55
90	130	0.63 or 0.58

From Bernstein, Yang, and Langer 1987b.

The diffusivity of heparin is 1.2×10^{-6} cm^2/sec. The porosity of the agarose support was 0.92 and the tortuosity was unity. The radius of a heparin molecule is 1.5 nm and the pores in the agarose support have a radius of 35 nm. The partition coefficient, K, was found to be equal to 0.36. The agarose support had a radius of 0.0112 cm. The flow rate through the reactor was 120 ml/min. The external mass-transfer coefficient was estimated to be 0.154 cm/min. Using the kinetic parameters obtained in Example 6.5, use the mathematical model for a well-mixed immobilized enzyme reactor to predict the substrate conversions. Compare the results to the experimental conversions in the above table.

SOLUTION Mathcad^R Chart 6.6 summarizes the solution to this problem. We see that the reaction-diffusion model in combination with the well-mixed reactor model and the enzyme kinetics from Example 6.5 provides a remarkably good prediction of the exiting substrate conversions. It can also be shown that for the most part, the external mass-transfer resistance is negligible.

Chart 6.6 for **Example 6.7:** Prediction of heparin conversion in a well-mixed reactor.

$k_{cat} := .891$ mg/U/hr, rate constant from Example 6.5

$k_{cat} := \frac{k_{cat}}{3600}$

$K_m := .078$ mg/ml, Michaelis constant from Example 6.5

$D := 1.2 \times 10^{-6}$ cm^2/sec, heparin diffusivity

$\varepsilon := .92$ porosity of the enzyme particle

$r_0 := .0112$ cm, enzyme particle radius

$r_m := 1.5$ nm, heparin molecular radius

$r_p := 35$ nm, pore radius in the enzyme particle

$K := .36$ partition coefficient

$Q_b := 120$ ml/min, flow rate through the reactor

$Q_b := \frac{Q_b}{60}$

$k_{ext} := .154$ cm/min, external particle mass transfer coefficient

$k_{ext} := \frac{k_{ext}}{60}$

We have two inlet compositions and 4 enzyme loadings; let index i = inlet composition and index j := enzyme loading.

$i := 1 \ldots 2$ $\quad$ $j := 1 \ldots 4$

mg/ml, inlet heparin composition U/ml, enzyme loading

$s_{bin_i} :=$

.2
.5

$E_j :=$

100
150
120
130

ml, total volume of the enzyme particles

$V_{b_j} :=$

80
85
96
90

Now calculate the effective diffusivity.

$\lambda := \frac{r_m}{r_p}$ $\quad\omega_r := 1 - 2.1 \cdot \lambda + 2.09 \cdot \lambda^3 - .95 \cdot \lambda^5$ $\quad D_e := \frac{\varepsilon \cdot D}{1} \cdot \omega_r$

$D_e = 1.005 \times 10^{-6}$ $\quad$ cm^2 sec, effective diffusivity

Now calculate the Thiele modulus for each of the 4 enzyme loadings.

$$\phi_j := \frac{r_0}{3} \cdot \left[\frac{k_{cat} \cdot E_j}{(K_m \cdot D_e)}\right]^{.5}$$

the Thiele moduli for the 4 enzyme loadings

ϕ_i

2.098
2.569
2.298
2.392

At this point we need to calculate the effectiveness factors which depend on β and the unknown particle surface concentration that is related to the bulk or exit concentration which is not known yet; we can estimate the values of η by assuming the reaction is first order which gives the values of η by Equation 6.58.

$$\eta_{(1,j)} := \frac{1}{\phi_j} \cdot \left[\frac{1}{\tanh(3 \cdot \phi_j)} - \frac{1}{(3 \cdot \phi_j)}\right]$$

Initial estimates of the values of the effectiveness factors for each enzyme loading assuming the reaction is first-order; these values will be adjusted using Chart 6.5 until convergence.

$\eta_{(2,j)} := \eta_{(1,j)}$

After using Chart 6.5 and iterating several times we get:

$\eta_{(1,j)} :=$	$\eta_{(2,j)} :=$
.52	.62
.43	.51
.47	.56
.46	.54

We can now calculate the surface concentration for each inlet concentration and enzyme loading.

$s_{bs} := .1$ mg/ml, initial guess of the particle surface concentration

Equation 6.64 can be solved for the exiting bulk substrate concentration which can be substituted into Equation 6.60 to provide the following relationship between the inlet and surface concentrations.

$$f(s_{bs}) := (s_{b\,in_i} - s_{bs}) - \left[\frac{\eta_{(i,\,j)} \cdot k_{cat} \cdot E_j \cdot K \cdot s_{bs}}{(K_m + K \cdot s_{bs})}\right] \cdot \left[\frac{V_{b_j}}{Q_b} + \frac{r_0}{(3 \cdot k_{ext})}\right]$$

$s_{bs_{(i,\,j)}} := \mathrm{root}(f(s_{bs}), s_{bs})$

$s_{bs_{(i,\,j)}} :=$ mg/ml, the surface concentration for feed concentration = 0.2 mg/ml

0.07
0.056
0.057
0.057

$s_{bs_{(2,\,j)}}$ mg/ml, the surface concentration for feed concentration = 0.5 mg/ml

0.197
0.154
0.156
0.159

the substrate concentration leaving the reactor can be found from Equation 6.64

$$s_{b\,out_{(i,\,j)}} := s_{b\,in_i} - \frac{V_{b_j} \cdot \eta_{(i,\,j)} \cdot k_{cat} \cdot E_j \cdot K \cdot s_{bs_{(i,\,j)}}}{[Q_b \cdot [K_m + K \cdot s_{bs_{(i,\,j)}}]]}$$

$s_{b\,out_{(1,\,j)}}$ mg/ml, exiting concentrations for feed concentration of 0.2 mg/ml

0.074
0.061
0.061
0.061

$s_{b\,out_{(2,\,j)}}$ mg/ml, exiting concentrations for feed concentration of 0.5 mg/ml

0.208
0.166
0.166
0.169

$X_{s_{(i,\,j)}} := 1 - \dfrac{s_{b\,out_{i,\,j}}}{s_{b\,in_i}}$ calculates the substrate conversion

Enter table of data for the experimental substrate conversions.

$X_{s\,exp_{(i,\,j)}} :=$

.58
.64
.69
.63
.54
.62
.55
.58

Compare the predicted and experimental conversions for each inlet substrate concentration.

$s_{b\,in_1} = 0.2$ mg/ml

$X_{s_{(i,\,j)}}$	$X_{s\,exp_{(1,\,j)}}$
0.628	0.58
0.696	0.64
0.695	0.69
0.693	0.63

$s_{b\,in_2} = 0.5$ mg/ml

$X_{s_{(2,\,j)}}$	$X_{s\,exp_{(2,\,j)}}$
0.585	0.54
0.669	0.62
0.668	0.55
0.661	0.58

Present conversions as a parity plot,

$x = 0, .01 \ldots 1 \qquad y(x) := x$

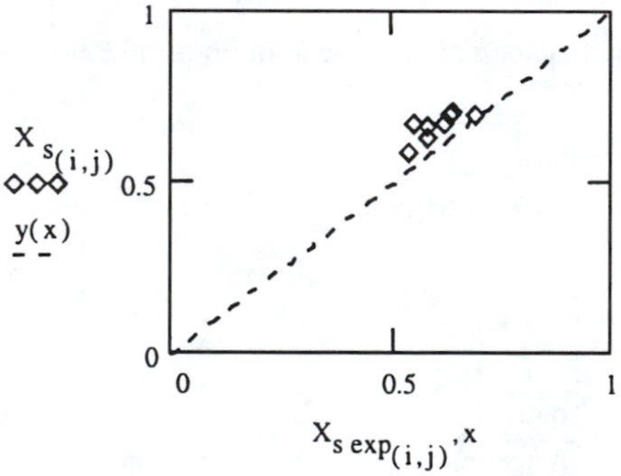

We see that the model predicted conversions compare rather well with the experimental conversions.

Calculation of the dimensionless Michaelis constant, since we need to go back to Chart 6.5 and calculate the values of η based on the Michaelis-Menten model.

$$\beta_{(i,\,j)} = \frac{K \cdot s_{bs_{(i,\,j)}}}{K_m}$$

$\beta_{(1,\,j)}$ for inlet concentration of 0.2 mg/ml

0.323
0.258
0.262
0.263

$\beta_{(2,j)}$ for inlet concentration of 0.5 mg/ml

0.909
0.711
0.719
0.733

These values of β for the given values of ϕ have pretty much converged and we conclude we have correct estimates for the values of η obtained from Chart 6.5.

PROBLEMS

1. Estimate the clearance and fractional % removal of urea from a flat-plate dialyzer for both cocurrent and countercurrent flow under the following conditions:

urea diffusivity = 2×10^{-5} cm^2/sec
4 flat plate membranes each with the following dimensions:
dialyzer length = 85 cm
dialyzer width = 15 cm
blood flow rate = 200 ml/min (total)
dialyzer flow rate = 500 ml/min (total)
mass transfer resistances
membrane = 25 min/cm
blood-side = 8 min/cm
dialysate side = 35 min/cm

2. Estimate the urea flux [$(NH_2)_2CO$] (g/m^2/hr) through a cellulosic membrane that is 0.001 inches in thickness. The blood at the membrane surface has a urea concentration of 1 mg/ml. Assume the dialysis fluid on the other side of the membrane surface is free of urea. The void volume of the cellulosic membrane is 0.40 and the membrane tortuosity is 2.0. The pores in the membrane have a diameter such that they exclude all molecules larger than a molecular weight of 20,000 dalton.

3. The following data for the mass-transfer coefficient of water flowing within hollow fibers was reported by Yang and Cussler (1986). Compare their results to the correlation given by Equation 2.5.

$Pe = \dfrac{V D^2_{\text{hollow fiber}}}{DL}$	$Sh = \dfrac{k_m D_{\text{hollow fiber}}}{D}$
30	4.8
50	6.0
100	7.6
300	9.8
500	13.0
1000	18.0
3000	30.0

4. The following data for the antibiotic imipenem were obtained from six patients suffering multiorgan failure (Hashimoto et al. 1997. *ASAIO Journal*. 43: 84–88, see also Problem 10 from Chapter 5). All patients were anuric because of complete renal failure and were on continuous venovenous hemodialysis. The antibiotic was first delivered by an intravenous infusion pump for 30 minutes resulting in an initial plasma concentration of 32.37 μg/ml. Following the infusion, blood samples were taken over the next 12 hours. In addition to the blood samples, drug concentrations in the blood exiting the hemodialysis unit were also taken. The dialysate entering the hemodialyzer was free of drug and the concentration of the drug in the dialysate leaving the dialyzer was also measured. The blood flow rate through the hemodialyzer was set at 60 ml/min and the dialysate flow was set at 20 ml/min. The hemodialyzer had a total surface area of 0.5 m^2. Using this data, develop a model that describes the drug removal process by the patient's body and the hemodialyzer. Carefully state all of your assumptions. Determine key parameters such as the total body clearance, the dialyzer clearance, and the overall mass-transfer coefficient for the dialyzer. Compare your model to the actual data. The table below summarizes the drug concentrations in the blood entering the dialyzer, the blood leaving the dialyzer, and the exiting dialysate.

Time, minutes	Blood drug concentration entering dialyzer, μg/ml	Blood drug concentration leaving dialyzer, μg/ml	Drug concentration in dialysate leaving dialyzer, μg/ml
0	32.47	23	21
30	24	20.5	14
60	18.5	16	12
120	13	11	8.8
180	10	8	6.4
360	4.6	3.8	2
540	2	1.6	1.4
720	1.12	0.9	0.6

5. At the start of CAPD using a 2.5 wt% glucose solution, the filtration rate of water from the plasma was determined to be about 6 ml/min. Estimate the hydraulic conductance (L_P in ml/hr m^2 mm Hg) of the peritoneal membrane.

6. The following data were presented by Lysaght and Farrell (1989) for urea (60 daltons) and inulin (5200 daltons) transport during CAPD. The initial glucose concentration in the dialysate solution was equal to 1.5 wt%. Assuming a constant peritoneal fluid volume of 2.4 liters, estimate the permeability of the peritoneal membrane for each solute. What thickness of a stagnant film of water would be required to give an equivalent permeability for urea and inulin? Explain the difference in the observed peritoneal membrane permeabilities based on your understanding of solute diffusion through hydrogels.

Time, minutes	Ratio of dialysate to plasma concentration for urea	Ratio of dialysate to plasma concentration for inulin
0	0.03	0
50	0.33	0.085
100	0.57	0.16
150	0.69	0.21
200	0.77	0.27
300	0.91	0.36
400	0.96	0.42

7. The following data were presented by Lysaght and Farrell (1989) for urea (60 daltons) transport during CAPD. The urea concentration in blood was 100 mg/dL. From this data, estimate the urea permeability of the peritoneal membrane.

Time, minutes	Peritoneal volume, ml	Dialysate urea concentration, mg/dl
0	1983	10
15	2304	30
30	2543	42
45	2938	55
60	3146	62

8. Analyze the performance of an oxygenator that uses an ultrathin (25 microns) membrane made of polyalkylsulfone cast on microporous polypropylene. The oxygen permeability of this membrane is reported (Gray, 1981) to be 1100 ml(STP)/min m^2 atm and that for carbon dioxide is reported to be 4600 ml(STP)/min m^2 atm. What membrane area is need to deliver 250 ml/min of oxygen and remove 200 ml/min of carbon dioxide? Carefully state your assumptions.

9. Analyze the performance of an oxygenator in which the hollow fibers have a length of 20 cm. Consider whether or not the assumption of fully developed flow is valid for these conditions. What membrane area is need to deliver 250 ml/min of oxygen and remove 200 ml/min of carbon dioxide? Carefully state your assumptions.

10. Consider an oxygenator where the gas flows through the lumen of the hollow fibers and the blood flows on the outside across the hollow fibers. Assume the that the gas flow is sufficiently high that a reasonable assumption is that the gas concentrations are constant. Develop a mathematical model that describes the performance of this oxygenator. Carefully state your assumptions. What membrane area is need to deliver 250 ml/min of oxygen and remove 200 ml/min of carbon dioxide? The external mass transfer coefficient for blood flowing across a bank of hollow fibers is given by the following expression (Yang and Cussler, 1986).

$$Sh = 1.38 Re^{0.34} Sc^{0.33}$$

The characteristic diameter for this equation is the external diameter of the hollow fibers. The velocity of the blood (V_b) between the fibers may be calculated by the following equation.

$$V_b = \frac{Q_b}{\varepsilon_{\text{unit}} A_{\text{unit}}}$$

Here Q_b is the volumetric flow rate of blood, $\varepsilon_{\text{unit}}$ is the void volume in the hollow fiber unit, and A_{unit} is the total cross-sectional area of the unit that is perpendicular to the blood flow.

11. Yang and Cussler (1986) considered the design of a human gill using hollow fibers. To support a man, they assumed the human gill would have to provide up to 2,500 ml/min of oxygen at 1 atm and 37°C for moderate levels of activity. Develop your design for the human gill and determine the hollow fiber membrane area that is required. How big would your human gill be ? What other issues need to be considered for the design of the human gill?

12. Calculate the effectiveness factor for an enzyme reaction that follows Michaelis-Menten kinetics. Assume the Thiele modulus is 4 and $\beta = 2$. Compare your result to the value obtained assuming the reaction is first order.

13. The following kinetic data (Sung et al. 1986) were obtained for the enzyme bilirubin oxidase at a pH of 7.4 and a temperature of 37°C. The substrate was bilirubin. The enzyme loading in the reactor was equivalent to 0.15 units/liter, where a "unit" of enzyme activity is defined as the amount of enzyme that catalyzes the conversion of 1 μmol/min of bilirubin. Assuming the data follows the Michaelis-Menten rate law, determine the values of V_{max}, K_m, and k_{cat}.

Bilirubin concentration, μM	Reaction rate, μM bilirubin/min
1	0.015
5	0.055
10	0.089
15	0.130
25	0.185
40	0.230
50	0.244

14. Suppose the bilirubin oxidase is immobilized within an agarose gel (Sung et al. 1986). The enzyme loading in the gel equals 10 Units/ml of wet gel. If external and internal mass-transfer effects are negligible, calculate the conversion of bilirubin in a well-mixed reactor containing a total gel volume of 15 ml. The void fraction in the reactor is 70%. Assume the flow rate through the reactor is equal to 1.2 ml/min and that the entering serum bilirubin concentration is 200 μM. The value of k_{cat} for these conditions is equal to 4.67×10^{-4} μmol/min/unit. Assume the value of K_m is unaffected by immobilization or other substances found in serum.

CHAPTER

SEVEN

TISSUE ENGINEERING

7.1 INTRODUCTION

Over eight million surgical procedures are performed each year in the United States for the treatment of lost tissue or organ function as a result of disease or injury (Holder et al. 1997). Organ transplantation has also become a routine and successful treatment method for the replacement of a failed organ. Organs routinely transplanted include the kidneys, heart, lungs, liver, and pancreas. Part of this success is due to better therapies to prevent rejection episodes as well as improved techniques for the procurement and storage of whole organs prior to transplant. Organ transplantation has saved and improved the quality of many lives, however, its widespread application is severely limited by a shortage of whole-organ donors. In the United States, only a few thousand organ donors are available each year. Yet tens of thousands are on waiting lists for an organ transplant and most of these will die waiting for an organ. Certainly, extracorporeal devices such as the hemodialyzer, bioartificial liver (to be discussed in Chapter 8), and artificial heart (Jarvik 1981; Rosenberg 1995; Spotnitz 1987) can be used until an organ is available. These devices therefore serve as a "bridge" to a transplant. The continued development and improvement of these devices will help alleviate the pain and suffering while awaiting an organ transplant. However, these devices cannot be used indefinitely and by themselves are not a perfect solution.

Many diseases do not involve an entire organ but only certain tissues or cell types that have lost their function. In these cases, it may be simpler to restore the lost function of the tissue or cells than to replace an entire organ. A good example would be the treatment of insulin-dependent diabetes mellitus. The islets of Langerhans are a specialized cluster of cells scattered throughout the pancreas. Their mass represents only about 1–2% of the pancreas. They secrete a variety of hormones and one of these, insulin, is responsible for the regulation of blood glucose levels. Loss of insulin secretion through an autoimmune disease process (Atkinson and MacLaren 1990; Notkins 1979) results in diabetes. It is estimated that perhaps only about 500 thousand to 1 million islets (Warnock and Rajotte 1988; Friedman 1989; Smith, Sarver, and Fournier 1991) would be needed to treat a patient with diabetes. With an islet averaging about 150 microns in diameter, this translates to an islet volume of 1 to 2 cm^3, a volume considerably less than that of a whole pancreas. Transplanting just the islets would therefore result in a far simpler surgical procedure.

7.2 CELL TRANSPLANTATION

The selective transplantation of just those cells necessary to replace the lost function of a tissue or an organ is receiving considerable attention (Cima et al. 1991b; Langer and Vacanti 1993). Selective cell transplantation is being developed for the treatment of a variety of diseases and medical problems, some of which are outlined in Table 7.1. The Suggested Readings section at the end of the book are a good resource for additional information concerning specific applications of cell transplantation.

Selective cell transplantation forms the basis of the relatively new fields of *tissue engineering* and *bioartificial organs*. Tissue engineering is defined as a developing multidisciplinary field that utilizes life science and engineering principles to construct biological substitutes containing viable and functioning cells for the restoration, maintenance, or improvement of tissue function.

The transplanted cells can be harvested from a variety of sources. For example, the cells can be obtained from a human donor (*allograft*), animals (*xenograft*), the patient themselves (*autograft*), or from a genetically engineered cell line (Morgan, Tompkins, and Yarmush 1994; Chang 1997). The donor tissue is normally dissociated using proteolytic enzymes (Ricordi et al. 1988) into single cells or small cellular aggregates that are then infused as is, incorporated into a support matrix or scaffold, or encapsulated

Table 7.1 Opportunities for cell transplantation and tissue engineering

Liver	Plastic/reconstructive surgery
Diabetes	Dental
Parkinson's disease	Nerve regeneration
Kidney	Site-specific delivery of growth factors
Chronic pain	Cornea
Esophagus	Skin
Ureter	Intestine
Bladder	Trachea
Cartilage	Bone
Muscle regeneration	Blood
Artificial blood vessels	Destruction of tumors
Gene therapy	Parathyroid
Cell production	In vitro hematopoiesis
Alzheimer's disease	Neurological diseases
Heart valves	Lymphocytes

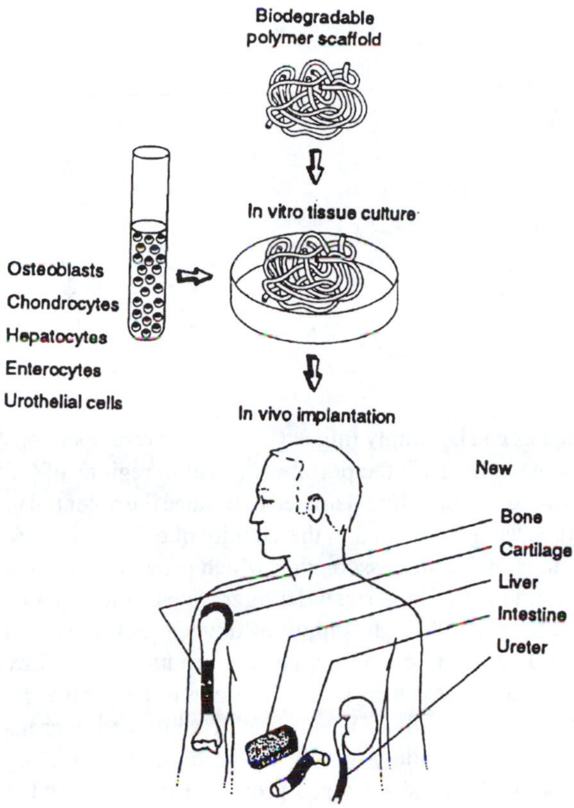

Figure 7.1 The tissue engineering process. (From Langer et al. [1995], with permission.)

within a device. The supporting structure or device containing the cells is then implanted within the patient's body. Figure 7.1 illustrates the basic steps of the tissue engineering process.

There is considerable overlap between the two emerging fields of bioartificial organs and tissue engineering. In some cases they are synonymous. For our purposes, we will categorize the field of bioartificial organs as a subset of the larger field of tissue engineering. The major distinction between the two being that in a bioartificial organ, the goal is to use a cellular transplant to replace a lost organ function, for example the use of the islets of Langerhans for the treatment of diabetes or *hepatocytes* for treatment of liver failure. Bioartificial organs also isolate the cells from the host's immune system through the use of an *immunoisolation membrane* (Colton 1995). This membrane is permeable to small molecules, such as required nutrients and the therapeutic agent released by the cells, but impermeable to the larger molecules (antibodies and complement) and cells of the host's immune system. The immunoisolation concept is shown in Figure 7.2 (Lanza and Chick 1994b). Because of this immunoprotection, cells from animals (xenografts) can also be used, vastly increasing the availability of donor tissue. Several examples of bioartificial organs will be examined in greater detail in Chapter 8.

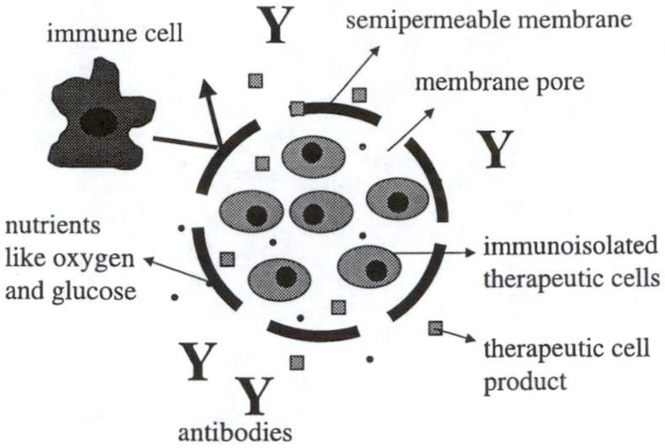

Figure 7.2 Concept of immunoisolation.

Isolated cells or cellular aggregates can be simply infused as is into several locations within the body. Typical sites include the liver via the portal vein, within regions of fat, the spleen, under the kidney capsule, and the omentum. Isolated cells have the potential to grow and form larger tissue structures when implanted in the vicinity of existing mature tissue. However, a potential problem is their loose association which provides no guide for attachment, restructuring, and formation of larger cellular aggregates. Additionally, cells can not survive if they do not have an adequate supply of oxygen, nutrients, and a means for waste-product removal. Furthermore, any therapeutic product released by the transplanted cells needs to have access to the host's vasculature in order for it to be effective. This requires that each transplanted cell needs to be within a hundred microns or so of a capillary. Initially, the existing capillary supply in the tissue surrounding the cellular transplant may not be optimal for the sustained growth of the transplanted cells. This severely limits the size of the cellular aggregates that can form, unless a means is provided for the formation of a vascular capillary bed (angiogenesis) within the transplanted cells. The goal of tissue engineering is therefore to develop methods and techniques than can enhance the success of cellular transplants.

7.3 THE EXTRACELLULAR MATRIX

Most of the effort in tissue engineering is now focusing on the use of polymeric support structures for guiding the growth and organization of the transplanted cells (Cima et al. 1991b). To better understand how polymeric support structures can enhance the growth of transplanted cells, it is necessary to first examine how cells organize themselves within natural tissue structures.

Tissues are not made up entirely of cells, but also consist of an extracellular gel-like fluid containing a variety of macromolecules collectively referred to as the *extracellular matrix* (ECM) (Alberts et al. 1989; Rubin and Farber 1994; Long 1995; Naughton 1995; Anderson 1994; Hubbell 1997; Parsons-Wingerter and Sage 1997). The ECM consists of glycosaminoglycans (GAGs) and fibrous proteins that are in contact with the cells and hold them together forming an organized cross-linked mesh-like structure. This is shown in Figure 7.3 for the case of *epithelial cells* that are supported by a layer of

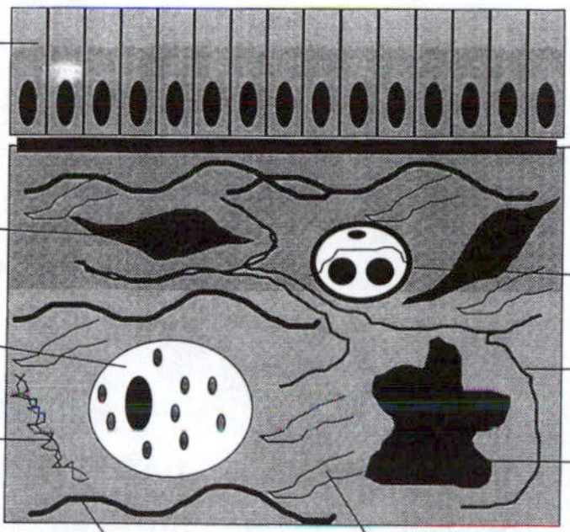

Figure 7.3. Connective tissue

connective tissue. Connective tissues (also called the *stroma*) provide the framework for the formation and organization of most of the larger structures found within the body. The specific functional cells that define a tissue, for example the hepatocytes found in the liver or the islets of Langerhans in the pancreas, are called *parenchymal cells*.

The ECM not only serves to provide three-dimensional organization of cells and adjacent layers of tissue, but also provides a mechanism for intercellular communication and controls such cellular processes as proliferation, cell migration, attachment, differentiation, and repair. The major components of the ECM are the GAGs, proteoglycans, collagens, elastic fibers, structural glycoproteins, and the basement membrane.

Most mammalian cells are anchorage dependent and possess cell-surface receptors for a variety of the macromolecules or so-called *attachment factors* found within the ECM that are responsible for the development, growth, and metabolic functions of cells. The ECM macromolecules are secreted locally for the most part by specialized cells called *fibroblasts*. Several of the most important types of macromolecules found within the ECM are summarized in Table 7.2.

7.3.1 Glycosaminoglycans

GAGs form from long linear polymers consisting of a repeating disaccharide unit. Molecular weights range from several thousand to over a million daltons. GAGs tend to have a large number of negative charges because of the presence of carboxyl and sulfate groups. They are highly extended, forming random coil-like structures. Because they are also hydrophilic, they form, in combination with the interstitial fluid the hydrated gel-like

Table 7.2 Major macromolecules of the extracellular matrix

Glycosaminoglycans (GAGs)—forms the ECM gel
Collagen—provides strength and organization to the ECM
Elastin—provides resilience or elasticity
Fibronectin—adhesion of fibroblasts and other cells
Laminin—adhesion of cells to the basal lamina

material that contains the other macromolecules of the ECM. There are four types of GAGs, depending on what types of sugars they are made from. They are *hyaluronic acid, chondroitin sulfate* and *dermatan sulfate, heparin sulfate* and *heparin*, and *keratin sulfate*. With the exception of hyaluronic acid, they also form *proteoglycans* through covalent bonds with proteins.

7.3.2 Collagens

Collagens are fibrous proteins that are made from three polypeptide chains that form the tough, triple-stranded helical structure of the collagen molecule. Tensile strength in tissues is provided by collagen. Although about ten types of collagen molecules have been identified, the most important types of collagen molecules are the *fibrillar collagens* (type I, type II, and type III) and type IV collagen. The first three types of collagen molecules assemble themselves into much larger structures called collagen fibrils that aggregate further to form collagen fibers. Type I collagen makes up about 90% of the collagens found in the body. It is found mostly in the skin, tendons, ligaments, various internal organs, and bone. Type II collagen is primarily found in cartilage and type III collagen is found in blood vessels as well as the skin. Type IV collagen organizes itself to form sheets within the basal lamina of the basement membrane. Fibroblasts have the ability to organize the collagen fibrils they secrete forming sheets or rope-like structures, therefore they can affect the spatial organization of the matrix they produce.

7.3.3 Elastin

Elastin is a hydrophobic glycoprotein molecule that through cross-links with other elastin molecules, can form a network of sheets and filaments with the unique property of being elastic, allowing for recoil after periods of stretch. The elasticity is a result of the random coil-like structure of the elastin molecule. The elastic nature of elastin is important in blood vessles, skin, the lungs, and the uterus. Inelastic collagen fibrils are interwoven with elastin to limit and control the degree of stretching.

7.3.4 Fibronectin

Fibronectin is the principal adhesive glycoprotein found within the ECM. Fibronectin binds to other ECM molecules such as collagen and to cell-surface receptors. It therefore has a principal role in the attachment of cells to the ECM. Fibronectin is a dimer made up of two subunit chains that are bound together at one end by a pair of disulfide bonds. Along the length of the chains are a series of functional domains that can bind to specific types of molecules (for example collagen or heparin) or cell-surface receptors. The cell binding domain has a specific tripeptide sequence consisting of the amino acids: arginine (R), glycine (G), and aspartic acid (D), often referred to as the *RGD sequence*. Peptides containing this sequence will inhibit the attachment of cells to fibronectin through their competition for the RGD binding site on the cell surface.

In addition to fibronection, there are several other ECM adhesion proteins that express the RGD sequence. These include vitronectin found primarily in blood cells. Thrombospondin is secreted by a variety of cells involved in the development of the

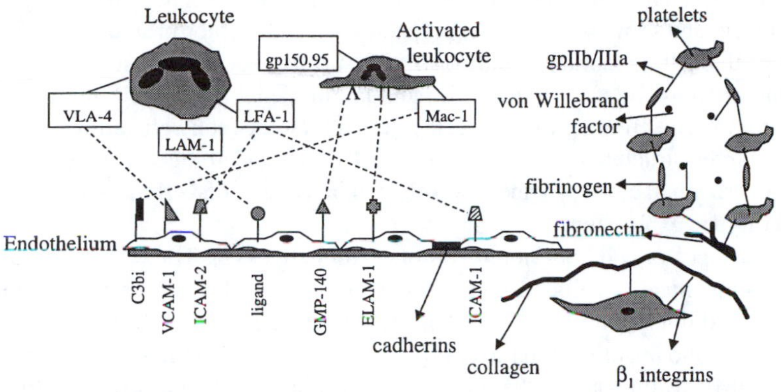

Figure 7.4 Cellular interactions.

ECM. It serves to bind together other components of the ECM like fibronectin. The von Willebrand factor is made by megakaryocytes (platelet-generating cells found in the bone marrow). It is stored in circulating platelets. von Willebrand factor is released by platelets as a result of injury to blood vessels. Fibrinogen is the clotting protein found in blood. During blood-clot formation, fibrinogen is converted to fibrin that forms a mesh-like structure that traps red blood cells and platelets.

7.3.5 Basement Membrane

The *basal lamina* is a continuous, mat-like structure of ECM materials that separates specific cells; such as epithelial, endothelial, or muscle, from the underlying layer of connective tissue. The basal lamina consists of two distinct layers, the *lamina rara*[1] directly beneath the basal membrane of the specific cells above, and the *lamina densa*[2] just below the lamina rara. Below the two layers of the basal lamina is the collagen-containing *lamina reticularis* that connects the basal lamina to connective tissue that lies below it. All three of these layers together constitute what is known as the *basement membrane* (Figure 7.3). The basal lamina consists primarily of type IV collagen, proteoglycans such as those formed from heparin sulfate, and the glycoprotein laminin. Laminin is an extremely large protein (850,000 daltons) made from three long polypeptide chains that form the shape of a cross. It also contains a number of functional domains throughout its structure with sites for binding to type IV collagen, heparin sulfate, and laminin cell-surface receptors.

7.4 CELLULAR INTERACTIONS

There are three types of interactions involving cells. These interactions may be classified as cell-ECM, cell-cell, and cell-growth factor (Long 1995). Figure 7.4 illustrates the

[1]Under the electron microscope this layer is translucent to electrons.

[2]Under the electron microscope this layer is somewhat opaque to electrons.

various interactions of cells with their surrounding environment. Cells interact with other cells, the ECM, and growth factors through cell-surface receptor proteins that are an integral part of the cell membrane. Binding of cell-surface receptors to surrounding ECM components such as collagen, fibronectin, and laminin enables the cell to link up to the surrounding matrix. Binding of cell surface receptors between cells allow them to organize into larger structures such as tissues and organs. Cell surface receptor's interactions with growth factors provide control over a variety of cellular functions through control of gene expression.

There are four types of cell adhesion receptors (Hubbell 1997). Three of these receptors are mainly involved in cell-cell interactions and the fourth is involved in both cell-cell and cell-ECM interactions. The adhesion receptors used in cell-cell interactions include the *cadherins*, the *selectins*, and the *CAMs* (cell adhesion molecules). The receptors involved in cell-ECM interactions belong to a general family of adhesion receptors known as *integrins* (Horwitz 1997).

7.4.1 Cadherins

Cadherins take part in what is known as homophilic binding, that is, the binding of a cadherin molecule on one cell with an identical cadherin on another cell of the same type. They do not bind with ECM components. Cadherin molecule binding is strongly dependent on the presence of extracellular Ca^{++}.

7.4.2 Selectins

The selectins are a family of cell-adhesion molecules that are found on the surfaces of endothelial cells, leukocytes, and platelets. They exhibit heterophilic binding between the blood cells and the endothelial cells that line the blood vessels. They are important in the localization of leuokocytes to sites of inflammation and tissue injury. The three most important selectins are called GMP-140, endothelial leukocyte adherence molecule-1 (ELAM-1), and leuokocyte adhesion molecule-1 (LAM-1). Like the cadherins, their activity is dependent on the presence of extracellular calcium.

7.4.3 Cell Adhesion Molecules

The CAMs are members of the immunoglobulin gene superfamily. These receptors bind independently of extracellular calcium levels. CAMs are involved in recognition of antigens (foreign agents) and cell-cell interactions. Antigen recognition involves both the T-lymphocyte receptor and two special antigen-presenting molecules called major histocompatibility class I and II (MHC I and MHC II). These specialized CAMs will be discussed further in Chapter 8. Two other CAMs are important factors in localizing leuokocytes to regions of tissue injury. For example, intercellular adhesion molecule-1 (ICAM-1) is found on the surfaces of both endothelial cells and leukocytes and binds to the integrin called leukocyte function antigen-1 (LFA-1) that is found on macrophages and neutrophils. ICAM-2 found on endothelial cells also binds to LFA-1 and helps to localize neutrophils to sites of injured tissue. VCAM-1 (vascular cell adhesion molecule) is expressed by endothelial cells as a response to injury or inflammation and also promotes adherence of leukocytes to endothelial cells.

7.4.4 Integrins

The integrins are a family of adhesion molecules that are involved in both cell-cell interactions as well as binding with molecules found in the extracellular matrix. Integrin molecules are heterodimers consisting of two protein subunits, the α chain and the β chain. There are at least 15 variants of the α subunit and 8 variants of the β subunit. Overall, there are at least 20 different heterodimer $\alpha\beta$ integrins. The β subunits provide the functional aspects of the dimeric integrin molecule, hence integrins are classified according to the type of their β chain. The subclasses having the β_1, β_2, and β_3 chains are the most common. The β_1 and β_3 subclasses are primarily involved in interactions between the cell and the ECM. The β_2 subclass involves mostly cell-cell interactions involving leukocytes.

β_1 integrins are also referred to as VLA (very late after) antigens (Margiotta, Benton, and Greco 1994). VLA antigens are numbered according to the number of the alpha subunit that forms the heterodimer with the β_1 subunit. These subclasses bind with a variety of ECM proteins such as collagen, fibronectin, and laminin. The primary recognition site for β_1 integrins is the RGD sequence described earlier. Therefore this group of integrins are important for the overall organization and function of a variety of cell types. VLA-4 is also a receptor found on leukocytes for VCAM-1 expressed by endothelial cells.

Cell-cell interactions, mainly leuokocyte-to-leuokocyte, involve mostly the β_2 group of integrins. Hence this subclass is known as the "leukocyte" integrins. The common β_2 subunit is also known as CD18 (cluster designation), a 95-kD protein. The most important alpha subunits, in combination with β_2, form the leukocyte integrins known as LFA-1 (leuokocyte function antigen-1), Mac-1 (macrophage antigen-1), and gp150,95 (a glycoprotein with an alpha subunit of molecular weight 150 kD along with the β_2 subunit protein of molecular weight 95 kD). These alpha subunits are also respectively known by the cluster designations CD11a, CD11b, and CD11c. The expression of these integrins on leukocytes is particularly important for defense against bacterial infections.

The β_3 integrins comprise the platelet glycoprotein receptor known as gpIIb/IIIa and the vitronectin receptor. Both of these receptors also recognize the RGD sequence of ECM matrix proteins such as vitronectin, fibronectin, thrombospondin, fibrinogen, and von Willebrand factor. Aggregation of platelets is enhanced by the release of von Willebrand factor. gpIIb/IIIa is calcium dependent and found on the surface of activated platelets. This integrin is important for platelet aggregation and adherence of platelets to the subendothelium. This occurs through binding of the gpIIb/IIIa receptor to fibrinogen, fibronectin, vitronectin, and von Willebrand factor.

7.4.5 Cytokines and Growth Factors

A variety of cells also secrete soluble proteins known as *cytokines*. Cytokines serve as intercellular chemical messengers. Many of these proteins are required for the normal development, growth, and proliferation of cells. They are also involved in such processes as inflammation, wound healing, and the varied responses of the immune system. Many

Table 7.3 Growth factors

Growth factor	Property
Platelet derived growth factor (PDGF)	Mitogen for smooth muscle cells, fibroblasts, and neuroglial cells, chemotactic signal for immune system cells
Epidermal growth factor	Mitogen for a variety of cells types, accelerates wound healing
Fibroblast growth factor (aFGF, bFGF) and vascular endothelial cell growth factor (VEGF)	Stimulates growth of blood vessels, mitogen for endothelial cells, fibroblasts, smooth muscle cells
Tumor necrosis factor (TNF)	Stimulates growth of blood vessels
Insulin-like growth factor (INF-I)	Stimulates growth of blood vessels
Transforming growth factor (TGF-β)	Controls cellular response to other growth factors
Nerve growth factor (NGF)	Stimulates axon growth
Interleukin-2 (IL-2)	Stimulates proliferation of T lymphocytes
Interleukin-6 (IL-6)	Stimulates proliferation and differentiation of B lymphocytes

are mitogens, that is, they stimulate the proliferation of specific types of cells. They are often then referred to as growth factors.

Growth factors bind to specific receptors on the surface of their target cell and have the ability to induce or direct the action of specific genes in the targeted cell. A variety of growth factors have been identified and some of the more important of these are summarized in Table 7.3. All of these growth factors play a key role during the wound healing process and many are involved in the growth of new blood vessels. Growth factors can also be used to stimulate and accelerate the development of blood vessels in polymeric scaffolds used in tissue engineering applications.

Platelet derived growth factor (PDGF) is a mitogen for such cells as smooth muscle, fibroblasts, and neuroglial cells. PDGF is stored in platelets and is released after platelet aggregation at the site of injury to blood vessels. Damaged endothelial and smooth muscle cells also secrete PDGF. PDGF is also a chemotactic signal for such inflammatory cells as macrophages and neutrophils. Epidermal growth factor (EGF) stimulates the growth of a variety of cell types. EGF accelerates the healing of many wounds and stimulates fibroblasts to deposit collagen.

Fibroblast growth factor (FGF) comes in two forms, acidic (aFGF) and basic (bFGF). bFGF is an order of magnitude more potent than aFGF. FGF and vascular endothelial cell growth factor (VEGF) stimulates the growth of capillaries, a process called angiogenesis (Folkman 1985; Folkman and Klagsbrun 1987). FGF also promotes the growth of fibroblasts, endothelial cells, and smooth muscle cells. These processes affected by FGF also accelerate wound healing. Another angiogenic growth factor is tumor necrosis factor-α (TNF-α) which is secreted by activated macrophages during the initial stages of inflammation. Insulin-like growth factor-I (IGF-I) not only plays a major role in growth, but also stimulates angiogenesis. Transforming growth factor-β (TGF-β) can either promote or inhibit the growth of cells by controlling their response to other growth factors. Several other important growth factors include nerve growth factor (NGF) that enhances function of neurons and stimulates axon growth, interleukin-2 (IL-2) that stimulates the proliferation of T lymphocytes, and interleukin-6 (IL-6) that stimulates the proliferation

Table 7.4 Desirable properties of support structures for transplanted cells

Biocompatible
Non-immunogenic
Negligible toxicity
Locally and systemically
Polymer and any degradation products
Chemically and mechanically stable
Processable into a variety of shapes
Hollow tubes, sheets, arbitrary 3-D shapes
Open foam or sponges
Woven or non-woven mesh-like structures
High porosity
High internal surface area to volume ratio
Controllable pore size
Promotes cell attachment
Promotes angiogenesis
Favorable interaction/mobilization of host cells
Ability to release active compounds such as growth factors
Ability of the cells and ECM materials to interact with the support structure
Ability to obtain the desired cellular response

of B lymphocytes and their differentiation to antibody secreting plasma cells. There are also several hematopoietic[3] growth factors that are specific to the development and differentiation of blood cells.

7.5 POLYMERIC SUPPORT STRUCTURES

With this background on how cells organize themselves within the body, we can now address the properties required for an artificial polymeric support structure for transplanted cells. Table 7.4 summarizes some of the characteristics that are important to consider in selecting materials to serve as support structures for transplanted cells.

A variety of biological and synthetic polymers are used as biomaterials and many of these can be used to provide the support structure for transplanted cells. Table 7.5 summarizes some of the materials that are being used or considered for use as support structures in tissue engineering.

The biomaterials listed in Table 7.5 are grouped as to whether or not they are biodegradable. Long-term implantation of nonbiodegradable biomaterials poses the risk of localized infection, chronic inflammation, and development of fibrous encapsulation of the implant. This can compromise the proper functioning of transplanted cells if they were to reside within the porous structure of these materials. Nonbiodegradable materials such as PTFE, PU, and Dacron, are preferred for those applications where their structural integrity is more important than the requirement for them to serve as scaffolds for transplanted cells. For example, as vascular grafts or structural components

[3]Hematopoiesis is the process of blood-cell development and differentiation.

Table 7.5 Materials for tissue engineering and other biomedical applications

Non-Biodegradable polymer	*Applications*
Polydimethylsiloxane or silicone (PDMS)	Breast, penile, testicular prostheses catheters, drug delivery, heart valves, membrane oxygenators, shunts, tubing, orthopedics
Ceramics	Bone repair
Polyurethanes (PU)	Artificial hearts and ventricular assist devices, catheters, intraaortic balloons, wound dressings
Polytetrafluorethylene (PTFE)	Heart valves, vascular grafts, reconstruction, shunts, membrane oxygenators, catheters, sutures, coatings
Polyethylene (PE)	Artificial hips, catheters, shunts, syringes, tubing
Polysulfone (PS)	Heart valves, penile prostheses, artificial heart
Polycarbonate	Hard contact lenses
Poly(methyl methacrylate) (PMMA)	Bone cement, fracture fixation, intraocular lenses, dentures, plasmapharesis membranes
Poly(2-hydroxyethylmethacrylate) (PHEMA)	Controlled drug release, contact lenses, catheters, coatings, artificial organs
Polyacrylonitrile (PAN)	Hemodialysis membranes
Polyamides (nylon)	Hemodialysis membranes, sutures
Polyethylene terephthalate (Dacron)	Vascular grafts, tissue patches, shunts
Polypropylene (PP)	Valve structures, plasmapheresis membranes, sutures
Polyvinyl chloride (PVC)	Tubing, plasmapheresis membranes, blood bags
Poly(ethylene-co-vinyl acetate)	Drug delivery devices
Polystyrene (PS)	Tissue culture flasks
Poly(vinyl pyrrolidone) (PVP)	Blood plasma extender
Polyvinyl alcohol (PVA)	Dental, tissue repair, scaffolds for bioartificial organs
Biodegradable polymers	*Applications*
Poly(L-lactic acid), poly(glycolic acid), and poly(lactide-co-glycolide) (PLLA, PGA, and PLGA)	Controlled release drug delivery devices, sutures, scaffolds for cell transplantation and tissue engineering
Collagen	Artificial skin, hemostasis, tissue regeneration scaffold
GAGs (hyaluronan)	Tissue repair, viscoelastic fluids, wound care
Chitosan	Scaffolds for tissue engineering, inhibitors of blood coagulation, cell encapsulation, membrane barriers, contact lens materials
Polyhydroxyalkanoates (PHA)	Controlled drug release, sutures, artificial skin
Poly(ε-caprolactone) (PCL)	Implantable contraceptive devices, controlled drug release, surgical staples

From Friedman, Orland, and Greco 1994; Marchant and Wang 1994; Saltzman 1997; Pachence and Kohn 1997.

of artificial organs. These materials can also be made into a variety of three-dimensional configurations.

Biodegradable materials are now attracting significant interest for tissue engineering applications. Biodegradable materials allow the transplanted cells sufficient time to organize the desired three-dimensional structure and develop their own blood supply. The biodegradable polymers are easily hydrolyzed by the body's fluids and slowly disappear wthout leaving behind any foreign residues. Examples of biologically derived biodegradable polymers include collagen, GAGs, and chitosan. Collagen is one of the ECM materials and can be prepared as fibers (Cavallaro, Kemp, and Kraus 1994) or as a gel. Collagen is being used in such applications as tissue repair and artificial skin (Morgan and Yarmush 1997). PLLA, PGA, and PLGA are naturally occurring hydroxy acids. They are biodegradable and have been approved for use by the Food and Drug Administration (FDA) for use in sutures, controlled drug release, and surgical support fabrics (Cima et al. 1991a, 1991b). A great deal of effort is focusing on these materials for tissue engineering applications (Holder et al. 1997; Mooney et al. 1996; Mooney et al. 1994; Mooney et al. 1995a, 1995b; Kaufmann et al. 1997). They degrade by hydrolysis and form natural byproducts. Their absorption rate can also be controlled from months to years (Agrawal, Niederauer, and Athanasiou 1995; Mooney et al. 1994, 1995a, 1995b). The degradation rate of these materials is dependent on their initial molecular weight, the surface area that is exposed, degree of crystallinity, and for the case of co-polymer blends, the ratio of the hydroxy acid monomers used (Pachence and Kohn 1997).

The resulting structure formed from a biomaterial must be mechanically strong enough to support the growth of the transplanted cells and chemically compatible with the intended duration of use. For example, cells secrete a variety of enzymes that may degrade the polymer used. For a biodegradable material, this cell-induced degradation rate must be considered during formulation of the biomaterial to achieve the desired degradation rate. In the case of nonbiodegradable materials, they must withstand the cellular attack and not form byproducts that are inflammatory or immunogenic and therefore capable of compromising the function of the implant. The material used to form the support structure will also generally be in direct contact with the host's immune system and connective tissue such as fibroblasts. For any biomaterial, activation of the host's immune system, as well as a fibrotic connective tissue response, will tend to wall off and isolate the transplanted support structure, blocking its intended function. The host reaction and formation of a fibrotic capsule around the implant is a function of not only the materials used, but also on the shape and microstructure of the implant.

Unlike in vitro tissue culture of cells wherein the cells are usually grown in only two dimensions, the transplanted support structure containing the cells will be three-dimensional. Furthermore, the resulting three-dimensional polymer scaffold must in general have high porosity, significant internal surface area, and a controlled pore size distribution. All of these properties must not compromise the structural strength of the material.

Polymer scaffolds used in tissue engineering applications must be processable into whatever shape is required for the implant. A variety of techniques may be used to process polymer scaffolds. These include such methods as fiber bonding, solvent casting, particulate leaching, membrane lamination, melt molding, polymer/ceramic fiber composite foams, phase separation, and in situ polymerization (Lu and Mikos 1996; Thomson, Yaszemski, and Mikos 1997).

For example, the biodegradable open sponge or mesh-like structures shown in Figure 7.5 exhibit high porosity (>80%) and large pores. The poly(DTE carbonate) shown on the left in the figure was prepared by a salt leaching technique. The resulting structure has pores in the range of 200 to 500 microns and is being used for bone regeneration. The PGA nonwoven mesh (on right) has fibers 10 to 50 microns in diameter and is

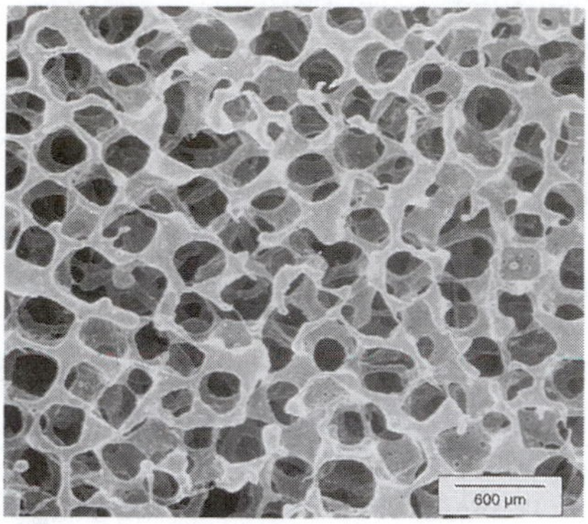

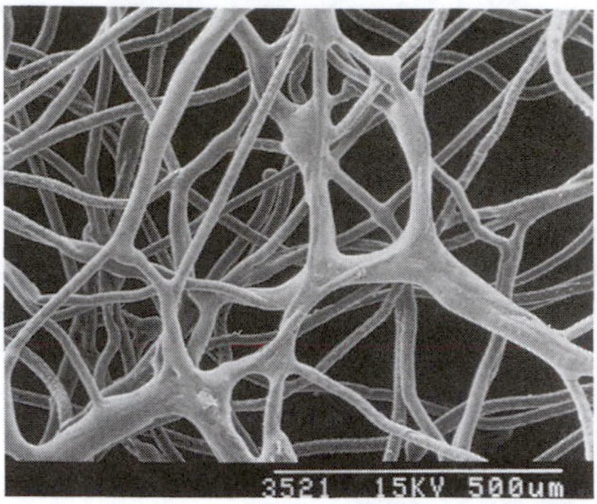

Figure 7.5 Scanning electron micrographs of polymeric scaffolds for tissue engineering. (From James and Kohn [1996] and Lu and Mikos [1996], with permission.)

being considered for use in organ regeneration. These materials interact favorably with the surrounding tissue, allowing for ingrowth of tissue and the formation of a vascular network throughout the support structure. The high porosity of these structures is also important for nutrient and product transport as well. To avoid the presence of nonvascular regions and to optimize implant size, it may be important to be able to control the pore size distribution. A high internal surface area to volume ratio will also provide numerous sites for cell adhesion to the polymer support structure. A high internal surface area also allows for surface treatment of the polymer structure with ECM materials and/or growth factors in order to enhance interaction of the support with the transplanted cells and promote their proliferation and differentiation.

In the best of situations, the implanted support structure becomes vascularized by the host during the period of time the transplanted cells are increasing in number. However, the metabolic demands of the transplanted cells in many cases cannot be met by the vasculature that is also developing at the same time within the implant. It may be better to first implant the support structure, allowing it to become vascularized, and then to transplant or seed the cells at a later point in time (Takeda et al. 1995). This gives the transplanted cells the opportunity to start in a well-vascularized region thus eliminating any transport limitations due to the metabolic requirements of the transplanted cells. An example of this is shown later in Section 7.11.

Prevascularization of the implant can also be enhanced by first seeding the implant with the host's endothelial cells (Holder et al. 1997). The endothelial cell–loaded support matrices are believed to enhance vascularization by any or all of the following mechanisms. The endothelial cells can form new capillaries, they may provide chemical signals (ECM and growth factors) for growth of blood vessels, and they may merge with the host's own vascular ingrowth. Immobilization of growth factors on the support material, or within controlled-release microspheres, can also be used to facilitate vascularization and improve survival of the transplanted cells (Thompson et al. 1988, 1989; Mooney et al. 1996). Figure 7.6 shows the tissue region in a polyester support matrix 28 days after implantation that was pretreated with collagen and aFGF. We see a well-vascularized

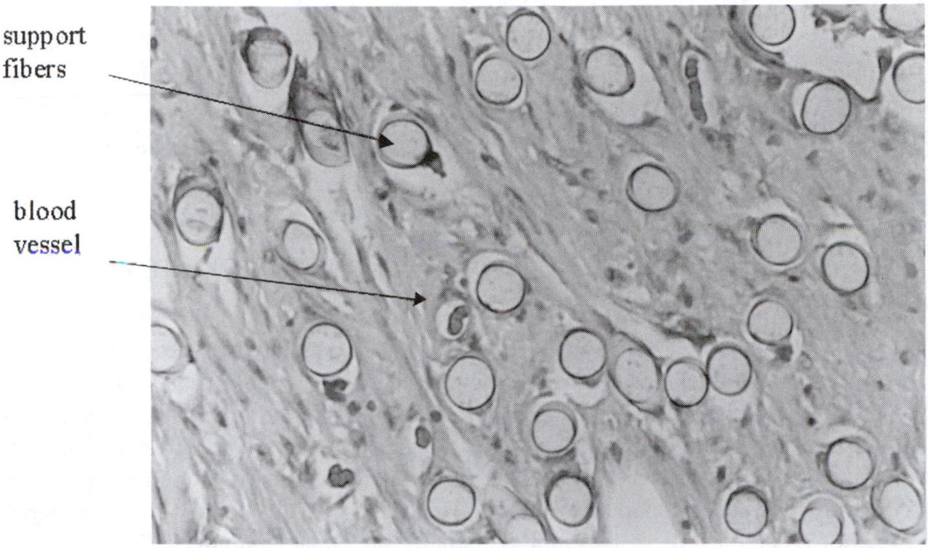

Figure 7.6 Vascularized tissue within a polyester support matrix 28 days after implantation.

tissue region within the fiber space as evidenced by the presence of numerous capillaries some of which contain red blood cells.

7.6 INITIAL RESPONSE TO AN IMPLANT

The first response after implant of the support structure will be the infiltration of the support by a variety of cell types. This process is very similar to what occurs during wound healing and involves three somewhat overlapping phases: *inflammation*, *proliferation*, and *maturation* (Hammar 1993; Arnold and West 1991).

The *inflammation period* lasts for several days and involves the arrival of *platelets* and *neutrophils*. Activation of the clotting process and release of growth factors (Alberts et al. 1982) (PDGF, EGF, TGF) leads to the formation of a collagen-free fibrin network that serves as a scaffold for the inflammatory cells. The neutrophils release factors that attract *monocytes* and also ingest foreign material by phagocytosis. The monocytes also enter the site and are transformed into *macrophages* and when fused together, they become *foreign body giant cells*. They continue to clean up the site, removing dead tissue, removing bacteria, and walling off large debris.

During the *proliferative phase*, the macrophages release a variety of factors (PDGF, TGF, EGF, FGF) that activate the migration and proliferation of fibroblasts and endothelial cells. The fibroblasts release a variety of ECM materials and begin to form a collagen network. The generally low oxygen levels within the site also stimulates the movement and growth of capillary sprouts from the surrounding vasculature. These capillary sprouts are formed from endothelial cells and invade the site, forming a vascular bed. Over a period of weeks, the entire site becomes vascularized and this marks the end of the proliferative phase.

The *maturation phase* involves final remodeling of the site, resulting in contraction of the wound and organization of the collagen matrix. This is also a critical time for tissue engineering applications, since the reduced oxygen demand of the cells involved in wound healing can result in regression of the vascular supply. Seeding of the transplanted cells at this time and their need for oxygen may promote the continued growth of the vascular supply.

7.7 TISSUE INGROWTH IN POROUS POLYMERIC STRUCTURES

The rate of tissue ingrowth is an important factor in the development of tissue engineered structures. Presently, the slow rate of ingrowth of vascularized tissue from the host severely limits the thickness of most tissue engineered structures (Holder et al. 1997). Much effort is therefore being focused on optimizing the rate of tissue ingrowth into support materials so that the thickness of these devices can be increased from millimeters to centimeters. These studies on the rate of ingrowth of new tissue in the support structure are being performed in laboratory animals. For example, Mikos et al. (1993) prepared highly porous polymer disks 13.5 mm in diameter and about 5 mm thick. The polymers used were made either from PLLA, PGA, or PLGA. In forming the polymer disks, NaCl particles on the order of several hundred microns, were incorporated in the polymer solution at 80–90 wt %, to form a composite membrane. By leaching out the salt particles, a highly porous foam-like support structure was formed. These structures typically had porosities in the range of 64–90% with mean pore sizes ranging from as low as 36 to 179 microns. Porosity and pore size was controllable by adjusting respectively the relative amount and particle size of the salt used. The devices were then implanted in the *mesentary* of male rats.

Figure 7.7 illustrates some typical results for a prewetted PLLA implant (83% porosity and 166 micron mean pore size) at two sites in the mesentary. Tissue ingrowth was determined from histological sections of implants removed at 5, 15, and 25 days. The optimal time for subsequent cell transplantation corresponds to the time required for 100% tissue ingrowth and assumes that vascularization has also occurred. For these results, it appears that 25 days is sufficient to fill the void space of the support structure with vascularized tissue. However, the regions occupied by tissue were not completely filled. The percent tissue area provides a measure of the device total volume fraction occupied by tissue. At day 25, this volume fraction of the tissue is 79%. Since the device total porosity (ε_p, recall it is 83%) is the sum of the tissue volume fraction ε_T and the remaining void volume (ε_v), we conclude that the remaining void fraction is 4% at day 25. This would be the volume remaining for the transplanted cells.

Also determined from the histological sections of the implant were the number of capillaries and their average cross-sectional surface area within a $1 \times 1\ \text{mm}^2$ field of view. The number of capillaries (N_c) after 25 days is about 5 per field, whereas the average capillary cross sectional area ($A_{\text{capillaries}}$) is about 1000 μm^2. This gives a mean capillary radius of 8 microns. The volume fraction of capillaries within the tissue (v) would then be given by the next equation:

$$v = \frac{A_{\text{capillaries}}}{\varepsilon_T \times 1\ \text{mm}^2} \tag{7.1}$$

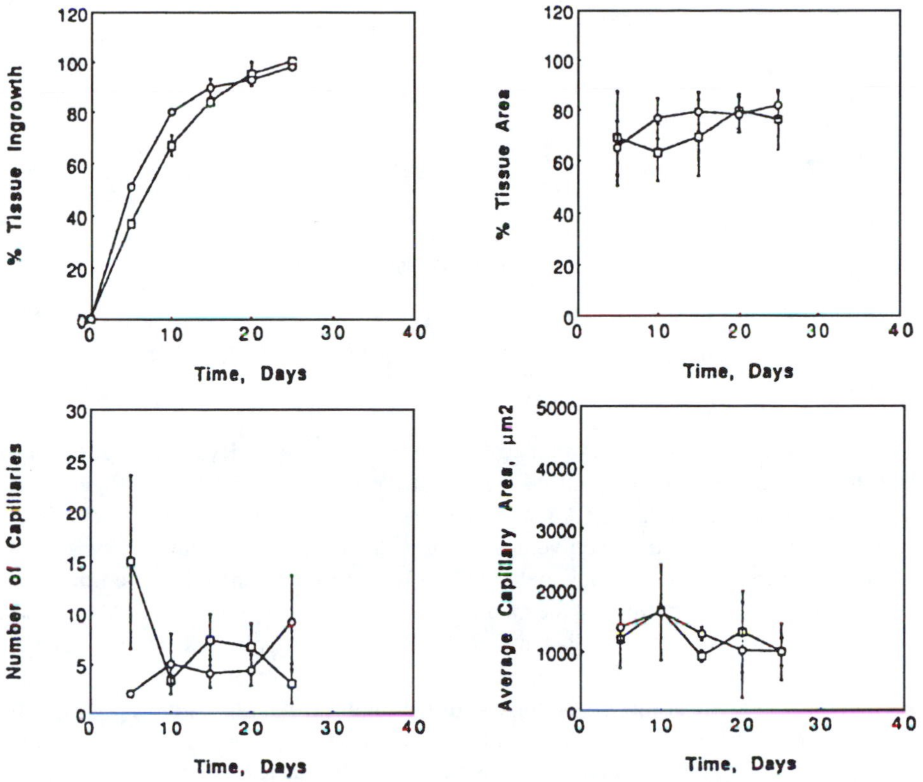

Figure 7.7 Normalized tissue ingrowth, percent tissue area, number of capillaries, and capillary area for prewetted Poly-L-Lactic Acid (PLLA) devices. (From Mikos et al. [1993], with permission.)

This gives a capillary tissue volume fraction of 0.13%. Assuming the idealized Krogh tissue cylinder model shown in Figure 2.10, we can calculate the Krogh tissue cylinder radius using Equation 2.56.

$$v = \left(\frac{r_c}{r_T}\right)^2$$

With a capillary radius (r_c) determined above of 8 microns, we obtain a Krogh tissue cylinder radius of about 220 microns.

In a subsequent study, Wake, Patrick, and Mikos (1994) focused on disks with the same geometry as the earlier Mikos et al. (1993) study, forming an amorphous PLLA device with pore sizes ranging from 91 to 500 microns. Figure 7.8 illustrates the tissue ingrowth for various implantation times out to 35 days. Once again the device is ready for cell transplantation once 100% tissue ingrowth has been achieved. Here we see the strong effect that mean pore size has on the rate of tissue ingrowth. A 500-micron pore size provides 100% ingrowth after only 5 days, compared to a requirement of 25 days for the 179-micron mean pore size. For the 91-micron pore size device, after 25 days, tissue ingrowth is only about 90%. Once again, however, the void space available for cell transplant after tissue ingrowth was quite small.

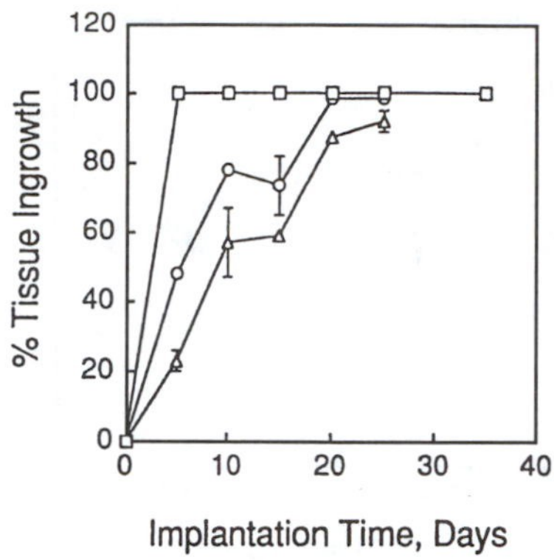

Figure 7.8 Normalized tissue ingrowth in PLLA devices of different pore sizes: 500 microns, square; 179 microns, circle; 91 microns, triangle. (From Wake, Patrick, and Mikos [1994], with permission.)

Sarver et al. (1995) have examined the tissue ingrowth and vascularization of a polyester surgical support felt placed within the test chamber shown in Figure 7.9. The support material itself was 1 cm in diameter and consisted of two layers for a total thickness of 3 mm (δ_T). The void fraction of the support material (ε_m) was 0.84. The polymer support material was pretreated with type I collagen and acidic fibroblast growth factor (aFGF) according to the method outlined by Thompson et al. (1988, 1989). The combination of the type I collagen and aFGF is a potent promoter of angiogenesis (Folkman 1985; Folkman and Klagsbrun 1987). The test chambers were implanted within the abdominal cavities of rats for periods of time ranging from 1 to 28 days.

Figure 7.10 (Sarver et al. 1995) shows the increase in capillary volume fraction of the support matrix as a function of time. Also shown is the fraction of the support matrix

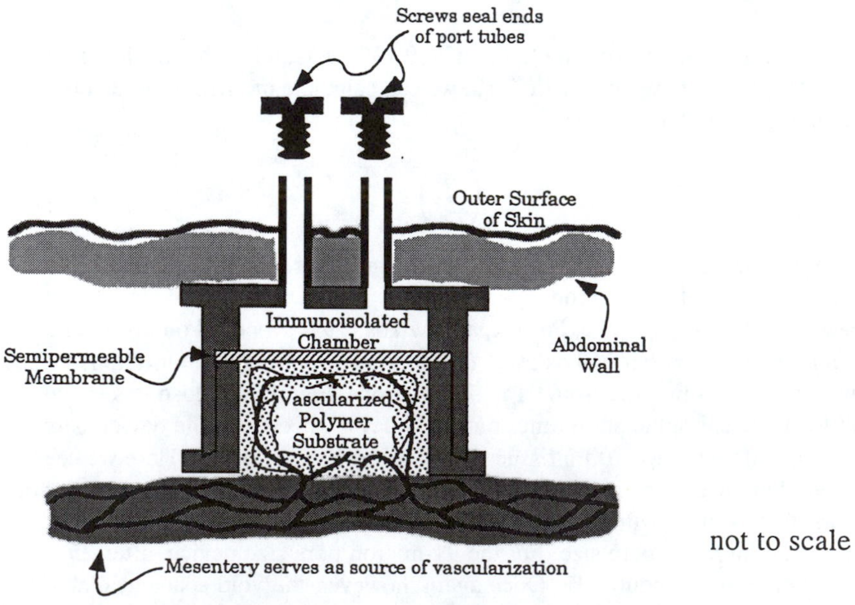

Figure 7.9 A device for cell transplantation. (From Sarver et al. [1995], with permission.)

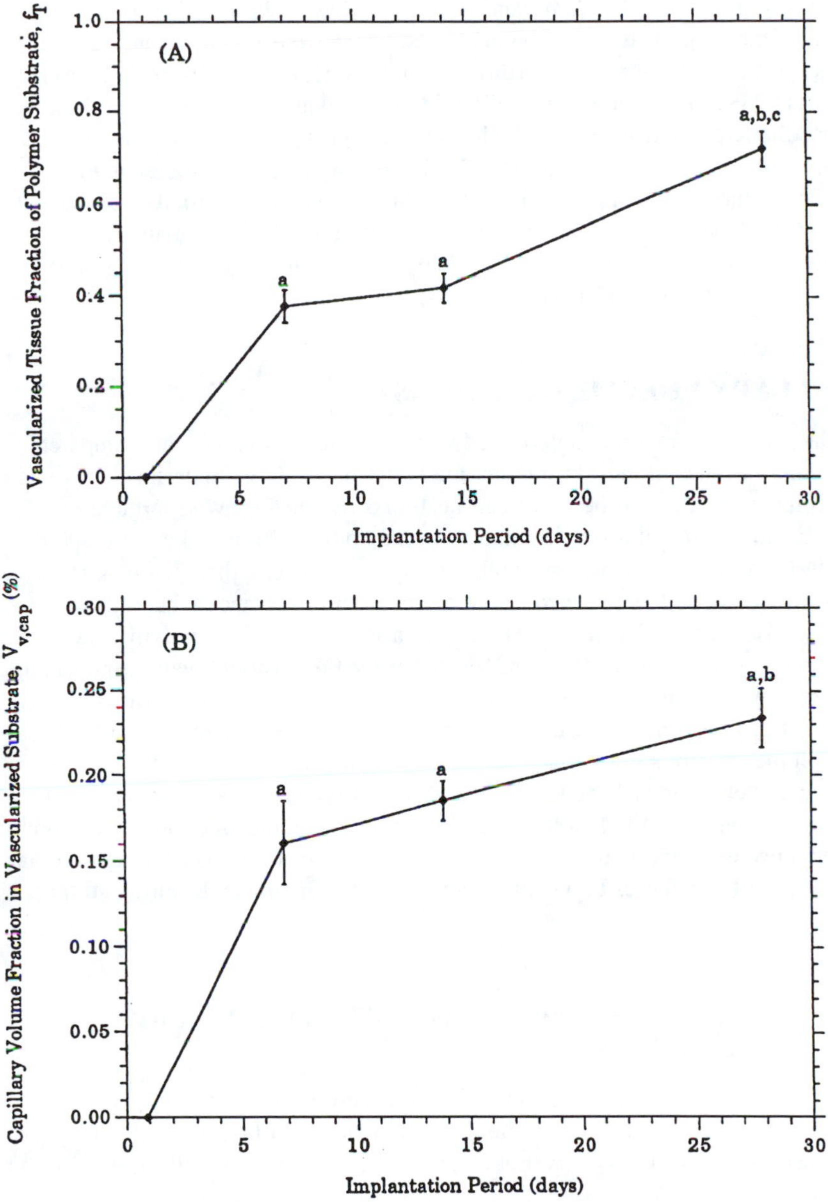

Figure 7.10 (A) Vascularized tissue fraction of polymer matrix for each period of implantation. (B) Capillary volume fraction within vascularized region of polymer matrix for each period of implantation. (From Sarver et al. [1995], with permission.)

(f_T) that has vascularized tissue which, in this case, is equal to the fractional depth of tissue penetration. We see that after 28 days, the capillary volume fraction (based on the total matrix volume) is nearly 0.23% and that about 72% (0.72×3 mm $= 2.2$ mm of penetration) of the support structure contained vascularized tissue. The tissue ingrowth in this case occurs for the most part in one direction only and is comparable to that observed by Mikos et al. (1993). The average capillary diameter was also found to be about 7.2 microns and the Krogh tissue cylinder radius after 28 days was about 68 microns.

Similar studies by Holder et al. (1997) showed significantly higher numbers of new capillaries during the first 3 weeks for implants that were initially seeded with endothelial cells. They also reported cellular ingrowth rates into their matrices of about 0.5–1 mm/week. These rates of ingrowth are comparable to the values obtained by Mikos et al. (1993) and Sarver et al. (1995).

7.8 CAPILLARY VOLUME FRACTIONS

The capillary volume fractions determined in the studies above can be compared to what are found in other tissues throughout the body. A rough estimate for the average capillary volume fraction in the body can be found by the following argument. The total cross-sectional area of all the capillaries in the body is estimated to be 2,500 cm^2 (Guyton 1991). Assuming an average capillary length of 0.05 cm, then in a 70-kg human (with density of about 1 g/cm^3), the volume fraction of capillaries would be: ($2{,}500\ cm^2 \times 0.05\ cm/70{,}000\ cm^3$) $\times 100 = 0.18\%$. Therefore, in a sense, the above polymeric support structures have capillary volume fractions comparable to the "average" value for the body. Certainly there will be quite a bit of variation around this average considering the many different functions the various tissues in the body perform. For example, the brain has capillary volume fractions in the range of 1–4% , cardiac muscle is on the order of 7–10% , whereas skeletal muscle is around 1% . Tumors typically have capillary volume fractions in the range of 0.1–1%. The higher capillary volume fractions of the brain and cardiac muscle reflect their much higher demand for oxygen. Skeletal muscle has a reserve supply of capillaries to meet the increased oxygen demands during strenuous exercise.

7.9 MEASURING THE BLOOD FLOW WITHIN POLYMERIC SUPPORT STRUCTURES

Another way to monitor the ingrowth of tissue, and the development of a vascular supply within the support structure, is through the use of radioactive or fluorescent microsphere techniques (Selman et al. 1984; Sarver et al. 1995). This technique is illustrated for the rat in Figure 7.11.

In this technique, a syringe pump, which we will see later serves as a reference flow rate, is connected to the left femoral artery and blood is removed at a constant rate. After about 1 minute, 15-micron-diameter radioactive or fluorescent microspheres are injected into the left ventricle of the heart via a cannula in the right common carotid artery. The microspheres are then allowed to distribute throughout the animal's body for 1 minute.

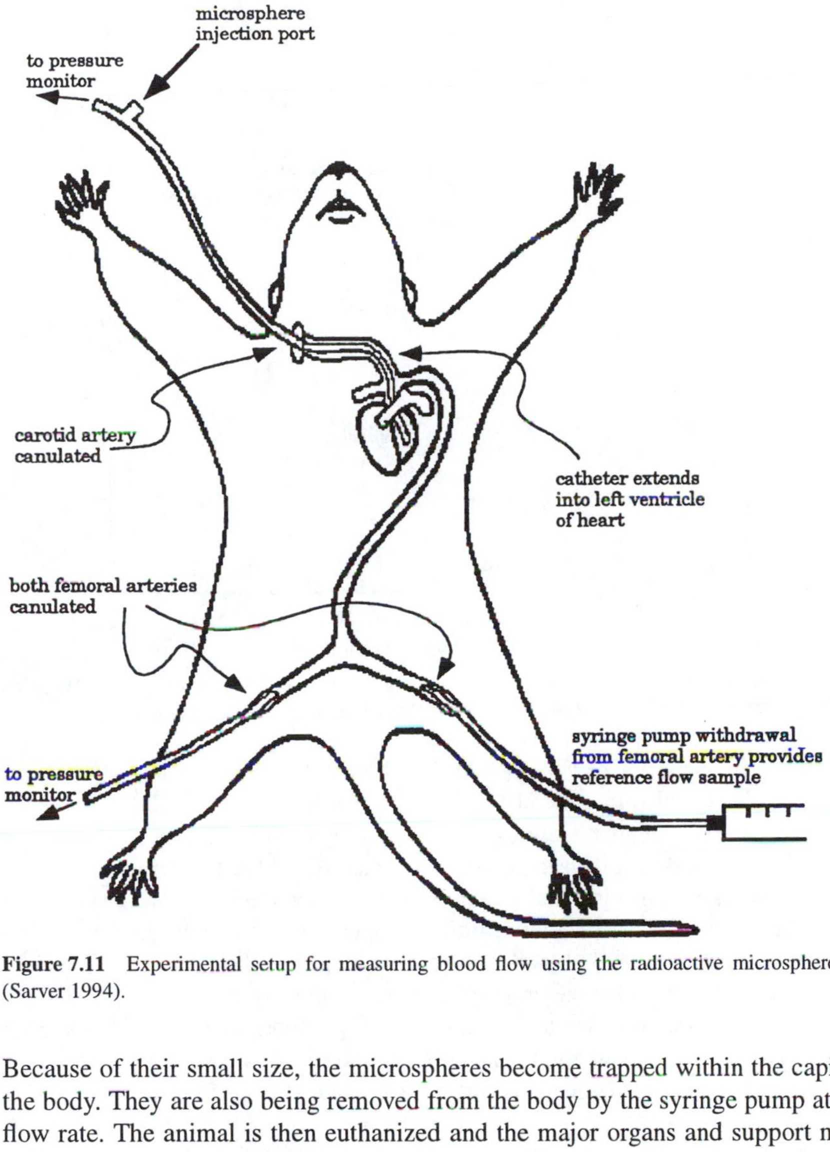

Figure 7.11 Experimental setup for measuring blood flow using the radioactive microsphere technique (Sarver 1994).

Because of their small size, the microspheres become trapped within the capillaries in the body. They are also being removed from the body by the syringe pump at a known flow rate. The animal is then euthanized and the major organs and support matrix are removed from the animal. The radioactivity or fluorescent levels in samples from the organs, the support matrix, and the syringe pump are determined. The blood flow rate to each of these samples is then proportional to their level of radioactivity or fluorescence, recognizing that the amount of radioactivity or fluorescence in the syringe pump sample is used to calibrate the flow rate.

Sarver et al. (1995) used this technique to also determine the blood flow rate within the polyester support matrix for various times after implantation. The results are shown in Figure 7.12. Recall that two layers of the support material were used in the device shown in Figure 7.9. The layer adjacent to the mesentary clearly shows the progressive

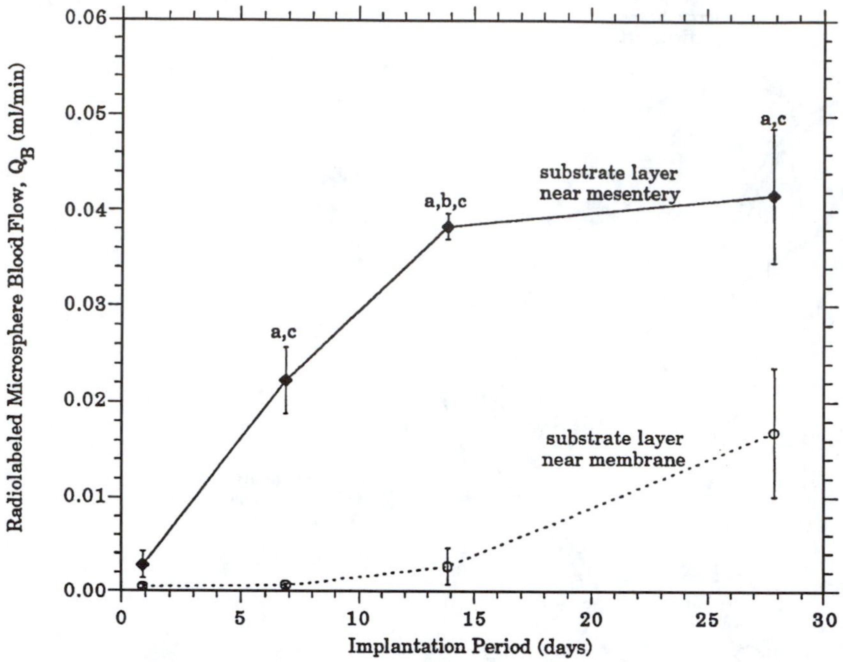

Figure 7.12 Blood flow measuring in polymer matrix for each implantation period. (From Sarver et al. [1995], with permission.)

development of the capillary bed as reflected by the increase in support matrix blood flow. These results are also in general agreement with the capillary volume fractions determined from the histological sections of the support material that were shown earlier in Figure 7.10. After about 14 days, a plateau is reached with little further improvement in the blood flow rate, indicating that perhaps 14 days after implantation would be an appropriate time for cell transplantation into a support matrix of this thickness. The innermost layer shows a delay in blood flow rate development that would be expected, recognizing the time needed for the tissue front to reach this layer.

Focusing on the first layer, we see that the maximum blood flow rate is about 0.042 ml/min after 28 days of implantation. The support material has a volume of $\pi/4 \times 1$ $cm^2 \times 0.15$ cm $= 0.118$ cm^3, of which the tissue could occupy 84% of this volume, giving a potential tissue volume of 0.1 cm^3. The blood perfusion rate for the tissue in this layer is then 0.042 ml/min $\times$ 1/0.1 cm^3 $=$ 0.42 ml/cm^3/min or 42 ml/min/100 g of tissue. This value of the tissue blood perfusion rate within the support structure compares favorably with values obtained for other tissues within the body as summarized in Table 4.2.

7.10 MEASURING MASS-TRANSFER RATES

The implant device shown in Figure 7.9 can also be used to determine the mass transfer characteristics of the capillary bed (Sarver et al. 1995). Note the presence of a small chamber in the region above the vascularized support structure. This chamber can also be

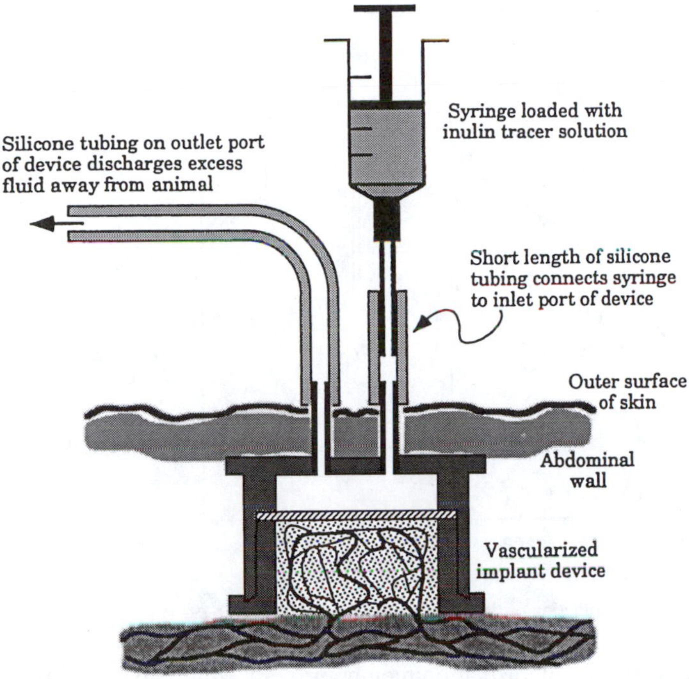

Figure 7.13 The radiolabeled inulin tracer technique. (From Sarver et al. [1995], with permission.)

immunoisolated from the vascularized support structure by a semipermeable membrane that typically has a nominal molecular weight cutoff of around 50,000 daltons. This nominal pore size will stop components of the host's immune system (immune cells and antibodies, which have molecular weights >150,000) from entering the chamber region of the device.

If the chamber region is loaded with a suitable tracer compound as shown in Figure 7.13, then the tracer's uptake and elimination from the body can be followed and analyzed using the pharmacokinetic techniques developed in Chapter 5. In this way, the transport properties of the capillary bed can be determined.

An ideal tracer compound for this type of study is radioactive inulin. Inulin is a polysaccharide molecule with a molecular weight of around 5,500 daltons. As mentioned in Chapter 5, it is only removed from the blood through the glomerulus and it is not secreted or reabsorbed within the renal tubules. Furthermore, it is not metabolized and is nontoxic. Therefore, if it is placed within the chamber of the device as shown in Figure 7.13, its only elimination pathway from the body is through the kidneys, and is equivalent at any time to the product of the glomerular flow rate (GFR) and the inulin plasma concentration. If the GFR and volume of distribution are known for the test subject, for example from an inulin infusion study as in Example 5.1, then a pharmacokinetic model can be readily developed to correlate the transient plasma inulin levels (Sarver 1994; Sarver et al. 1995). Correlation of the pharmacokinetic model to the plasma inulin levels then provides a variety of transport information such as the matrix blood

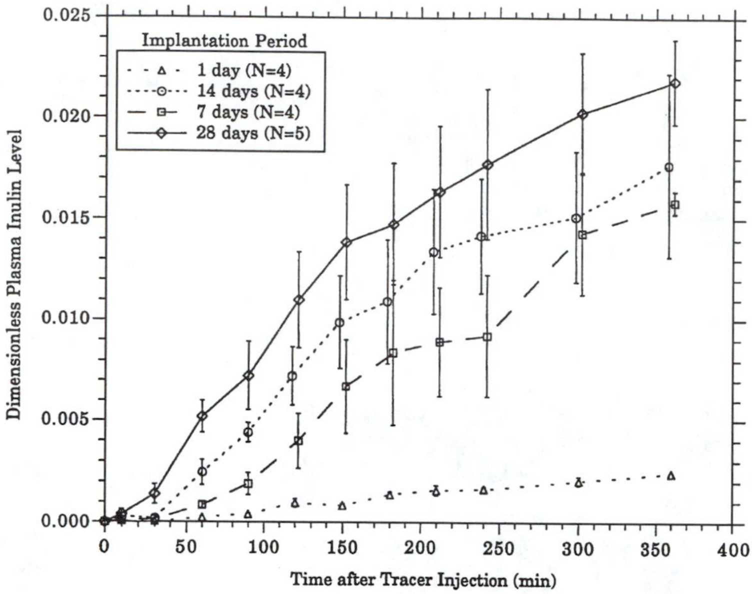

Figure 7.14 Plasma inulin level as a function of time after tracer injection (Sarver 1994).

flow and the capillary wall and immunoisolation membrane permeability to inulin. With this information, one can then predict the permeabilities of other compounds of interest using the solute permeability relationships developed in Chapter 2. This type of information can then be used to perform scaleup studies for bioartificial organs as discussed in Chapter 8.

Figure 7.14 shows the dimensionless plasma inulin levels for the period of time after injection of inulin into the device chamber. The time since device implantation is shown as a parameter. The devices once again contained two layers of the polyester surgical support felt and were implanted in the abdominal cavity of laboratory rats. The dimensionless plasma inulin concentration (X) is defined by the following equation.

$$X = \frac{C}{\left(\frac{D}{V_B}\right)} \tag{7.2}$$

In this equation, C is the measured plasma inulin concentration and the quantity (D/V_B) represents the ratio of the amount of inulin injected into the device chamber to the inulin distribution volume. This ratio represents the maximum possible plasma inulin concentration that would occur if the inulin dose were instantly mixed throughout its distribution volume after injection.

For a given time period since device implantation, we see from Figure 7.14 that the plasma inulin concentration increases steadily with time following its introduction into the chamber of the device. We also observe that at any time following addition of the inulin tracer to the device, the plasma inulin levels are higher for longer implantation periods. The rate of increase of the plasma inulin levels also increases for longer times since device implantation. This reflects the increase in capillary surface area and blood flow with time since implantation, and serves as another measure of tissue ingrowth and vascularization in the support matrix material.

7.11 PHARMACOKINETIC MODELING OF INULIN TRANSPORT IN A POLYMERIC SUPPORT STRUCTURE

We can develop a pharmacokinetic model to analyze the plasma inulin levels for the period of time after injection. A compartmental model approach can be used as outlined in Figure 7.15. Four separate well-mixed compartments are used to define the distribution of inulin within the device and the body. The device itself is assumed to consist of three compartments. The first compartment is the implantation chamber where inulin is initially injected. Inulin is then assumed to diffuse out of this chamber and across the semipermeable membrane and an adjacent layer of yet to be vascularized matrix. The inulin tracer then enters the interstitial fluid space of the tissue region that occupies the void space within the remaining portion of the support structure. This matrix tissue region is then the second compartment. The third compartment is the capillary bed itself that lies within the matrix tissue. These capillaries have blood flowing through them at the rate Q_{CB}. Inulin then diffuses from the tissue region into these capillaries where it is carried to the body represented by a single compartment having an inulin concentration of C_B. Inulin is then eliminated from the fourth or body compartment (V_B) by simple filtration through the kidneys.

Unsteady mass balances for the dimensionless inulin concentration can then be written as follows for each of the compartments.

$$V_I \frac{dX_I}{dt} = -P_{IT} S_{IT} (X_I - X_T) \quad (7.3)$$

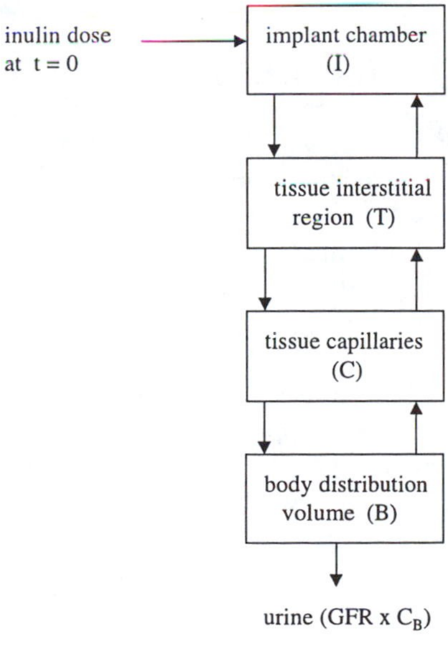

Figure 7.15 Compartmental model for analysis of plasma inulin concentrations.

$$V_T \frac{dX_T}{dt} = P_{IT} S_{IT}(X_I - X_T) - P_{TC} S_{TC}(X_T - X_C) \tag{7.4}$$

$$V_C \frac{dX_C}{dt} = Q_{CB}(X_B - X_C) + P_{TC} S_{TC}(X_T - X_C) \tag{7.5}$$

$$V_B \frac{dX_B}{dt} = -Q_{CB}(X_B - X_C) - \text{GFR} \times X_B \tag{7.6}$$

Four initial conditions are needed to solve these differential equations. Therefore, at $t = 0$, we have that $X_I = V_B / V_I$ and $X_T = X_C = X_B = 0$.

In Equations 7.3 to 7.6, the P's represent the solute permeability between the various compartments. P_{IT} is the apparent permeability between the implantation chamber and the tissue region. This value times the bulk concentration difference between the implantation chamber and the tissue region, $(\frac{D}{V_B})(X_I - X_T)$, provides the solute transport flux.

P_{IT} can be related to the membrane permeability (P_m). P_m can be estimated using the equations outlined in Chapter 2 or measured by in vitro experiments (Baker et al. 1997; Sarver et al. 1995; Dionne et al. 1996; Iwata et al. 1996). However, in vitro estimates of the membrane permeability may be significantly higher than the actual in vivo permeability of the membrane because of protein adsorption.

The additional resistance presented by the avascular matrix (P_{AV}) adjacent to the membrane must also be considered and can be estimated as discussed in Chapter 2. The membrane resistance, and the resistance of the adjacent avascular matrix, may then be combined to give the apparent membrane permeability $\frac{1}{\bar{P}_m} = \frac{1}{P_m} + \frac{1}{P_{AV}}$. As we will soon show in Example 7.1, experimental data such as that shown in Figure 7.14 can be used to obtain the apparent membrane permeability and an estimate of the in vivo membrane permeability. Therefore P_{IT} is given by the following equation:

$$P_{IT} = \frac{1}{\frac{1}{P_m} + \frac{1}{P_{AV}}} = \bar{P}_m \tag{7.7}$$

P_{TC} represents the solute permeability of the capillary wall (P_{CW}) which is assumed to be the controlling resistance for transport from the tissue region and into the capillary. S_{IT} is the surface area available for transport between the implantation chamber and the tissue region. This would be equal to the immunoisolation membrane surface area S_M. S_{TC} represents the surface area available for solute transport between the tissue region and the capillaries. This is equivalent to the total capillary surface area S_{CW} and is given by this relationship, $\varepsilon_m S_M \delta_T f_T s$, where s is the capillary surface area per volume of tissue and is given by Equation 2.55. Recall that f_T is the fraction of the support structure volume that contains tissue.

Q_{CB} is the blood flow rate within the tissue region corrected for the extracellular volume fraction of blood, ζ_C, which is about 0.63 for inulin. This accounts for the fact that the inulin concentration measurements were based on plasma volume. If the inulin concentration was based on the total blood volume, then ζ_C would be equal to one. V_I is the volume of the implantation chamber and is based on the given device dimensions. V_T is the volume of the tissue region extracellular fluid and is equal to $f_T \zeta_T S_M \delta_T \varepsilon_M$, where

ζ_T is the total extracellular volume fraction of the tissue within the support structure. Inulin only distributes within the extracellular fluid space in the tissue region and the extracellular volume fraction (ζ_T) of most tissues is about 0.15. This fluid surrounding the cells is also assumed to be well-mixed and the inulin concentration homogeneous. The capillary volume in the tissue region (V_C) is given by $S_M \delta_T \varepsilon_M f_T v$, where v is the capillary volume fraction given by Equation 2.56.

Equations 7.3 through 7.6 can be simplified somewhat by recognizing that the capillary volume, V_C, is much smaller than the other volumes. Since V_C is so small, it has the effect of making the left hand side of Equation 7.5 very nearly equal to zero. This creates a pseudo-steady-state situation for the solute concentration in the capillaries and means that we can replace the differential Equation 7.5 with the following algebraic equation.

$$Q_{CB}(X_B - X_C) + P_{TC}S_{TC}(X_T - X_C) = 0 \tag{7.8}$$

This equation can be simplified further by recognizing that the body compartment solute concentration (X_B) is much smaller than the solute concentration within the tissue capillaries of the implant device (X_C). This makes sense because the volume of the body compartment is several orders of magnitude larger than that of the capillaries within the device's support structure. Once the solute enters the body compartment, it becomes significantly diluted.

Equation 7.8 can then be rearranged to provide the solute capillary concentration within the tissue region in terms of the solute concentration in the tissue compartment only.

$$X_C = \left[\frac{P_{TC}S_{TC}}{Q_{CB} + P_{TC}S_{TC}}\right] X_T = K_{TC}X_T \tag{7.9}$$

This equation also shows that the capillary solute concentration is essentially at equilibrium with the tissue solute concentration as given by the constant distribution coefficient, K_{TC}. We can now rewrite Equations 7.3, 7.4, and 7.6 , using Equation 7.9 for the capillary concentration, to obtain the following set of equations.

$$V_I \frac{\mathrm{d}X_I}{\mathrm{d}t} = -P_{IT}S_m(X_I - X_T) \tag{7.10}$$

$$V_T \frac{\mathrm{d}X_T}{\mathrm{d}t} = P_{IT}S_m(X_I - X_T) - P_{CW}S_{CW}(1 - K_{TC})X_T \tag{7.11}$$

$$V_B \frac{\mathrm{d}X_B}{\mathrm{d}t} = Q_{CB}K_{TC}X_T - (\mathrm{GFR})X_B \tag{7.12}$$

These equations may be solved analytically using Laplace transforms (Sarver 1994; Gibaldi and Perrier 1982) to provide the following results for the dimensionless solute concentrations in the implantation chamber, the tissue region, and within the body compartment. Recall that Equation 7.9 provides the corresponding value of the concentration within the capillary compartment.

$$X_I = \left[\frac{V_B(\lambda_1 - k_{TI} - k_{CW})}{V_I(\lambda_1 - \lambda_2)}\right] e^{-\lambda_1 t} + \left[\frac{V_B(\lambda_2 - k_{TI} - k_{CW})}{V_I(\lambda_2 - \lambda_1)}\right] e^{-\lambda_2 t} \tag{7.13}$$

$$X_T = \left[\frac{V_B k_{IT}}{V_T(\lambda_1 - \lambda_2)}\right](e^{-\lambda_2 t} - e^{-\lambda_1 t}) \tag{7.14}$$

$$X_B = \left[\frac{k_{IT}k_{CW}}{(\lambda_1 - \lambda_2)(\lambda_1 - k_{TE})}\right]e^{-\lambda_1 t} + \left[\frac{k_{IT}k_{CW}}{(\lambda_2 - \lambda_1)(\lambda_2 - k_{TE})}\right]e^{-\lambda_2 t} + \left[\frac{k_{IT}k_{CW}}{(k_{TE} - \lambda_1)(k_{TE} - \lambda_2)}\right]e^{-k_{TE}t} \tag{7.15}$$

The rate constants (k's) are given by the following set of equations.

$$k_{IT} = \frac{P_{IT}S_m}{V_I} \tag{7.16}$$

$$k_{TI} = \frac{P_{IT}S_m}{V_T} \tag{7.17}$$

$$k_{CW} = -\frac{P_{CW}S_{CW}(K_{TC} - 1)}{V_T} = \frac{Q_{CB}K_{TC}}{V_T} \tag{7.18}$$

$$k_{TE} = \frac{\text{GFR}}{V_B} \tag{7.19}$$

The constants λ_1 and λ_2 are given by the equations below.

$$\lambda_1 = \frac{1}{2}\left[k_{IT} + k_{TI} + k_{CW} + \sqrt{(k_{IT} + k_{TI} + k_{CW})^2 - 4k_{IT}k_{CW}}\right] \tag{7.20}$$

$$\lambda_2 = \frac{1}{2}\left[k_{IT} + k_{TI} + k_{CW} - \sqrt{(k_{IT} + k_{TI} + k_{CW})^2 - 4k_{IT}k_{CW}}\right] \tag{7.21}$$

Solution of the above equations then provides a description of the solute concentrations in the various compartments following the injection of the solute in the implantation chamber. These equations can also be used to determine the membrane permeability and the capillary wall permeability using data such as that shown in Figure 7.14.

Example 7.1 Use the pharmacokinetic model for inulin distribution developed above and perform a nonlinear regression analysis of the 28-day data shown in Figure 7.14. Obtain best estimates of the glomerular filtration rate (GFR), the in vivo membrane permeability, and the capillary wall permeability $(PS)_{CW}$ using these data that were obtained from a total of five rats. The measured in vitro membrane permeability for inulin is 7.8×10^{-6} cm/sec. Also recall that the blood perfusion rate for the vascularized matrix was 0.42 ml/cm^3/min (see Figure 7.12) and the vascularized tissue fraction of the matrix was 0.72 (see Figure 7.10).

SOLUTION Additional data needed to solve this problem and not already discussed above are as follows. The depth of the implantation chamber (δ_I) is 0.07 cm and the volume of this chamber (V_I) is 0.052 ml. The inulin diffusivity in the bulk isotonic solution (D_I) at 37°C can be estimated from Figure 2.1 to be about 2.3×10^{-6} cm^2/sec. D_T is the effective diffusivity in the tissue region and can be found using the data presented earlier in Figure 2.6. For a molecule the size of inulin, the diffusivity will be reduced by about 85%, giving a value for D_T of 3.45×10^{-7} cm^2/sec. The inulin distribution volume for a rat was found in Example 5.1. The same example also provides an initial estimate of the GFR. The GFR will then be determined from the inulin data since GFR has a significant effect on the inulin clearance. The rats used in these experiments also had an average body weight of 481 grams.

Chart 7.1 shows the Mathcad® solution to this problem. We see that a nonlinear regression on the unknown apparent membrane permeability ($\bar{P}_m$), the capillary permeability × capillary wall surface area $(PS)_{CW}$, and the glomerular filtration rate (GFR) results in an excellent fit of the inulin data.

Chart 7.1 for **Example 7.1:** Pharmacokinetic model of inulin distribution following injection into the device shown in Figure 7.9.

$i := 0 \ldots 10$ data points, corresponds to the 28 day curve shown in Figure 7.14

minutes, time after inulin injection $t_i =$	dimensionless inulin levels after injection $x_i :=$
10	.0005
30	.0015
60	.0052
90	.0071
120	.011
150	.0138
180	.0146
210	.0162
240	.0177
300	.022
360	.0218

$TOL := .0001$ solution tolerance, default = .001

$r_m := .5$ cm, membrane diameter

$\delta_T := .3$ cm, support structure thickness

$f_T := .72$ fraction of matrix structure void volume containing tissue

$\delta_I := .07$ cm, depth of the implantation chamber

$V_I := .052$ ml, volume of the implantation chamber

$r_c := .00036$ cm, capillary radius

$\varepsilon_m := .84$ matrix structure void volume

$q_B := .42$ ml/cm^3/min, tissue region blood perfusion rate

$Q_B := q_B \pi r_m^2 \delta_T \varepsilon_m f_T$

$Q_B := \dfrac{Q_B}{60}$

$PS_{CW} := .01$ cm^3/sec/100 g, capillary wall inulin permeability × capillary surface area; these are guesses for P_m & PS_{CW}

$P_m := 1 \times 10^{-8}$ cm/sec, membrane inulin permeability

$D_I := 2.3 \times 10^{-6}$ cm^2/sec, inulin diffusivity in the implantation chamber

$D_T := 3.45 \times 10^{-7}$ cm²/sec, effective diffusivity of inulin in the tissue region

$\xi_C := .63$ extracellular void fraction of the capillary blood

$gfr := .48$ ml/hr/g BW, glomerular filtration rate, see Example 5.1

$gfr := \frac{gfr}{3600}$

$v_B := .21$ ml/g BW, inulin body distribution volume, see Example 5.1

$BW := 481$ g, body mass

$\xi_T := .15$

$V_T := \xi_T \varepsilon_m \pi r_m^2 \delta_T f_T$

$V_T = 0.021$ cm³, volume of interstitial fluid in the tissue region

$V_B := BW v_B$

$V_B = 101.01$ cm³, inulin body distribution volume

$S_m := \pi r_m^2$

$S_m = 0.785$ cm², membrane surface area

$PS_{CW} := \frac{PS_{CW}}{100} \varepsilon_m \pi r_m^2 \delta_T f_T$

$PS_{CW} = 1.425 \times 10^{-5}$ cm³/sec, based on actual tissue volume

$P_{IT}(P_m) := P_m$

$P_{IT}(P_m) = 1 \times 10^{-8}$ cm/sec, effective permeability of the membrane

$Q_{cb} := Q_B \xi_C$

$Q_{cb} = 6.284 \times 10^{-4}$ ml/sec, fluid flowrate in the support structure capillaries

$k_{te} := \frac{gfr}{v_B}$

$k_{te} = 6.349 \times 10^{-4}$ 1/sec, elimination rate constant

$k_{IT}(P_m) := \frac{P_{IT}(P_m) S_m}{V_I}$

$k_{TI}(P_m) := \frac{P_{IT}(P_m) S_m}{V_T}$

$K_{TC}(PS_{CW}) := \frac{PS_{CW}}{(Q_{cb} + PS_{CW})}$

$K_{TC}(PS_{CW}) = 0.022$ equillibrium constant = Xc/XT

$k_{CW}(PS_{CW}) := -PS_{CW} \frac{K_{TC}(PS_{CW}) - 1}{V_T}$

$$\lambda_1(P_m, PS_{CW}) := .5\Big[k_{IT}(P_m) + k_{TI}(P_m) + k_{CW}(PS_{CW}) \cdots + \sqrt{(k_{IT}(P_m) + k_{TI}(P_m) + k_{CW}(PS_{CW}))^2 - 4k_{IT}(P_m)k_{CW}(PS_{CW})}\Big]$$

$$\lambda_2(P_m, PS_{CW}) := .5\Big[k_{IT}(P_m) + k_{TI}(P_m) + k_{CW}(PS_{CW}) +- \sqrt{(k_{IT}(P_m) + k_{TI}(P_m) + k_{CW}(PS_{CW}))^2 - 4k_{IT}(P_m)k_{CW}(PS_{CW})}\Big]$$

$$A(t, P_m, PS_{CW}, k_{te}) := \left[\frac{k_{IT}(P_m)k_{CW}(PS_{CW})}{(\lambda_1(P_m, PS_{CW}) - \lambda_2(P_m, PS_{CW}))(\lambda_1(P_m, PS_{CW}) - k_{te})}\right] e^{-\lambda_1(P_m, PS_{CW})t60}$$

$$B(t, P_m, PS_{CW}, k_{te}) := \left[\frac{k_{IT}(P_m)k_{CW}(PS_{CW})}{(\lambda_2(P_m, PS_{CW}) - \lambda_1(P_m, PS_{CW}))(\lambda_2(P_m, PS_{CW}) - k_{te})}\right] e^{-\lambda_2(P_m, PS_{CW})t60}$$

$$C(t, P_m, PS_{CW}, k_{te}) := \left[\frac{k_{IT}(P_m)k_{CW}(PS_{CW})}{k_{te} - \lambda_1(P_m, PS_{CW}))(k_{te} - \lambda_2(P_m, PS_{CW}))}\right] e^{-k_{te}t60}$$

$$x_B(t, P_m, PS_{CW}, k_{te}) := A(t, P_m, PS_{CW}, k_{te}) + B(t, P_m, PS_{CW}, k_{te}) + C(t, P_m, PS_{CW}, k_{te})$$

sum of the squares of the error between the model and data that are to be minimized

$$\mathrm{SSE}(P_m, PS_{CW}, k_{te}) := \sum_{i=0}^{10} (x_i - x_B(t_i, P_m, PS_{CW}, k_{te}))^2$$

given

$\mathrm{SSE}(P_m, PS_{CW}, k_{te}) = 0$

$1 = 1$

$2 = 2$

dummy equations

$$\begin{bmatrix} P_m \\ PS_{CW} \\ k_{te} \end{bmatrix} := \mathrm{minerr}(P_m, PS_{CW}, k_{te})$$

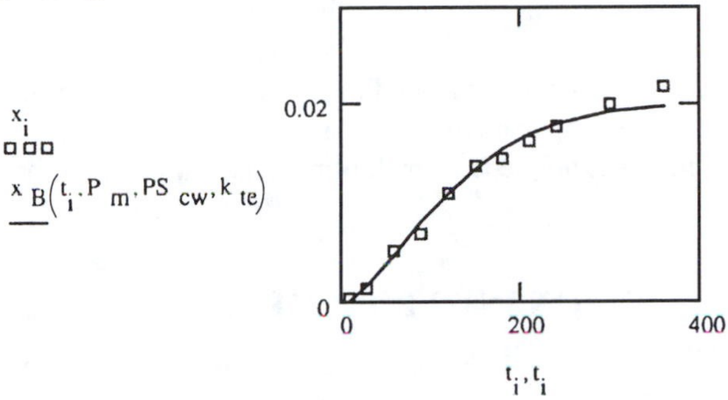

$P_m = 6.441 \times 10^{-7}$	cm/sec, membrane permeability
$PS_{CW} = 2.312 \times 10^{-6}$	cm^3/sec, capillary wall permeability × capillary surface area
$PS_{CW} := \dfrac{100 \times PS_{CW}}{\left(\varepsilon_m \pi r_m^2 \delta_T f_T\right)}$	
$PS_{CW} = 1.623 \times 10^{-3}$	cm^3/sec/100 grams of tissue
$k_{te} = 2.96 \times 10^{-4}$	1/sec
$gfr := v_B k_{te} 3600$	
$gfr = 0.224$	glomerular filtration rate, ml/hr/g BW

The calculated apparent permeability of the membrane includes the resistance of the adjacent unvascularized layer of tissue. This apparent membrane permeability ($\bar{P}_m$) is found to be 6.44×10^{-7} cm/sec. The apparent membrane permeability can be written in terms of the actual in vivo membrane permeability and the permeability of the avascular tissue. In the following equation, we simply assume that the permeability of the adjacent avascular matrix is given by the ratio of the inulin diffusivity corrected for matrix void tissue volume, divided by the thickness of the avascular region. The thickness of the avascular region is equal to $(1 - f_T)\delta_T$ and is 0.084 cm.

$$\bar{P}_m = \frac{1}{\dfrac{1}{P_m} + \dfrac{1}{\dfrac{\varepsilon_m D_T}{\delta_{\text{avascular region}}}}}$$

Using the above equation, we can calculate an in vivo membrane permeability of 7.92×10^{-7} cm^2/sec. The in vitro inulin permeability of this membrane (P_m) was found to be 7.8×10^{-6} cm/sec. The considerably lower in vivo value is most likely a result of protein adsorption within the 250 micron thick fibrous backing material used in this type of membrane (Opong and Zydney 1991; Robertson and Zydney 1990 a,b).

The capillary wall inulin permeability and surface area product (PS_{CW}) was found to be 0.0016 cm^3/sec/100 grams of tissue. Based on the data shown in Figure 2.7, a molecule the size of inulin (1.5 nm) should be expected to have a PS_{CW} value on the order of 0.01 cm^3/sec/100 grams of tissue. The observed reduction in the value of PS_{CW} could be a result of reduced capillary surface area and/or capillary wall permeability. The Krogh tissue cylinder radius was obtained from analysis of histological sections and found to equal 68 microns. This would give a capillary surface area per volume of tissue equal to 15.6 1/cm. Renkin (1977) presents estimates of the capillary surface area in tissues like that developing in this example and are on the order of 70 1/cm with inulin permeabilities on the order of 10^{-6} cm/sec, thus giving a predicted PS_{CW} on the order of 0.007 cm^3/sec/100 g of tissue. Therefore, it appears that the less than expected value of PS_{CW} obtained from the data used in this example is a result of reduced capillary surface area.

7.12 CELL TRANSPLANTATION INTO POLYMERIC SUPPORT STRUCTURES

The previous discussion shows that porous polymeric support structures are capable of developing within their confines a neovascularized tissue region. Furthermore, we find that this new tissue region has blood flow rates and capillary wall permeabilities comparable to those found in other tissues. We may now ask the following questions related to cell transplantation into these vascularized support structures. When is the best time to transplant cells? Is it possible to maintain the viability of transplanted cells and can they continue to perform their normal function? To answer these questions consider the results reported by Thompson et al. (1988, 1989). They induced the formation of *organoid* neovascular structures in rats using polytetrafluoroethylene (PTFE) fibers coated with collagen and the angiogenesis initiator acidic fibroblast growth factor (aFGF). The fibers are about 20 microns in diameter and a bundle or cotton ball-like mass of these fibers was implanted within the abdominal cavity adjacent to the liver.

To demonstrate the efficacy of a cellular transplant using the vascularized PTFE support structure, they used as a host for the implant the Gunn rat. *Homozygous* Gunn rats lack the liver enzyme needed for the conjugation of bilirubin. Recall that bilirubin is a product obtained from hemoglobin at the end of a red blood cell's lifespan. The conjugated form of bilirubin is readily passed from the liver via the bile and then removed from the body in the feces. Since the Gunn rat cannot conjugate bilirubin, its plasma bilirubin levels are increased significantly from the normal level of <1 mg/dl. The Wistar rat is genetically identical to the Gunn rat except that it has the ability to conjugate bilirubin. In their initial set of experiments, they obtained hepatocytes from the livers of the syngeneic Wistar rats. These hepatocytes were then seeded onto collagen coated PTFE fibers that did not contain the angiogenesis promoter aFGF. This structure was then implanted adjacent to the liver of the Gunn rat. Because the Gunn and Wistar rats are genetically the same there is no need for immunosuppressive agents or immunoisolation of the transplanted cells. For the first 10 days after implantation, the plasma bilirubin

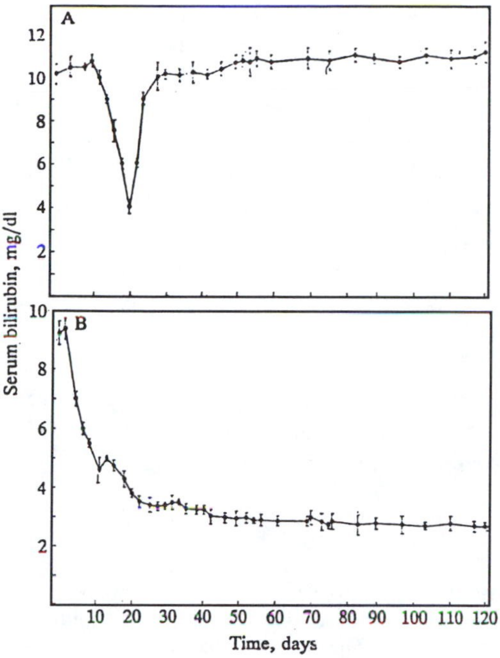

Figure 7.16 Plasma bilibrubin levels following transplantation of hepatocytes into Gunn rats (A) collagen-coated PTFE support structure with hepatocytes added at time of implantation; (B) collagen-coated PTFE support structure with aFGF. Hepatocytes added 28 days after implantation. (From Thompson et al. [1989], with permission.)

levels remained unchanged as shown in Figure 7.16a. By the 20th day, the plasma bilirubin levels had decreased by about 50%. Shortly thereafter, the plasma bilirubin levels returned to their original high levels and remained there for the remaining days of the experiment.

The lack of long-term function of the hepatocytes in these preliminary studies can be attributed to several effects. First the angiogenesis factor, aFGF, was not used in these first experiments. This would result in less vascularization of the support structure. Furthermore, the cells were seeded into the support structure at the time of implant. Lack of a good vascular supply at the time of cell seeding is clearly not optimal for cell growth and the transport of nutrients and bilirubin. This would account for the delayed response with regard to any effect on the plasma bilirubin levels. For the first 10 days, the implant was still in the process of tissue ingrowth and the development of a blood supply that is needed for efficient mass transport. It also allowed for the accumulation of bile acids that led to the death of the cells after about 20 days. Clearly these results show that seeding the cells at the time the support structure is implanted is not the best approach.

In their second set of experiments, collagen-coated PTFE fibers that contain adsorbed aFGF were first implanted. After 28 days, a suspension of Wistar rat hepatocytes was seeded into the network of the now vascularized fibers. As shown in Figure 7.16b, we see that the plasma bilirubin levels began to decrease within 1 day. After about 10 days, the plasma bilirubin levels had decreased by about 50% and at the end of the 120 day-experiment this reduction was >60%.

These results strongly suggest that polymeric support structures, through their ability to form their own vascular supply, can be used to sustain the long-term function of transplanted cells. Additionally, it appears better to first vascularize the support structure for several weeks before seeding the implant with the transplanted cells. Furthermore, these results demonstrate that transplanted cells carrying a normal gene are able to restore the function lost by the host's own genetically compromised cells. If the host's genetically deficient cells can be removed, the appropriate gene inserted, and their numbers expanded in tissue culture prior to their transplantation, then this approach becomes a powerful treatment method for a variety of diseases. The only limitation that needs to be addressed is the problem of immunorejection of the transplanted cells if they are not genetically identical to those of the host. The next chapter therefore addresses how immunoisolation of the transplanted cells can be incorporated to create bioartificial organs.

PROBLEMS

1. Consider the steady-state diffusion of glucose from a well-stirred bulk liquid solution across a thin hydrogel layer and then reaction and diffusion of the glucose within a multilayer of epithelium which lies on top of the hydrogel layer (Perez et al. 1995). The epithelium consists of N layers of cells, each cell h_c in thickness. The cells consume glucose at a first-order volumetric rate given by $k_1 C$, where C is the local concentration of glucose and k_1 is the first-order rate constant. The surface of the cells exposed to the gas phase above is a no-flux boundary for glucose, and C will therefore attain its minimum value at that interface ($C_{\min}$). Calculate the fractional drop ($C_b/C_{\min}$) in the glucose concentration from the bulk liquid medium (C_b) to the outermost epithelial surface ($C_{\min}$). Assume there is no mass-transfer resistance between the bulk solution and the surface of the hydrogel. Some additional information needed to solve this problem is shown in the table below.

Parameter	Value	Description
h_c	15 microns	Thickness of single cell layer
k_1	1.5×10^{-3} 1/sec	Glucose rate constant
C_b	3.2 mg/ml	Bulk glucose concentration
D_G	3×10^{-6} cm²/sec	Diffusivity of glucose in the hydrogel
K_p	2	Glucose hydrogel partition coefficient, bulk-gel or gel-epithelium
L	0.05 cm	Hydrogel thickness
N	5	Layers of epithelial cells
D_e	6×10^{-6} cm²/sec	Glucose diffusivity in epithelium

2. The following dimensionless inulin concentrations were obtained for a device like that shown in Figure 7.9. In this case, the immunoisolation membrane was that developed by Baker et al. (1997). This membrane consisted of a polyvinyl alcohol hydrogel imbedded within the pores of a 0.2-micron polyethersulfone microporous membrane. Using this data, determine the apparent membrane permeability, the value of PS_{CW} for the capillaries, and the GFR. The device used in these experiments had a membrane radius of 0.65 cm. The vascularized matrix was 0.15 cm thick. The chamber containing the inulin was 0.07 cm thick. The rat body weight was 345 grams. All other properties are as in Example 7.1. How does the value of PS_{CW} for inulin compare with the data shown in Figure 2.7?

Time, minutes	Dimensionless inulin concentration, X
0	0.0
10	0.0188
30	0.0714
60	0.1433
90	0.1210
120	0.0943
150	0.1148
180	0.1044
210	0.0773

3. Krewson, Klarman, and Saltzman (1995) presented the steady-state data in the table below for the distribution of nerve growth factor (NGF) in the vicinity of a thin, cylindrical, controlled–drug release device implanted into the brain of a rat. The polymeric disks containing radiolabeled I^{125}-NGF were 2 mm in diameter and 0.8 mm in thickness. NGF has a molecular weight of 28,000 daltons. Assuming that the NGF is eliminated from the brain tissue by a first-order process (i.e., the rate of elimination is proportional to the concentration of nerve growth factor, i.e., $k_{\text{apparent}} C_{\text{NGF}}$), develop a steady-state reaction-diffusion model to analyze this data. k_{apparent} is the apparent first-order elimination rate constant for NGF and accounts for such processes as metabolism, cellular internalization, or uptake by the brain's systemic circulation. The boundary conditions assuming the origin of the Cartesian coordinate system to be the midline of the polymeric disk are as follows:

$$C_{\text{NGF}} = C_0 \quad \text{at} \quad x = a \quad \text{and} \quad C_{\text{NGF}} = 0 \quad x = \infty$$

where a is the half-thickness of the polymeric disk. Estimate the value of the Thiele modulus ($a\sqrt{k_{\text{apparent}}/D}$) that provides the best estimate of the data. D represents the diffusivity of NGF in the brain tissue. After estimating a value of D for NGF, what is the value of k_{apparent}? Using Equation 5.16, calculate the half-life of NGF in the brain tissue. How does this value of the NGF half-life compare to the reported half-life of NGF in brain tissue of about 1 hour?

Distance from polymer/tissue interface, mm	NGF concentration, μg/ml
0	37.0
0.1	31.5
0.2	20.0
0.3	15.0
0.4	10.5
0.5	8.5
0.6	6.5
0.8	3.5
1.0	2.2

4. Estimate the value of k_{apparent} in Problem 3 assuming the only elimination pathway for NGF is uptake by the brain's systemic circulation. The capillary surface area for the brain is about 240 1/cm. Assume the concentration of NGF is zero in the blood because of dilution.

5. Select from Table 7.1 a subject for tissue engineering. Prepare a poster for presentation to your class that summarizes recent advancements in the area you selected. Topics covered in your poster should include a description of the clinical need, the methodology used, results that have been obtained (in vitro, in animals, and the status of any human clinical trials), the potential clinical impact, potential market impact, safety issues, and additional development needs.

6. Design a tissue engineered system for the delivery to the systemic circulation of human growth hormone (hGH). hGH has a molecular weight of 22 kD. Assume that the hGH has an apparent distribution volume of 30 liters. Autologous cells were transfected with a recombinant gene for hGH. The production rate of hGH from these cells is about 2,500 ng/10^6 cells/24 hours. The plasma concentration of hGH should be maintained at about 10 ng/ml. Assume hGH has a half-life in the body of about 2.3 hours. Describe your system for delivering hGH and carefully state all assumptions. How many cells will be required for the delivery of the hGH?

CHAPTER

EIGHT

BIOARTIFICIAL ORGANS

8.1 BACKGROUND

Tissue engineering is a very promising approach for the treatment of a variety of medical conditions. Treatment of a disease or medical problem using this approach requires the availability of the appropriate cells and the creation of an artificial support structure to contain them. The cells may be obtained from a variety of sources. For example, from an expanded population of the host's own (perhaps genetically modified) cells, from other compatible human donors, perhaps from animal sources, or even genetically engineered cell lines. The cells to be transplanted are then seeded at the appropriate time into the prevascularized polymeric support structure as discussed in the previous chapter. However, with the exception of *autologous cells*, one of the major obstacles that must be overcome is rejection of the transplanted cells by the host's immune system. The transplanted cells will be destroyed quickly by the host's immune system unless they are immunologically similar to the host's own cells (Benjamin and Leskowitz 1991).

Immunosuppressive drugs can be used to suppress the host's immune system and prolong the function of transplanted cells that are a relatively close match to the host. However, immunosuppressive drugs have potent side effects, and in the case for the example of islet transplants to treat diabetes, these drugs may provide a situation where the cure is worse than the disease. Genetically engineered cell lines, in addition to their possibly being rejected by the host's immune system, also pose additional risks that need to be considered. For example, direct implantation of an immortal (usually of tumor origin) cell line could lead to the unchecked growth of the implanted cells, they could move to and proliferate at a site other than the desired site, and they could change to a potentially hazardous form with a loss of their original therapeutic function. The full potential of tissue engineering will therefore be limited unless techniques can be developed to either restrict the host's immune response, or somehow modify the transplanted cells to make them more acceptable to the host's immune system, i.e., *immunomodification* (Lanza and Chick 1994b).

8.2 SOME IMMUNOLOGY

The immune response to foreign materials (*antigens*) such as transplanted cells consists primarily of a *cell-mediated component* and a *humoral component* (Benjamin and Leskowitz 1991). The major cellular components are the *B-lymphocytes* and the *T-lymphocytes*. The B-lymphocytes or B cells form in the *bone marrow*, and when properly activated by antigens, form proteins called *antibodies* that comprise the active agents of the humoral component of the immune system. The T-lymphocytes originate in the *thymus* and come in two basic types, the *CD4$^+$ (helper) T cells* and the *CD8$^+$ (killer or cytotoxic) T cells*.

The immune system is activated by its intimate contact with foreign molecules called *antigens* (*anti*body *gen*erating). Antigens must possess "foreignness", meaning the antigen is unlike anything the immune system has seen before. Immature B and T lymphocytes that react against *self-antigens* (the body's own proteins for example) are eliminated during their maturation phase in the bone marrow or the thymus. This is the so-called *clonal deletion theory*. Thus the mature B and T cells are *self-tolerant* and do not normally react against the body's own tissues.

Antigens generally have molecular weights $>$6,000 daltons and possess some degree of molecular complexity. In fact there can be many sites along the surface of such large molecules that are *immunogenic*[1]. These sites are referred to as *epitopes*. Fortunately, many synthetic polymers, although of high molecular weight, do not possess a sufficient amount of molecular complexity to provoke an immune response. Lower-molecular-weight materials, although not immunogenic by themselves, can associate with larger carrier molecules, for example a protein, to give an immune response directed at the lower-molecular-weight compound. In this case, the lower-molecular-weight compound is referred to as a *hapten*.

8.2.1 B-Lymphocytes

Each B cell has a unique antibody receptor on its surface that only recognizes a specific antigen. The B-cell inventory within an individual is capable of producing as many as 100 million distinct antibodies. This allows the immune system to respond to almost any known or unknown antigen. If a specific B cell comes into contact with its corresponding antigen, it becomes activated through a process, discussed later, that also involves the CD4$^+$ or helper T cells. The activated B cell then begins to reproduce, rapidly increasing the number of B cells that are specific for a given antigen. This expanded collection

[1]An immunogen is a molecule that can induce a specific response of the immune system. An antigen refers to the ability of a given molecule to react with the products of an immune response. For example, the binding of an antigen to an antibody. In this discussion we will assume that antigen and immunogen are synonymous.

of lymphocytes with a given specificity is referred to as a *clone of lymphocytes*. This selection by the antigen of a specific reactive clone of lymphocytes from a large pool of existing lymphocytes, each with their own unique antigen specificity, is called the *clonal selection theory*. These B cells then differentiate to form *plasma cells* that begin to secrete antibodies with the same antigen specificity.

8.2.2 Antibodies

Antibodies are proteins and are also called *gamma globulins* or *immunoglobulins*. They comprise about 20% of the total amount of protein found in plasma. There are five classes of antibodies that are called IgA, IgD, IgE, IgG, and IgM, where Ig stands for immunoglobulin and the letter designates the class. The basic structure of an antibody molecule consists of two light and two heavy polypeptide chains held together by disulfide bonds as shown in Figure 8.1.

Conceptually the antibody molecule is "Y-shaped", with the light chains forming the upper branches of the "Y" and the heavy chains forming the lower half of the "Y" structure. The molecule consists of three fragments. Two of these fragments are identical and reside at the top of the "Y" and each of these binds with antigen. These fragments are referred to as Fab, for fragment antigen binding. The base of the "Y" is called the Fc fragment, for fragment crystallizable. The Fc fragment does not bind with antigen but is responsible for the biological activity of the antibody after it binds with antigen. The Fc fragment is referred to as a *constant region* in terms of its amino acid sequence since it is the same for all antibodies within a given class. Since the Fab fragments bind to antigen, their structure is highly *variable* in order to provide the multitudinous shapes or conformations required for antigen specificity and recognition. The specific antigen-binding characteristics of the Fab fragments is determined by their unique sequence of amino acids. Although an activated B cell (or plasma cell) makes antibodies with only a single antigen specificity, it can switch to make a different class of antibody, while still retaining the same antigen specificity.

IgG is the major immunoglobulin in the body and is found in all the fluid spaces. It consists of a single "Y"-shaped molecule with a molecular weight of about 150,000

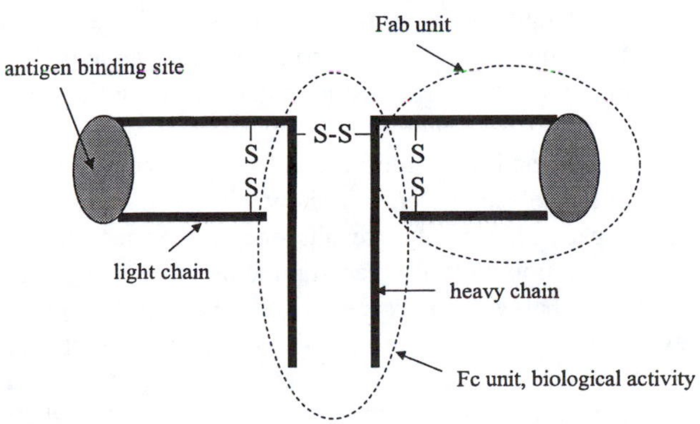

Figure 8.1 Structure of a antibody.

daltons. IgG has several important biological properties that are found in some of the other antibody classes as well. Because of its ability to bind two antigens per IgG molecule, IgG can cause the clumping or *agglutination* of particulate antigens such as those of invading microorganisms. These large antibody-antigen complexes are then readily phagocytized by phagocytic cells such as neutrophils and macrophages. In the case of smaller soluble antigens, this cross-linking of antigen and IgG leads to much larger complexes that become insoluble and *precipitate* out of solution and are then phagocytized.

IgG is also capable of binding to a variety of epitopes found on the surfaces of invading cells. The cell then becomes covered with IgG molecules that have their Fc fragments sticking out. Many phagocytic cells have surface receptors for the IgG Fc fragments allowing the phagocyte to bind to, engulf, and then destroy the invading organism. IgG is therefore an *opsonizing* antibody (opsonin is a Greek word that means to prepare for eating) since it facilitates the subsequent "eating" of the invading microorganism or cell by phagocytic cells.

The Fc receptor on an IgG molecule bound to a cellular antigen also binds to specific receptors on so-called *natural killer lymphocytes* (*NK cells*). The NK cells will then be attracted to and destroy the invading cell by the release of toxic substances called *perforins* (Ojcius, Liu, and Young 1998). Perforins form channels in the cell membrane of the target cell, making them leak and resulting in death of the cell. This process is called *antibody-dependent cell-mediated cytotoxicity* or ADCC.

IgG is also capable of neutralizing many toxins by binding to their active site. Viruses are also neutralized in a similar manner. In this case, the IgG molecule binds and blocks sites on the virus's surface coat that is used by the virus for attachment to the cells it normally invades. IgG, when bound to antigen, is also capable of activating the *complement system*. The complement system consists of a group of protein enzyme precursors that, when activated, undergo an amplifying cascade of reactions that generate cell membrane attack complexes (MACs) that literally punch holes through the cell membrane. This results in lysis and death of the targeted cell.

IgA is found primarily in the fluid secretions of the body, for example in saliva, tears, mucous, and in the gastrointestinal fluids. It is found in a dimeric form, that is, it consists of two "Y"-shaped antibodies linked at the base of the "Y". The molecular weight of this antibody is about 400,000 daltons. Its primary role is as a defense against local infections where it stops the invading microorganism from penetrating the body's epithelial surfaces that line the respiratory and gastrointestinal tract. It does not activate the complement system.

IgM forms a planar star-like pentamer and consists of five antibody molecules joined together at the base of the Fc regions. Its molecular weight is about 900,000 daltons and it is found mainly in the intravascular spaces. It has only five antigen binding sites because the restricted pentameric structure does not allow the Fab fragments to be open fully. Therefore, large antigens bound to one Fab will block the adjacent Fab. IgM is a very efficient agglutinating antibody since its large structure provides for binding between epitopes that are widely spaced on the antigen. IgM is also an excellent activator of the complement system. IgM is also found on the surface of mature B lymphocytes where it serves as the specific antigen receptor.

IgD is also present primarily on the cell surface of the B lymphocytes and is believed to be involved in the maturation of these cells. It exists as a single antibody molecule and therefore has a molecular weight of about 150,000 daltons. Not much is really known about this antibody.

Finally, *IgE* also exists as a single antibody molecule and has a molecular weight of about 200,000 daltons. The Fc region of the IgE molecule binds with high affinity to specific receptors found on *mast cells* and *basophils*. Mast cells are found outside of the capillaries within the connective tissue region (see Figure 7.3). They are involved in the process of inflammation and are responsible for the secretion of *heparin* into the blood as well as *histamine*, *bradykinin*, and *serotonin*. Basophils are a type of white blood cell or *granulocyte* that mediates the inflammation process through release of these same substances. When an antigen binds to two adjacent cell surface–bound IgE molecules, the cell (mast or basophil) becomes activated and releases a host of potent biologically active compounds such as histamine. These agents are responsible for the dilation and increased permeability of the blood vessels. In normal situations, these changes facilitate the movement of the immune system components, such as white blood cells, antibodies, and complement, into localized sites of inflammation and infection. However, in people with allergies, a particular antigen, in this case called an *allergen*, stimulates the IgE on the surface of mast cells and basophils leading to the unwanted effects of an allergic reaction.

8.2.3 T Lymphocytes

The T lymphocytes are the other major cell type that forms the basis of the cell-mediated immune response. The T cells are characterized by the presence of an antigen-specific T-cell receptor or TcR. The T-cell receptor consists of an antigen recognizing molecule called Ti in close association with another polypeptide complex called CD3. The T cells, unlike the B cells, are not, however, activated by free antigens. T-cell activation requires that the antigen be *presented* by other cells such as macrophages or other B cells. These cells that present antigens to the T cells are collectively called *accessory cells* or *antigen presenting cells* (APCs).

The APCs ingest and breakdown the polypeptide antigen into much smaller fragments. These fragments then become associated with special molecules called the *major histocompatibility complex* (MHC). There are two major types of MHC molecules, called *MHC class I* and *MHC class II*. Recall that these molecules belong to the class of cell adhesion molecules (CAMs). The MHC class I molecule is expressed by almost every nucleated cell found in the body. The MHC class II molecule is only found in specialized antigen-presenting cells such as B cells and macrophages. The MHC molecule forms an antigen-binding area in the shape of a cleft or pocket. This pocket can accept polypeptide antigens consisting of up to 20 amino acids. The complex of small antigen and MHC is then transported to the surface of the cell where the antigen is presented for recognition by the T cell receptor. The T cell therefore only recognizes antigen in combination with the MHC molecule.

T-cell antigen recognition is shown in Figure 8.2. On the left side of this figure we see a foreign material such as a dead virus, a cancer cell, a microorganism, or a large

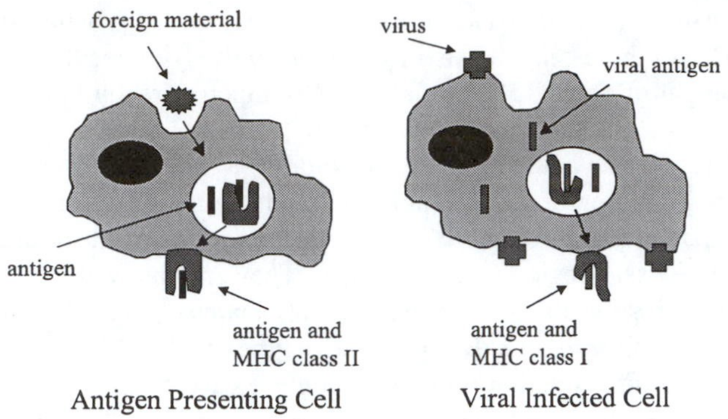

Figure 8.2 Presentation of antigen by MHC class I and II molecules.

polypeptide, being ingested and broken down by an APC. The antigen, in combination with an MHC class II molecule, is then transported to the cell surface for antigen presentation. On the right side of this figure, we see a live virus infecting the cell. As the virus takes over the operation of the cell to increase the number of viruses, many of the viral peptides that are produced are transported to the cell surface by MHC class I molecules for antigen presentation. The presence of antigens and MHC will then be recognized by the T cell.

However, in either case, the T-cell receptor by itself has a low affinity for antigen bound to MHC. To facilitate the recognition and binding of the T cell receptor with the antigen/MHC complex, there exists two important accessory molecules called CD4 and CD8 that also serve to distinguish between the two principal types of T cells.

T cells that have the CD4 molecule are known as $CD4^+$ or helper T cells, whereas those that have the CD8 molecule are known as $CD8^+$ or cytotoxic T cells. $CD4^+$ T cells only recognize antigen that is bound to the MHC class II molecule on the surface of an APC, see the left side of Figure 8.2. The $CD8^+$ T cell can only recognize antigen that is bound to the MHC class I molecule on the surface of all other nucleated cells found in the body as shown on the right side of Figure 8.2. This is called *MHC restriction*, that is, the response of $CD4^+$ T cells is restricted to only that antigen bound to MHC class II molecules, whereas the response of $CD8^+$ T cells is restricted to only that antigen bound to MHC class I molecules.

8.2.4 Interaction Between APCs, B Cells, and T Cells

With this background, we can now examine the cooperative relationship that exists between the APCs, B cells, and the T cells during the immune response. First let's consider the interaction between a resting $CD4^+$ T cell, an APC, and a B cell. This is illustrated in Figure 8.3. The first step involves ingestion by an APC of an antigen that, in this example, contains both a B-cell epitope and a T-cell epitope. The APC could be a macrophage that engulfs the antigen, or a B cell that internalizes the antibody-antigen complex. The T-cell epitope is then expressed on the surface of the APC in combination with an MHC class II molecule. A $CD4^+$ T cell that has a Ti-CD3 receptor, specific for this

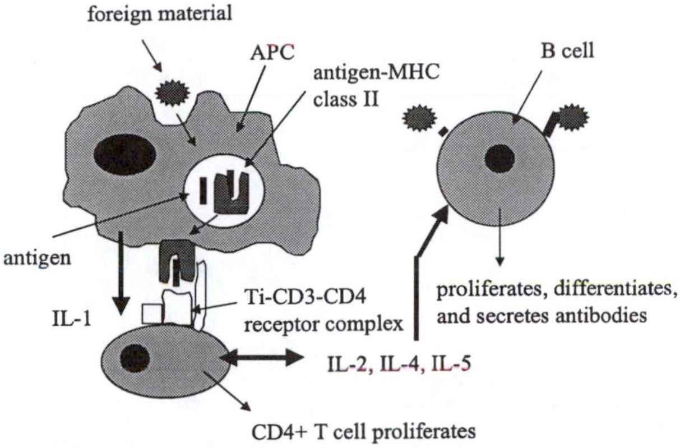

Figure 8.3 Interaction between APC, CD4$^+$ T cell, and the B cell.

particular antigen, then binds with the antigen/MHC class II complex. Note that the CD4 molecule stabilizes the binding of the antigen/MHC complex with the T-cell receptor.

This binding of the T cell and the antigen/MHC class II complex constitutes the first signal for activation of the T cell. This is then followed by the release of a second signal by the APC of a soluble substance called interleukin-1 (IL-1) that is also essential for T-cell activation. IL-1 is a small protein with a molecular weight of 15,000 daltons that belongs to a class of substances called *lymphokines*. Lymphokines are cellular messengers that have an effect on other lymphocytes and are a subcategory of a broader class of intercellular messengers called *cytokines*. The CD4$^+$ T cell then becomes activated and begins to secrete its own interleukins, specifically IL-2, IL-4, and IL-5. IL-2 induces the CD4$^+$ T cell to proliferate, rapidly forming a clone of CD4$^+$ T cells that are reactive to the specific antigen presented by the APC. IL-4 then activates the B cells and IL-5 induces the activated B cells to proliferate in number, forming a clone of B cells.

It is important to note that binding of antigen with the B-cell receptor is not sufficient for the B cell to become activated and then differentiate to antibody-secreting plasma cells. The B cell must become activated through the process shown in Figure 8.3 which also involves the CD4$^+$ T cell. Additionally, the B cell need not bind with the same epitope that activated the T cell. Hence the activated B cell may release antibodies with a different epitope specificity. Since the CD4$^+$ T cell is responsible for the activation of the B cell, it is referred to as the "helper" T cell. The activated CD4$^+$ T cell also secretes other lymphokines, such as γ *interferon*, that serves to attract and activate macrophages and natural (NK) killer cells and inhibits viral replication.

Figure 8.4 illustrates how the CD8$^+$ T cell becomes activated. Here we see that the CD8$^+$ T-cell receptor binds with the antigen/MHC class I complex presented, for example, by a virally infected cell. Once again the CD8 molecule stabilizes the interaction between the T-cell receptor and the antigen/MHC complex. This binding of the T-cell receptor and antigen/MHC complex represents the first signal. In order for the CD8$^+$ T cells to proliferate (i.e., form a clone of CD8$^+$ T cells) and become activated, it also must receive a second signal that is provided by IL-2 that is released as a result of the activation of the CD4$^+$ T cell shown in Figure 8.3. The activated CD8$^+$ T cell will then

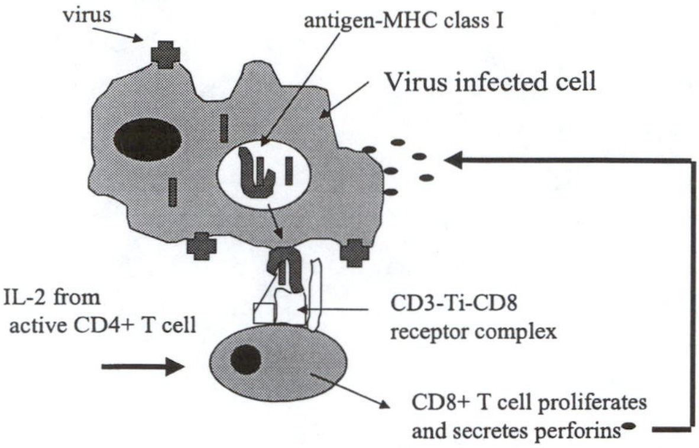

Figure 8.4 Activation of CD8$^+$ T cells.

kill any cell that expresses the appropriate combination of antigen and MHC class I. The activated CD8$^+$ T cell accomplishes this cellular destruction through the release of special molecules called *perforins* that destroy the cell membranes of the target cell. Because of their role in cell death, the CD8$^+$ T cells are also known as cytotoxic T cells or killer T cells.

8.2.5 The Immune System and Transplanted Cells

The above discussion illustrates the complexity as well as the coordination that exist between the different components of the immune system. The immune system, through an elegant process, is capable of recognizing "self" from "nonself". Our interest here is to prevent or restrict the action of the immune system towards the transplanted cells. The immune system response to transplanted cells involves a combination of effects resulting from antibodies, complement, macrophages, B cells, and T cells.

Antibodies recognize the foreign antigens presented by the transplanted cells and induce the destruction of the transplanted cells through activation of the complement system and NK cells by the process of ADCC. Rejection of the transplanted cells also occurs by a T-cell response against the MHC molecules that are expressed by the transplanted cells. The transplanted cells contain foreign MHC class I and class II molecules. The foreign MHC class II molecules are present on "passenger" leukocytes and macrophages that are present in the transplanted tissue. The foreign MHC class II molecules are sufficiently similar to the host's own combination of antigen/MHC class II complex to trigger the activation of the CD4$^+$ T cells. This is also true for the foreign MHC class I molecules, leading to the activation of the host's CD8$^+$ T cells. Activation of these T cells and the antibody-producing B cells, then leads to the destruction of the transplanted cells.

8.3 IMMUNOISOLATION

A promising method for restricting the host's immune response is to immunoisolate the transplanted cells (Lanza et al. 1995; Zielinski et al. 1997). This concept was shown earlier in Figure 7.2. Figure 8.5 presents a more detailed view of immunoisolation showing

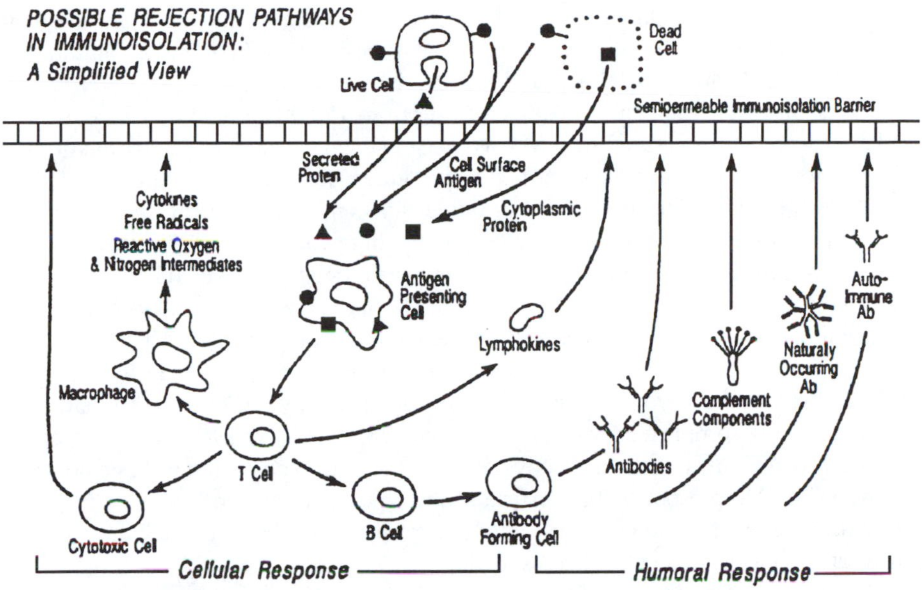

Figure 8.5 Pathways for rejection of immunoisolated cells. (From Colton [1995], with permission.)

the possible pathways for rejection of the transplanted cells (Colton 1995). Immunoisolation can be accomplished through the use of a polymeric membrane that prevents passage of the major components of the immune system, that is the immune cells, antibodies, and complement.

The immune response occurs as a result of antigens shed by the transplanted cells. These antigens may be products secreted by the functioning cells or materials released by the death of the transplanted cells. These antigens will cross the immunoisolation membrane and when recognized and presented by the host's immune system, lead to the cellular and humoral immune responses discussed earlier.

To restrict the passage of cellular components such as lymphocytes and macrophages, the pores or spaces within the membrane must be no larger than a micron or so. Recent evidence indicates that blocking the entry of the cellular component of the immune system is sufficient to prevent the rejection of allografts (Colton 1995). This is of special interest for applications involving allogeneic cells for gene therapy (Chang et al. 1993; Liu et al. 1993; Hughes et al. 1994; Al-Hendy et al. 1995; Chang 1997).

However, openings that restrict the entry of immune cells will still allow the passage of antibodies and complement, which must be blocked for successful use in a bioartificial organ of xenogeneic cells. Therefore, the solute permeability characteristics of the immunoisolation membrane are critical for achieving successful immunoisolation of xenogeneic transplanted cells. This requires that the membrane have significant permeability to essential small-molecular-weight solutes such as oxygen, glucose, growth factors, and molecular carriers such as albumin and transferrin[2]. The membrane also must be permeable to waste products and to the therapeutic product, for example in the

[2]Albumin carries fatty acids and transferrin carries iron.

case of diabetes, this would be insulin. On the other hand, the membrane must have negligible permeability to the soluble components of the immune system, i.e., antibodies, complement, and various cytokines and lymphokines.

Antibodies have molecular weights ranging from 150,000 to 900,000 daltons. Naturally occurring antibodies of the IgM class exist in the host and are reactive against the MHC of xenografts. In addition, antibodies produced from autoimmune diseases such as diabetes would be expected to bind to antigens on xenograft islets of Langerhans.

In the absence of immune cells, antibody binding to antigens on the surface of the transplanted cells is usually not sufficient to damage the cells. However, if both antibodies and complement are present, then the transplanted cells can be destroyed by complement activation and formation of the membrane attack complex (Iwata et al. 1995).

The complement system is activated by two different routes called the *classical pathway* and the *alternative pathway*. The classical pathway requires the formation of antigen-antibody complexes for its initiation, whereas the alternative pathway does not. The alternative pathway becomes activated in response to recognition of large polysaccharide molecules present on the surface of the cell membranes of invading microorganisms. As far as the immunoisolation of transplanted cells is concerned, the classical pathway is of most interest for this discussion.

Activation of the classical pathway involves nine protein components known as C1 through C9. C1 is the first component to become activated and requires the binding with antigen of either two or more adjacent IgG antibodies or a single IgM antibody. C1 consists of three subunits called C1q, C1r, and C1s. Their molecular weights are respectively 400,000, 95,000, and 85,000 daltons. The C1q subunit is the first subunit activated and it binds with the Fc region of the bound antibodies. Its activation then activates the C1r subunit which then activates the C1s subunit which activates C4 and so on.

As we discussed earlier, in the absence of immune cells, binding of antibody with antigen usually is not sufficient to damage the transplanted cells. Destruction of the antibody laden cell will require the presence and activation of complement. Since the C1q subunit must bind with either IgM or several IgG to start the cascade of complement reactions, the immunoisolation membrane must be capable of preventing the passage of the C1q molecule.

C1q has an interesting molecular shape as shown in Figure 8.6. The critical dimension of C1q is the span of the six projections from the cylindrical base, which is about 30 nm. Because of the presence of proteins coating the walls of the pores in the membrane, the maximum pore size allowable for blocking C1q is on the order of 50 nm (Colton and Avgoustiniatos 1991, 1997; Colton 1995; Zielinski et al. 1997).

Several other issues with regard to immunoisolation also need to be considered. The immunoisolation membrane may be capable of preventing the passage of immune cells and complement C1q. However, activation of the immune cells by antigens released by the transplanted cells will also result in the release of such lymphokines as IL-1 and other cytotoxic agents such as nitric oxide, peroxides, and free radicals. These substances can be toxic to the transplanted cells. Lymphokines have small molecular weights, typically around 20,000 daltons, and because of their small size, may not be retained by the immunoisolation membrane. Recent studies, however, have shown that immunoisolated

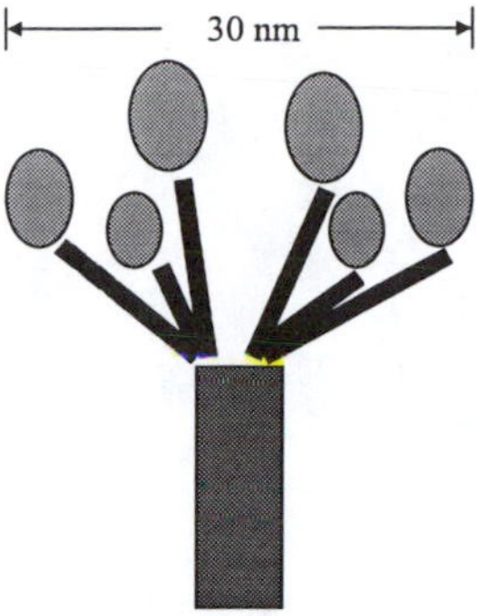

Figure 8.6 Complement C1q molecule.

cells were not affected by the presence of IL-1 (Zekorn et al. 1990). Substances such as IL-1, and other highly reactive species, may be consumed by other reactions before they can penetrate the immunoisolation membrane to any significant distance (Colton 1995). In addition, many bioartificial organs operate at very low tissue densities in order to provide for effective oxygenation of the transplanted cells. The reduced tissue density results in a lower concentration of shed antigens and hence a reduced concentration of soluble immune agents.

The shed antigens could also cause the host to undergo a life-threatening *anaphylactic reaction*[3] or to suffer from *antibody-mediated hypersensitivity reactions*[4]. Lanza, Beyer, and Chick (1994) examined these issues by transplanting immunoprotected canine and porcine islets into rats. The islets were encased within an acrylic hollow-fiber membrane with a nominal molecular weight cutoff of about 80,000 daltons. Their studies showed that the immunoprotected islet xenografts caused the host to generate antibodies to antigens given off by the transplanted islets. However, there was no evidence of any other pathological effects of these antibody-antigen immune complexes. Subsequent studies using immunoprotected human islets in human patients with diabetes (Scharp et al. 1994; Shiroki et al. 1995) have shown that the immunoisolation membrane can protect against not only the allogeneic immune response, but also the autoimmune components responsible for the development of insulin dependent diabetes mellitus.

8.4 PERMEABILITY OF IMMUNOISOLATION MEMBRANES

Figure 8.7 presents permeability data obtained on a hollow-fiber immunoisolation membrane for a variety of solutes of different molecular weight (Dionne et al., 1996). This particular membrane was made from a copolymer of acrylonitrile and vinyl chloride. The permselective membrane consists of a thin skin on either side of a much thicker spongy

[3] A result of a significant release of inflammatory mediators such as histamine, causing severe hypotension and bronchiolar constriction. Can cause death due to circulatory and respiratory failure.

[4] An inappropriate or exaggerated response of the immune system to an antigen. Binding of antibodies to the antigen starts the process.

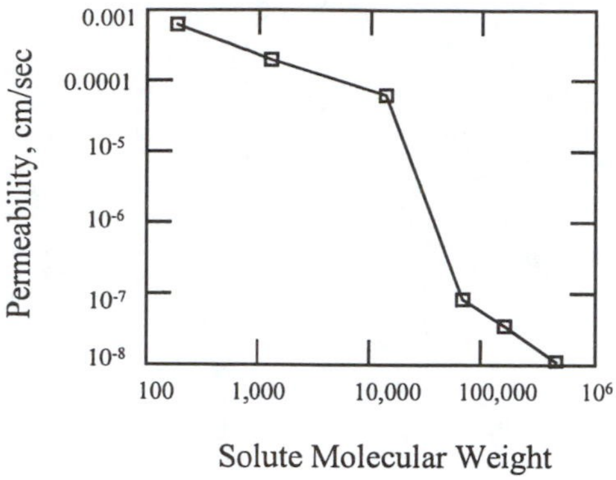

Figure 8.7 Immunoisolation membrane permeability. (From Dionne et al. [1996].)

wall region that provides overall structural strength. The thickness of the spongy wall was about 100 microns and these membranes were similar to those that were successfully used in preliminary tests to immunoisolate human islets in patients with diabetes (Scharp et al. 1994). These membranes had a reported nominal molecular weight cutoff of 65,000 daltons.

Notice how the permeability of this membrane at first decreases gradually as the solute molecular weight increases. As the solute molecular weight continues to increase, the permeability decreases much more rapidly. This can be explained by recalling our equation for solute permeability given by Equation 2.19. The size of low-molecular-weight solutes is much smaller than the pores in the membrane. For these solutes, steric exclusion (K) and solvent drag (ω_r) effects caused by the pore wall are negligible. The decrease in permeability is therefore directly related to the decrease in solute diffusivity, which according to Equations 2.9 and 2.10, would be inversely proportional to the solute size, or the one-third power of the solute molecular weight. As the solute size reaches a critical fraction of the membrane average pore size, steric exclusion and solvent drag become more important, and the solute permeability rapidly decreases with increasing solute molecular weight.

These data also clearly show the problem of defining the "molecular weight cutoff" for the membrane. There does not exist a molecular weight above which the solute permeability drops to zero. Surprisingly, we see that even very large molecules, here at the extreme represented by a molecular weight of about 440,000 daltons have a measurable permeability. Therefore, a membrane's molecular weight cutoff is very subjective and dependent on the definition that one chooses to apply. For example, some membrane manufacturers define the molecular weight cutoff on the basis of 90% retention of a given solute after so many hours of dialysis. Others base it on ultrafiltration of a solution in which a particular solute is 90% retained.

Although the exact nature of the pores or openings within the membrane is not known, descriptions such as that based on the hydrodynamic pore model developed in Chapter 2 (Equation 2.19) provides a useful framework for understanding the transport of various solutes, and also provides a means for correlating data. The data shown in Figure 8.7 can therefore be used to obtain an estimate of the average pore size in the membrane. This is shown in the following example.

Example 8.1 Using the solute permeability data shown in Figure 8.7, determine the average size of the pores in the membrane. Recall that the membrane is 100 microns thick. The table below summarizes the physical properties of the solutes that were used. Use the hindered diffusion model developed by Bungay and Brenner (1973) to describe the effect of the pore size on the diffusion of the solute. This equation is used, rather than the Renkin model given in Equation 2.11, because the Bungay and Brenner equation is generally valid for the entire range of the ratio of solute to pore size, i.e., $0 \leq \frac{a}{r} \leq 1$. The Renkin equation is only valid for $\frac{a}{r} < 0.4$ and may not provide a correct description of the diffusion of larger molecules in comparably sized pores. In the Bungay and Brenner model, $\omega_r = \frac{6\pi}{K_t}$ in the permeability equation given by Equation 2.19. K_t is given by the following equation. In this equation a is the molecular radius and r is the pore radius.

$$K_t = \frac{9}{4}\pi^2\sqrt{2}\left(1-\frac{a}{r}\right)^{-5/2}\left[1+\sum_{n=1}^{2} z_n\left(1-\frac{a}{r}\right)^n\right]+\sum_{n=0}^{4} z_{n+3}\left(\frac{a}{r}\right)^n$$

with $z_1 = -1.2167$; $z_2 = 1.5336$; $z_3 = -22.5083$; $z_4 = -5.6117$; $z_5 = -0.3363$; $z_6 = 1.216$; $z_7 = 1.647$.

Solute	Molecular weight	Stokes radius, nm	$D_{\text{water}} \times 10^6$ cm^2/sec, 37°C	$P_m \times 10^8$ cm/sec
Glucose	180	0.35	9.24	63,200
Vitamin B_{12}	1,300	0.77	5.00	20,000
Cytochrome C	13,400	1.65	1.78	6,160
Bovine serum albumin (BSA)	67,000	3.61	0.964	7.95
IgG	155,000	5.13	0.629	3.59
Apoferritin	440,000	5.93	0.611	1.07

From Dionne et al. (1996).

SOLUTION Mathcad Chart 8.1 summarizes the solution to the problem. Since the membrane consists of two permselective skins and a much thicker spongy wall region, the observed solute permeabilities must account for the resistances of all three regions. Therefore, the observed permeability is given by the following equation assuming the two skins are equivalent in their resistance to mass transfer.

$$\frac{1}{P_m} = \frac{2}{\dfrac{\varepsilon_{\text{skin}} D K \omega_r}{\tau_{\text{skin}} t_{\text{skin}}}} + \frac{1}{\dfrac{\varepsilon_{\text{sponge}} D}{\tau_{\text{sponge}} t_{\text{sponge}}}}$$

The porosity of the sponge is assumed to be about 0.8 and the tortuosity of the sponge is assumed to be unity. The regression analysis is based on two variables, the pore radius (r) and a paramter α. The parameter α accounts for the group $\dfrac{\varepsilon_{\text{skin}}}{\tau_{\text{skin}} t_{\text{skin}}}$ for which the individual values of the parameters are not known.

Since the permeability values vary over a considerable range, it is appropriate to define the objective function for the regression in terms of the natural logarithm of the permeabilities. The result of the regression analysis provides an estimated pore diameter of 14.2 nm.

Chart 8.1 for **Example 8.1:** Find the effective membrane pore size for the solute permeabilities shown in Figure 8.7.

$i := 0 \ldots 5$ number of data points

cm/sec, measured solute permeabilities $P_i :=$	solute molecular weights $mw_i :=$
6.32×10^{-4}	180
2×10^{-4}	1300
6.16×10^{-5}	13400
7.95×10^{-8}	67000
3.59×10^{-8}	155000
1.07×10^{-8}	440000

$t_m := .01$ cm, membrane thickness

$\varepsilon_{sponge} := 0.8$ porosity of the sponge region of the membrane

cm^2/sec, solute diffusivity $D_i :=$	nm, solute radius $a_i :=$
9.24×10^{-6}	.35
5.00×10^{-6}	.77
1.78×10^{-6}	1.65
9.64×10^{-7}	3.61
6.29×10^{-7}	5.13
6.11×10^{-7}	5.93

$r := 10$ nm, pore size estimate

$\alpha := 100$ 1/cm, estimate of $(\varepsilon/t/\tau)$ for the permselective skin of the membrane

$$K_t(i,r) := \frac{9}{4}\cdot\pi^2\cdot 2^{\frac{1}{2}}\cdot\left(1-\frac{a_i}{r}\right)^{-\frac{5}{2}}\cdot\left[1-1.2167\cdot\left(1-\frac{a_i}{r}\right)+1.5336\left(1-\frac{a_i}{r}\right)^2\right]$$
$$-22.5083-5.6117\cdot\left(\frac{a_i}{r}\right)+-.3363\cdot\left(\frac{a_i}{r}\right)^2-1.216\cdot\left(\frac{a_i}{r}\right)^3+1.647\cdot\left(\frac{a_i}{r}\right)^4$$

$$P_{calc}(i,\alpha,r) = \frac{1}{\left[\dfrac{2}{\alpha\cdot D_i\cdot\left(1-\frac{a_i}{r}\right)^2\cdot 6\cdot\pi\cdot\dfrac{1}{K_t(i,r)}}+\dfrac{1}{\left(\dfrac{\varepsilon_{sponge}\cdot D_i}{t_m}\right)}\right]}$$

sum of the squares of the error between the measured solute permeabilities and that predicted from the hydrodynamic pore model

$$SSE(\alpha,r) := \sum_{i=0}^{5}\left(\ln(P_i)-\ln(P_{calc}(i,\alpha,r))\right)^2$$

given

$SSE(\alpha,r)=0 \qquad 1=1$

$\binom{\alpha}{r} := \text{minerr}(\alpha,r)$

$r = 7.091$ nm, effective pore radius

$\alpha = 77.386$

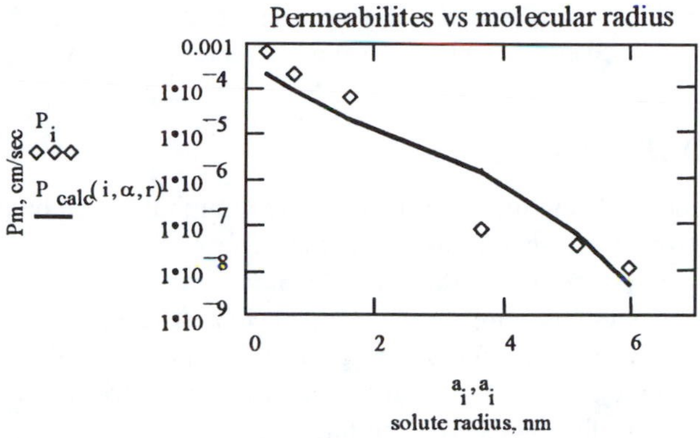

8.5 MEMBRANE SHERWOOD NUMBER

The solute permeabilities given in Figure 8.7 can also be compared to the permeability the solute would have in an aqueous layer that has the same thickness as the membrane itself. Therefore, we may write the following equation that expresses the ratio of these respective quantities.

$$\frac{P_m}{P_{\text{aqueous layer}}} = \frac{P_m}{D/t_m} = \frac{P_m t_m}{D} = \frac{D_{\text{effective}}}{D} = Sh_m \tag{8.1}$$

We see that this ratio is equivalent to defining a Sherwood number for the membrane, i.e., Sh_m, that represents the ratio of the effective diffusivity of the membrane to the aqueous diffusivity. This process has the effect of eliminating membrane thickness as a variable when correlating membrane permeability data. For a given solute and membrane structure, the Sh_m should be independent of the membrane thickness.

The data shown in Figure 8.7 are replotted as the membrane Sherwood number versus solute molecular weight in Figure 8.8. We see that for low-molecular-weight solutes, the membrane Sherwood number is relatively constant. In this example it is about 0.68. As

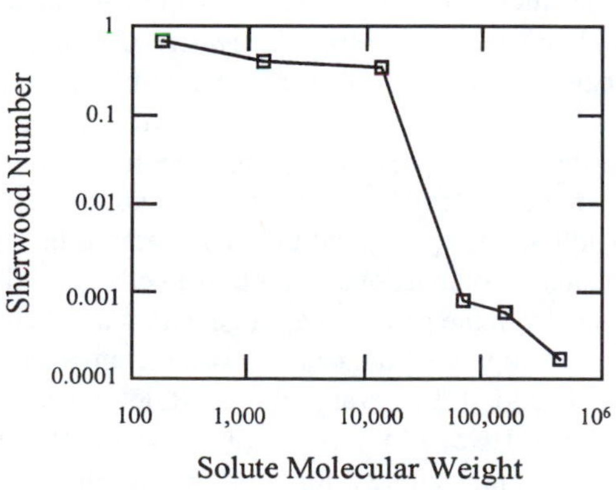

Figure 8.8 Immunoisolation membrane Sherwood number. (From Dionne et al. [1996].)

the solute molecular weight is increased, and the solute size becomes comparable to that of the pores, the membrane Sherwood number begins to decrease rapidly.

8.6 BIOARTIFICIAL ORGANS

Bioartificial organs contain living tissue or cells that are immunoisolated by polymeric membranes with the properties discussed above. These devices are also known as *hybrid artificial organs* since they consist of both artificial materials and living tissue or cells. Bioartificial organs are being considered for the treatment of a variety of diseases such as diabetes, liver failure, kidney failure; for neurological disorders such as Parkinson's or Alzheimer's disease; for the control of pain; and for delivery of a variety of therapeutic products secreted by genetically engineered cell lines.

For diabetes, the bioartificial organ contains cells that secrete minute-by-minute the appropriate amount of insulin needed to maintain blood glucose levels within a very narrow range. Bioartificial livers and kidneys attempt to replace all of the physiological functions of these complex organs. Treatment of neurological disorders involves the use of specialized genetically engineered cells that secrete neuroprotective products such as dopamine, nerve growth factor, glial cell-line–derived neurotrophic factor, or brain-derived neurotrophic factor. Immunoprotected cells that secrete pain-reducing neuroactive compounds such as β-endorphin are being considered for the treatment of pain resulting from cancer. Genetically engineered cell lines can also be used to treat a variety of other diseases. Examples include providing a specific hormone (for example growth hormone), replacing missing clotting factors in hemophilia (factor VIII or IX), or for providing needed enzymes in such diseases as adenosine deaminase deficiency that results in severe combined immunodeficiency disease (SCID). More details about these specific applications of bioartificial organs may be found in the list of suggested readings at the end of the book. In the following discussion, we will look at bioartificial organ replacements for the endocrine pancreas, the liver, and the kidney as representative systems.

8.6.1 The Bioartificial Pancreas

The focus of much of the research on bioartificial organs has been on the treatment of insulin dependent diabetes mellitus (IDDM) with a *bioartificial pancreas*. This is because IDDM is a chronic disease with many serious complications such as blindness, kidney disease, gangrene, heart disease, and stroke that lead to a poorer quality of life for those afflicted with this disease. IDDM therefore represents a major health problem in the United States. Over 1 million people in the United States have IDDM and the cost of their health-care exceeds several billion dollars each year. Additionally, methods have recently been developed that allow for the isolation of mass quantities of the cells needed to treat diabetes, i.e., the islets of Langerhans, from the pancreas of large mammals such as pigs, thus providing a potentially unlimited supply of donor islet tissue (Lanza and Chick 1994a; Ricordi et al. 1988, 1990a, 1990b, 1992; Warnock and Rajotte 1988; Warnock et al. 1989; Inoue et al. 1992; Lacy 1995; Maki et al. 1996; Lakey et al. 1996). Use of these xenogenic cells without immunosuppression requires that they be immunoisolated in a device such as a bioartificial organ. Research on the treatment of

IDDM with a bioartificial pancreas also provides a well-defined disease state[5] that can be used to explore alternative approaches for the design of a bioartificial organ that may have broader application in the treatment of other endocrine diseases, liver disease, and renal failure.

IDDM is believed to be caused by an autoimmune process (Notkins 1979; Atkinson and MacLaren 1990) that destroys the insulin-secreting cells found within the islets of Langerhans. Both environmental (viruses or chemicals) and heriditary factors are involved in its development. The islets of Langerhans are found scattered throughout the pancreas and represent about 1–2% of the pancreas mass.

Insulin is a small protein (6,000 daltons) and the key hormone involved in the regulation of the body's blood glucose levels. In the absence of insulin, glucose levels in the blood exceed normal values by several times. This is a result of the body's cells not being able to metabolize glucose. Without the ability to use glucose as an energy source, the body responds by metabolizing fats and proteins with a corresponding increase in the body fluids of ketoacids such as acetoacetic acid. This excess acid results in a condition known as ketoacidosis which if left untreated, can lead to death.

The conventional treatment for IDDM involves the daily administration of exogenous insulin to replace the insulin that is no longer produced by the patient's islets of Langerhans. This results in the almost normal metabolism of carbohydrates, fats, and proteins. However, even with exogenous insulin, the patient's blood glucose levels are not controlled as well as normal. The healthy islet cells are able to regulate the release of insulin in such a manner so as to maintain blood glucose levels within a narrow range, about 80–120 mg/dl. Insulin injections are not capable of providing such close control because of the difficulty of properly timing the injections and the inherent insulin transport delays from the site of injection. The abnormally high blood glucose levels, or *hyperglycemia*, in patients treated with exogenous insulin is now believed to be the primary cause of the long-term complications of diabetes (American Diabetes Association 1993). These elevated blood glucose levels lead to protein *glycosylation* that damages the microvasculature and other tissues in the body.

The goal of the bioartificial pancreas is therefore to utilize the glucose regulating capability of healthy donor islets in order to provide improved blood glucose control in patients with IDDM. This approach should minimize the complications of the disease and improve the quality of the patient's life. Transplanting the donor islets without immunoprotection is only warranted in those cases where the patient is already receiving immunosuppressive therapy, for example as a result of a kidney transplant to treat end-stage renal failure. In an otherwise healthy patient with diabetes, immunosuppressive drugs and their attendant problems, would not be needed if the islets were immunoisolated. Immunoisolation of the islets would not only protect them from the host's normal response to foreign tissue, but also from those agents of the immune system that caused the original autoimmune destruction of the patient's own cells. These immune system *memory cells* are still capable of responding to the antigens that resulted in the original destruction of the patient's own islets and could possibly destroy the donor islets.

[5]In that the transplanted cells (islets) respond to an easily measured stimulus (glucose) and respond by secretion of a therapeutic agent (insulin) with a measurable outcome, i.e., improved blood glucose.

8.6.1.1 Bioartificial pancreas approaches. A variety of approaches have been described over the years for the bioartificial pancreas and most of these have been tested extensively in laboratory animals. An excellent review of these devices through 1990 may be found in the paper by Colton and Avgoustiniatos (1991) and more recent reviews may be found in Mikos et al. (1994), Lanza Kuhtreiber, and Chick (1995), and Colton (1995). A series of books related to islet transplantation and the bioartificial pancreas have also been recently published (Ricordi 1992; Lanza and Chick 1994c).

These devices can be broadly classified into the following four categories: *intravascular*, *microencapsulation*, *macroencapsulation*, and *organoid*. These approaches are illustrated in Figure 8.9. These devices are for the most part general and are being considered for the immunoisolation of a variety of cell types.

Of concern in any of these approaches for a bioartificial pancreas would be the ability of the device to be readily fabricated and seeded with islets, their reliability over an extended period of time, and their cost. These issues also relate to the overall biocompatibility of the device, specifically in regard to device toxicity, immune system response, potential for fibrotic encapsulation; and for intravascular devices, the potential for thrombosis. Furthermore, the devices must be capable of maintaining long-term viablity of the donor tissue and have the proper mass-transfer characteristics to normalize blood glucose control. This requires proper understanding and integration during the design process of the islet glucose-insulin kinetics, the mass-transfer properties of the immunoisolation membrane, and the islet density within the device.

8.6.1.2 Intravascular devices. The intravascular devices (Figure 8.9A) (Chick et al. 1977; Moussy et al. 1989; Lepeintre et al. 1990; Lanza et al. 1992, 1993; Sarver and Fournier 1990; Maki et al. 1991, 1993; Petruzzo et al. 1991; Sullivan et al. 1991) involve a direct connection to the vascular system of the host via an arterio-venous shunt. Blood flows through the lumen of small polymeric hollow fibers or in the spaces between flat membrane sheets. Surrounding the blood flow path and protected by the membrane are the islets. An advantage of this type of device is that the convective flow of the blood provides the potential for better mass-transfer rates. In some cases, an ultrafiltration flow can cross the membrane at the arterial end and reenter at the venous end, much like the Starling flow found in capillaries. The close contact with a sizable flow of blood

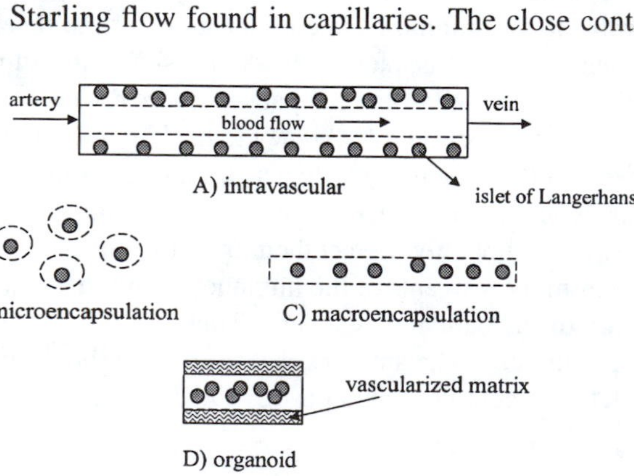

Figure 8.9 Approaches for a bioartificial pancreas.

also offers the potential for good tissue oxygenation. A disadvantage of the intravascular approach is the surgical risk associated with device implantation, connection to the host's vascular system, and removal at a later time.

One of the first experiments to demonstrate the feasibility of the intravascular approach, and for that matter, the overall concept of a bioartificial pancreas, was performed by Chick et al. (1977). A rat with chemically-induced diabetes was connected ex vivo to an intravascular device containing cells obtained from neonatal rats. The device used in these experiments consisted of a bundle of 100 hollow fibers with diameters less than 1 mm and a length of 11 cm. The fibers had a nominal molecular weight cutoff of about 50,000 daltons.

After connection of the animal to the ex vivo intravascular device, there followed a rapid decrease in blood glucose levels reaching the normalized values of 110–130 mg/dl (considered normal for a rat) after about 6 hours. After the removal of the device, blood glucose levels rapidly increased and returned to their diabetic levels. These results indicate the feasibility and the potential of the bioartificial pancreas for the treatment of IDDM.

A particularly serious problem with devices that contained numerous hollow-fiber tubules was the formation of blood clots in the entrance and exit header regions of the device. These regions of the device, where the fluid stream is either expanding or contracting, tend to form secondary flows or swirls that induce the formation of blood clots.

In an effort to minimize the formation of blood clots without the use of anticoagulation agents such as heparin, a device with a single coiled hollow-fiber membrane tube was recently described (Maki et al. 1991). This device is shown in Figure 8.10 This device consists of an annular-shaped acrylic chamber 9 cm in diameter and 2 cm thick.

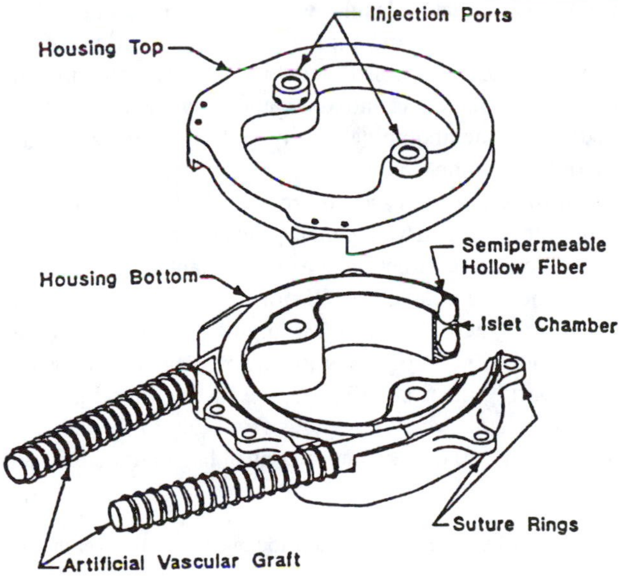

Figure 8.10 The biohybrid artificial pancreas. (From Maki et al. [1991], with permission.)

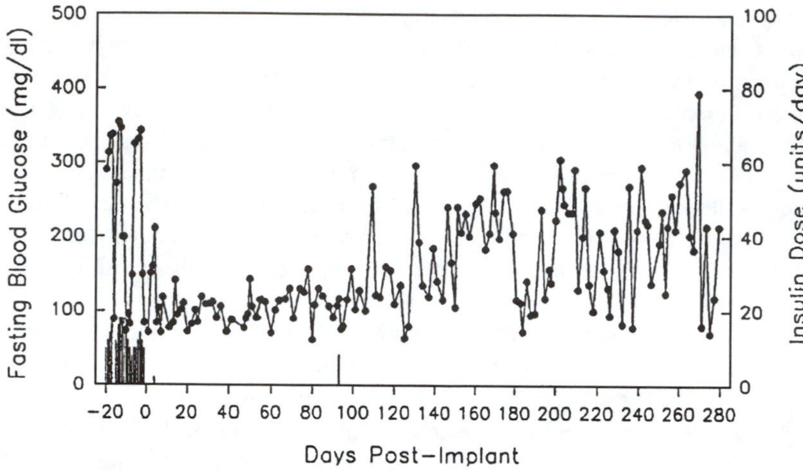

Figure 8.11 Fasting glucose and insulin requirements after implantation of device shown In figure 8.10. (From Lanza et al. [1992d], with permission.)

The chamber contains a 30–35 cm coiled hollow-fiber tubule with an inner diameter of 5–6 mm and a wall thickness of 120–140 microns. The hollow fiber tubule was connected at each end to polytetrafluoroethylene (PTFE) arterial grafts that provided connection of the device to the vascular system by an arteriovenous shunt. The larger bore size of the single hollow-fiber tubule eliminates the thrombosis observed in smaller bore hollow-fiber tubes and in the header regions of the device described earlier. Low-dose aspirin therapy was also found to contribute to long-term patency, presumably through aspirin's ability to prevent platelet activation.

The hollow-fiber membrane had a nominal molecular weight cutoff of about 80,000 daltons. A cavity surrounded the coiled tubular membrane and provided a 5-to 6-ml volume for the placement of the islets of Langerhans. This device was evaluated in pancreatectomized dogs using allogeneic canine islets and xenogeneic bovine and porcine islets. The device was implanted within the abdominal cavity and *anastomosed* to the left common iliac artery and the right common iliac vein.

This device was able to demonstrate in many cases improved blood glucose control for varying periods of time using both allo- and xenogeneic islets (Lanza et al. 1992). Figure 8.11 illustrates an example of long-term function in a dog implanted with a single device containing about 42,000 canine islet equivalents[6]. Prior to device implantation, this particular dog had a fasting blood glucose level of 230 mg/dl and required about 16 U/day of insulin[7]. After implantation, no exogenous insulin was provided for the 280-day period of the experiment. During the first 140 days, the fasting blood glucose was nearly normal and averaged 116 mg/dl, which compares favorably to the value of 91 mg/dl observed in normal dogs. After 140 days, glucose control began to deteriorate

[6]The equivalent islet number (EIN) is defined as the number of islets 150 microns in diameter that are equivalent in volume to a given sample of islets.

[7]Insulin dose is based on U(nits) of insulin where 1 U of insulin = 40 micrograms of insulin.

over the rest of the experiment. A higher fraction of animals was insulin free for longer periods of time when two devices were implanted as a result of the greater mass of islets.

Results obtained with xenogeneic implants containing porcine islets in this device were also very encouraging (Maki et al. 1996). Prior to receiving porcine islets, one particular dog had a fasting blood glucose level of 479 mg/dl and required 39 U/day of insulin. After this dog received 216,000 EIN of porcine islets, fasting blood glucose averaged 185 mg/dl and exogenous insulin averaged about 10 U/day. This device failed after 271 days of operation. Analysis of the porcine islets after device removal did not show any evidence of immune cell infiltration. These tests therefore presented clear evidence that xenogeneic porcine islets could be protected from the host's immune system using immunoisolation membranes. The only major complication in the use of these devices in dogs was vascular thrombosis. However, this clotting problem may be unique to dogs because of their known hypercoagulability.

8.6.1.3 Microencapsulation. The microencapsulation approach (see Figure 8.9B) involves the placement of one or several islets within small polymeric capsules (Altman et al. 1986; Gharapetian, Davis, and Sun 1986; Lum et al. 1991; Soon-Shiong et al. 1992a, 1992b, 1993; Sugamori and Sefton 1989; Sun et al. 1977, 1996; Lanza et al. 1995; Tun et al. 1996). The microcapsules are then injected within the peritoneal cavity. The diameter of the microcapsule is typically in the range of 300 to 600 microns. Immunoprotective surface coatings of the islets have also been described (Pathak, Sawhney, and Hubbell 1992).

Microencapsulation provides diffusion distances on the order of 100 to 200 microns and very high surface areas per volume of islet tissue. Accordingly, their small size provides good diffusion characteristics for nutrients and oxygen which improves islet viability. This also provides for a good glucose/insulin response that offers the potential for normalization of blood glucose levels.

The microcapsules usually consist of the islet immersed within a hydrogel material with another eggshell-like layer that provides the immunoisolation characteristics and mechanical strength. A wide variety of polymer chemistries have been described for the hydrogel and the immunoprotective layer and some of these are described in the following references (Douglas and Sefton 1990; Gharapetian et al. 1987; Gharapetian, Davis, and Sun 1986; Matthew et al. 1993; Sefton et al. 1987; Douglas and Sefton 1990, Sun et al. 1996, Lanza et al. 1995). Of particular concern in selecting the microcapsule chemistry is the formation of a fibrotic capsule around the encapsulated islet. Fibrotic capsule formation can severely limit the diffusion of nutrients and oxygen resulting in loss of islet function. The success of the microenapsulation approach is therefore strongly dependent on choosing membrane materials that minimize this fibrotic reaction.

A particularly promising microencapsulation approach has been recently described by Soon-Shiong (1994) and Soon-Shiong et al. (1993, 1992a, 1992b) and demonstrated in large animals and several human patients. They used an alginate-poly-L-lysine encapsulation system. Alginates are natural polymers composed of the polysaccharides mannuronic acid and guluronic acid. Their results showed that the high mannuronic acid residues in the alginate are responsible for the fibrotic response. Mannuronic acid was shown to induce the lymphokines IL-1 and TNF (tumor necrosis factor), which are known to promote the proliferation of fibroblasts and lead to fibrotic capsule formation. By reducing the alginate's mannuronic acid content, and increasing the guluronic content (>64%), they were able to minimize the fibrotic response. The higher guluronic acid content also provided another benefit. It was found that alginates with higher guluronic contents were mechanically stronger.

The modified alginate microcapsules were evaluated in a series of nine spontaneously diabetic dogs. Three of the dogs received free unencapsulated islets and the other six dogs received encapsulated islets. The donor islets were obtained from other dogs and each recipient received their quantity of islets by an intraperitoneal injection. The islets were provided at an average dose of about 20,000 EIN/kg of body weight.

Figure 8.12 summarizes the plasma glucose levels prior to and following the injection of the islets. Exogenous insulin was stopped four days prior to islet injection and the plasma glucose levels at that time averaged 312 mg/dl. The first day after receiving islets, the blood glucose levels were reduced to an average of 116 mg/dl in those animals that received encapsulated islets, and averaged 120 mg/dl in those receiving free islets. Rejection of the unprotected free islets occurred rapidly with hyperglycemia returning in about 6 days. The animals receiving encapsulated islets exhibited normoglycemia for periods of time ranging from 63 to 172 days, for a median period of 105 days. Failure of the encapsulated islets was attributed to membrane failure as a result of the water soluble nature of the alginate system.

Sun et al. (1996) developed a microencapsulation system for islets using alginate-poly-lysine-alginate microcapsules. Most microcapsules contained only a single islet and

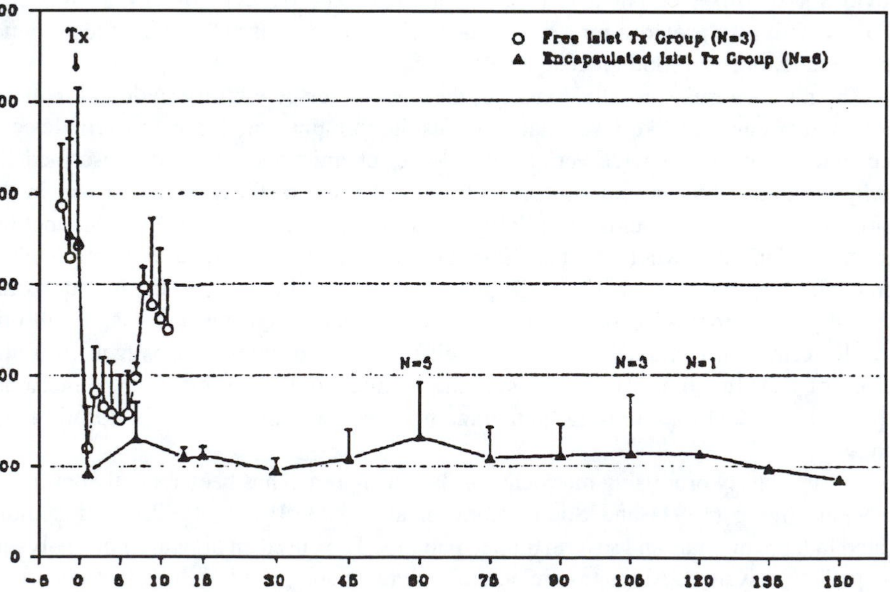

Figure 8.12 Fasting glucose levels following treatment with either free or microencapsulated islets. (From Soon-Shiong et al. [1992b], with permission.)

had diameters in the range of 250 to 350 microns. Preclinical studies were done using microencapsulated porcine islets implanted into spontaneously diabetic cynomologus monkeys. Seven monkeys receiving an average of 16,100 EIN islets/kg body weight were insulin independent for an average of 289 days. The shortest duration of control was 120 days and the longest 803 days. Prior to receiving the islets, fasting blood glucose levels averaged 353 mg/dl. Following the transplant, fasting blood glucose levels averaged 112 mg/dl. During the experiments, the sera of the animals was tested repeatedly for the presence of anti-porcine islet antibodies. No antibodies were detected, confirming effective immunoisolation of the porcine islets. Diabetic control animals also received unencapsulated islets. However, euglycemia was only maintained for about 9 days, after which the islets were destroyed by the host's immune system with a return to pretransplant diabetic levels of the blood glucose.

These results using the microencapsulation approach are very promising. However, there still are some technical difficulties with this approach that must be overcome. The strength and integrity of the microcapsules needs to be improved to provide long-term functioning and viability of the islets. Aggregation of the beads is also a problem, as this affects their mass-transfer characteristics and impacts glucose control and islet survival. The total volume for the smaller microcapsules of Sun et al. (1996) is modest. For example, using their monkey results, and assuming a 70-kg patient, a total of 1.1 million EIN would be required (16,100 EIN/kg body weight). With one islet per 300 micron diameter microcapsule, this amounts to a total bead volume of only 16 ml. Loss of islet function with time can be compensated for by injection of fresh microcapsules. Retrieval of the small microcapsules could be difficult if this should be necessary after their introduction.

8.6.1.4 Macroencapsulation. Macroencapsulation (see Figure 8.9C) involves placing numerous islets within the immunoisolation membrane structure. Typical approaches involve the use of multiple hollow fibers that encase the islets (Altman et al. 1986; Lanza et al. 1991, 1992a, 1992b, 1992c; Lacy et al. 1991) or a much larger flat bag-like structure (Gu et al. 1994; Inoue et al. 1992; Hayashi et al. 1996). Their larger size and smaller number also makes retrievel less of a problem in comparison to microcapsules. The membranes used for macroencapsulation must not only possess the needed permeability requirements, but because of their large size, must also provide sufficient mechanical strength to maintain their integrity. This is a difficult objective since the membrane wall must be thin for diffusional purposes and this may compromise overall strength. In many cases it has been found that thin-wall hollow fiber membranes were prone to breaking (Lanza, Kuhtreiber, and Chick 1995). Rupture of the membrane wall will result in rejection of the islet tissue and loss of function. As is the case for microcapsules, membrane chemistry is an important factor in minimizing the formation of fibrous tissue around the implant.

Most of the attention on the macroencapsulation technique has been focused on the use of hollow fibers. A trade-off exists on the selection of the fiber diameter. A large diameter fiber will result in a shorter overall length, but can lead to diffusional limitations. Lack of oxygen transport and the accumulation of waste products can therefore result in a central core of necrotic tissue. On the other hand, a small fiber diameter, while

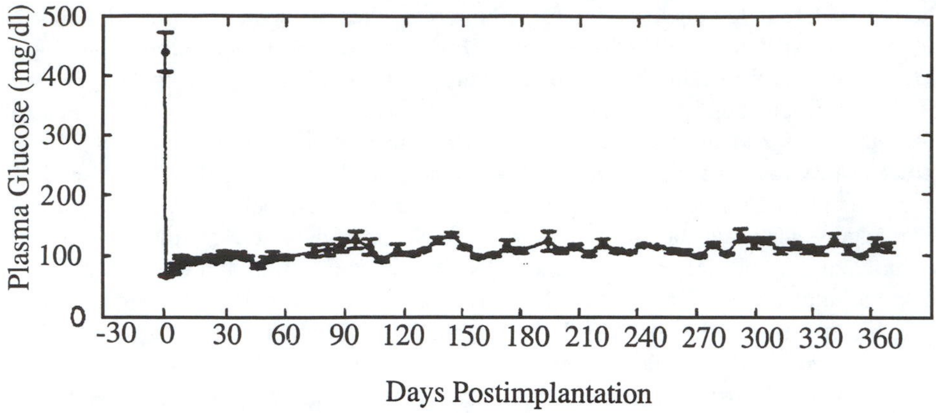

Figure 8.13 Fasting glucose levels in rats that received hollow fiber implants containing bovine islets. (From Lanza et al. [1993], with permission.)

improving the transport characteristics, can result in an incredibly long fiber length. This longer length increases the probability of breakage and makes implantation of the fibers more difficult.

An example of the use of hollow fibers as a vehicle for the macroencapsulation of the islets of Langerhans has recently been reported (Lanza et al. 1991; 1992a, 1992b, 1992c; Lanza, Kuhtreiber, and Chick 1995). This particular system has been studied in some detail in both rats and dogs using a variety of sources for the islets. The islets were enclosed within semipermeable hollow fiber membranes that were 2–3 cm in length and had an internal diameter of 1.8–4.8 mm. The membrane wall thickness ranged from 69–105 microns and the membrane had a nominal molecular weight cutoff of 50,000–80,000 daltons. Depending on the fiber diameter, approximately 9–50 fibers were implanted within the peritoneal cavity of each rat.

Figure 8.13 (Lanza 1993) summarizes fasting plasma glucose levels after fibers containing bovine islets (20,000 EIN) were implanted in the peritoneal cavity of rats whose diabetes was induced chemically using the drug, *streptozotocin*. In each case, the diabetic state was reversed within 24 hours after implantation. In some cases normoglycemia was maintained for 1 year. The wider bore membranes (4.5–4.8 mm internal diameter [ID]) retrieved after several months were found to contain a necrotic core. Viable islets were only found to exist within a 0.5–1 mm layer along the membrane wall. Some of the late failure of the implants was also attributable to membrane breakage as a result of the thin membrane wall.

Similar results were obtained using canine islets contained within the same hollow fibers and implanted within the peritoneal cavity of diabetic BB/Wor rats (Lanza et al. 1992 b). This type of rat develops autoimmune diabetes with a pathology similar to that found in humans. The canine islet xenografts provided normoglycemia in these rats for more than 8 months. This is an important result since it shows that the immunoisolation membrane protected the islet xenografts from the host's normal graft rejection response, as well as from the autoimmune disease process that produced the original diabetes.

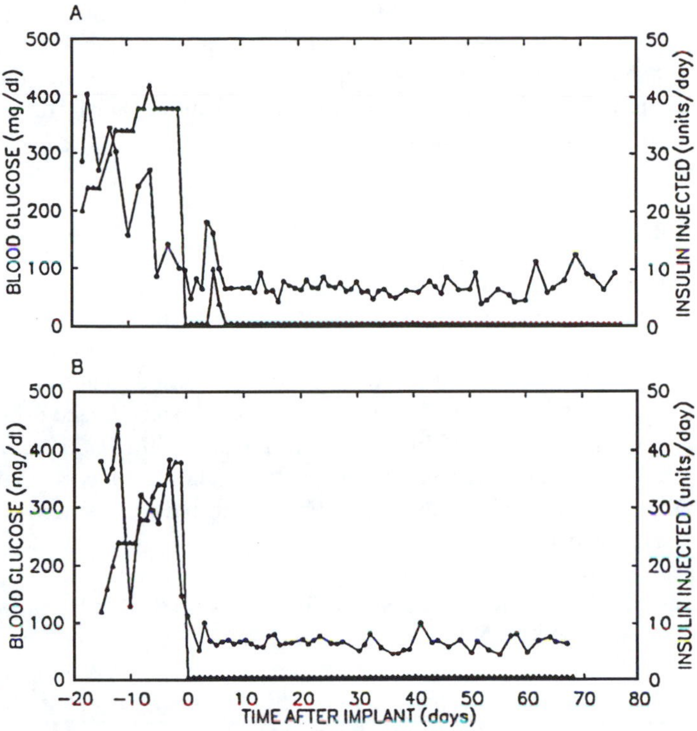

Figure 8.14 Fasting glucose levels in two dogs receiving hollow-fiber implants–containing canine islets. (From Lanza et al. [1992a], with permission.)

Protection of the islets from the autoimmune disease process is important in order to successfully treat human patients with diabetes.

The hollow-fiber macroencapsulation of islets was also tested in pancreatectomized dogs. Figure 8.14 (Lanza, Sullivan, and Chick 1992) summarizes results obtained for two dogs that were insulin-free for at least 70 days after receiving islet allografts contained within these fibers. Each dog received between 155 and 248 fibers containing a total of about 300,000 EIN. This would equate to a total fiber length of about 500 cm. The dogs had an average weight of about 17 kg, providing an islet loading of nearly 20,000 EIN/kg of body weight. For a 70-kg human, the total number of islets would therefore be about 1.4 million EIN, or a total fiber length on the order of 2100 cm (70 feet).

8.6.1.5 Organoid. The organoid approach (see Figure 8.9D) (Sarver et al. 1995; Hill et al. 1994; Suzuki et al. 1998) represents the union of the tissue engineering and macroencapsulation concepts. The bioartificial organoid consists of a thin disk-like compartment for the tissue that is enclosed by an immunoisolation membrane. Adjacent to the membrane is a thin layer of a porous scaffold material that becomes vascularized by the host. The use of tissue engineering allows control over the environment adjacent to the immunoisolation membrane. Therefore, the fibrotic response can be minimized, and through the growth of a vascular bed, intimate contact with the host's vasculature can be obtained. This has the potential to improve the viability and long-term functioning of the transplanted tissue. Additionally, these devices, unlike the microcapsule or hollow-fiber approaches, have ports that allow the device to be seeded with the islet tissue once the porous material has become sufficiently vascularized. The ports also allow for islet tissue to be replaced on a regular basis by a relatively minor procedure.

8.7 NUMBER OF ISLETS NEEDED

One issue confronting all of the devices discussed thus far concerns the mass of tissue required to achieve normoglycemia in a human patient with diabetes. The above studies in rats and dogs have used islet loadings on the order of 20,000 EIN/kg body weight. This is considerably higher than the benchmark study reported by Warnock and Rajotte (1988). They found that the threshold islet loading for treating pancreatectomized dogs corresponded to $\approx$4.3 μl of islet tissue/kg. This equates to about 2,500 EIN/kg. It is important to point out, however, that these were unencapsulated autologous islets that were implanted within the liver or the spleen. These islets had no artificial mass-transfer resistance to overcome and no host immune response that would lead to rejection. This result most likely defines the lower limit on islet loading. Certainly, improvements in the mass-transfer characteristics of the devices discussed so far could lead to islet loadings more on the order of 5,000 to 10,000 EIN/kg (Suzuki et al. 1998). Overall device size is dependent on the amount of tissue needed, as well as on the device mass-transfer characteristics for nutrients, oxygen, waste products and the therapeutic agent itself. The mass-transfer characteristics therefore are a critical factor in defining the islet tissue loading within the device itself.

The results presented so far demonstrate the potential of the bioartificial pancreas as a method for treating diabetes. All of the approaches, however, suffer from the problem of poor long-term islet survival. Islets have the potential to function for many years as evidenced by the fact that, unless they become diseased, they exist for as long as the lifespan of the animal they are found in. Therefore, islets should be capable of surviving within a bioartificial pancreas for many years.

Islet survival is a function of many variables including the amount of trauma they experience during their isolation and purification; the device mass-transfer characteristics for nutrients, oxygen, and waste products; the integrity of the membrane immunoisolation system; and finally the host's response to the implant. This latter factor determines the nature of the physiological environment in the vicinity of the implant. Certainly the formation of a region of under-vascularized fibrotic tissue would not provide the conditions needed for the longterm functioning and survival of the islets. Because of questions concerning the long term viability and functioning of the islet tissue, it may be necessary to periodically recharge the patients with islets. A bioartificial pancreas that has the ability to be reseeded would certainly offer an advantage.

8.8 ISLET INSULIN RELEASE MODEL

The design of a bioartificial pancreas requires an understanding of the insulin release rate from an islet and its dependence on plasma glucose levels. Islets that have been

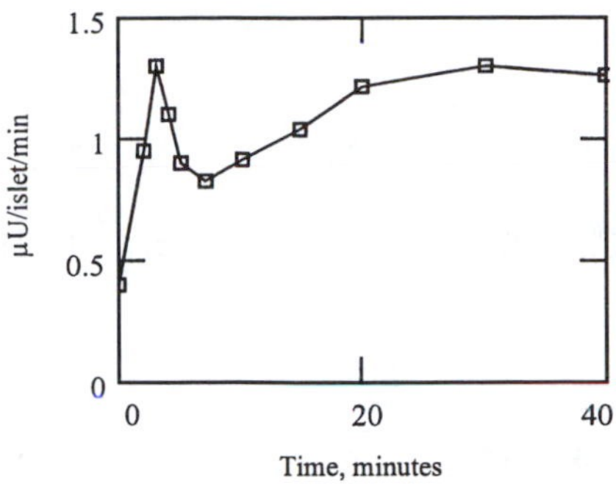

Figure 8.15 Rat islet insulin secretion rate. (From Nomura et al. [1984].)

isolated from a pancreas are usually challenged in culture with a step change in the glucose concentration. The glucose challenge is used to assess islet viability and glucose responsiveness. When challenged in this manner, islets release insulin in a biphasic fashion (Grodsky 1972).

Figure 8.15 (Nomura et al. 1984) illustrates the insulin response of rat islets to a rapid ramp increase in the glucose concentration from an initial value of 80 mg/dl to a final value of 200 mg/dl. Initially, the islets respond to the rate of change of the plasma glucose concentration, resulting in a maximum of the first-phase insulin release rate within a few minutes of the glucose challenge. Following this, the second-phase islet insulin release rate is proportional to the difference between the plasma glucose concentration and its fasting value of about 80 mg/dl. We see that this second phase release rate is relatively constant once the glucose level has achieved a steady-state value.

A mathematical description of the islet insulin release rate during a glucose challenge can be obtained through application of control theory. Nomura et al. (1984) have shown that the dynamics of the glucose induced secretion of insulin can be expressed as the sum of the proportional response to the glucose concentration, and a derivative response to the rate of change in the glucose concentration, each with a first order lag time, respectively given by T_1 and T_2. The Laplace transform of the islet insulin release rate can therefore be expressed as follows:

$$r_{\text{islet}}(s) = \left[\frac{K_p}{1+T_1 s} + \frac{T_d}{1+T_2 s}\right] C_G(s) \tag{8.2}$$

This equation may be inverted (in the Laplace transform sense) to give the following equation for the islet insulin release rate in terms of the glucose concentration and its rate of change.

$$r_{\text{islet}}(t) = \int_{-\infty}^{t} C_p C_G(z) e^{-(t-z)/T_1}\,\mathrm{d}z + \int_{-\infty}^{t} C_d \frac{\mathrm{d}C_G(z)}{\mathrm{d}z}\, e^{-(t-z)/T_2}\,\mathrm{d}z \tag{8.3}$$

The parameters in this equation; i.e., C_p $(=K_p/T_1)$, C_d $(=T_d/T_2)$, T_1, and T_2 may be obtained by performing a nonlinear regression of islet insulin release-rate data such as that shown in Figure 8.15. This can be accomplished by first representing the ramp glucose profile as follows:

$$C_G(t) = C_G^0, \qquad \text{for } t < 0$$

$$C_G(t) = \left(\frac{C_G^{SS} - C_G^0}{t_0}\right)t + C_G^0, \qquad \text{for } 0 \le t \le t_0 \tag{8.4}$$

$$C_G(t) = C_G^{SS}, \qquad \text{for } t \ge t_0$$

These equations may be substituted into Equation 8.3 to obtain algebraic expressions for the islet insulin release rate for a ramp change in the glucose concentration.

$$r_{\text{islet}}(t) = C_p T_1 C_G^0 e^{-t/T_1} + C_p T_1^2 \left(\frac{C_G^{SS} - C_G^0}{t_0}\right)\left[\frac{t}{T_1} - \left(1 - e^{-t/T_1}\right)\right]$$
$$+ C_p C_G^0 T_1\left(1 - e^{-t/T_1}\right) + C_d T_2 \left(\frac{C_G^{SS} - C_G^0}{t_0}\right)\left(1 - e^{-t/T_2}\right), \qquad \text{for } 0 \le t \le t_0 \tag{8.5}$$

$$r_{\text{islet}}(t) = C_p C_G^0 T_1 e^{-t/T_1} + C_p T_1^2 \left(\frac{C_G^{SS} - C_G^0}{t_0}\right)\left[e^{-(t-t_0)/T_1}\left(\frac{t_0}{T_1} - 1\right) + e^{-t/T_1}\right]$$
$$+ C_p C_G^0 T_1\left[e^{-(t-t_0)/T_1} - e^{-t/T_1}\right] + C_p C_G^{SS} T_1\left(1 - e^{-(t-t_0)/T_1}\right)$$
$$+ C_d \left(\frac{C_G^{SS} - C_G^0}{t_0}\right) T_2\left[e^{-(t-t_0)/T_2} - e^{-t/T_2}\right], \qquad \text{for } t > t_0 \tag{8.6}$$

The example below illustrates how the parameters in the Nomura et al. islet insulin release model may be determined for rat islets through the use of the above equations.

Example 8.2 Determine the constants in the Nomura et al. insulin release model for the rat islet perifusion data shown in Figure 8.15.

SOLUTION Mathcad® Chart 8.2 shows the nonlinear regression of the data in Figure 8.15. Note that we assumed the ramp period lasted until the first phase insulin release rate peaked. The figure shows that the model provides an excellent fit to the data.

Chart 8.2 for **Example 8.2:** Determine the parameters in the Nomura et al. model for the rat islet data shown in Figure 8.15 (Nomura et al. 1984).

$i := 0 \ldots 10$ number of data points

minutes $t_i :=$	μU/islet/min $r_{exp_i} :=$
0	.4
2	.95
3	1.3
4	1.1
5	.9
7	.82
10	.91
15	1.04
20	1.21
30	1.3
40	1.26

$t_0 := 3$ minutes, ramp period, corresponds to peak insulin rate
$C_{g_0} := 80$ mg/dl, initial glucose level
$C_{gss} := 200$ mg/dl, final glucose level
enter initial parameter estimates
$T_1 := 10$ minutes
$T_2 := 1$ minutes
$C_p := .0007$ μU dl/mg/min^2/islet
$C_d := .002$ μU dl/mg/min/islet
for $0 < t < t_0$
$j := 2$ index of data points corresponding to $t < t_0$
$k := 0 \ldots j$

$$r_1(k, T_1, T_2, C_p, C_d) := C_p \cdot T_1 \cdot C_{g_0} \cdot e^{-\frac{t_k}{T_1}} + C_p \cdot T_1^2 \cdot \left(\frac{C_{gss} - C_{g_0}}{t_0}\right) \cdot \left[\frac{t_k}{T_1} - \left(1 - e^{-\frac{t_k}{T_1}}\right)\right] \cdots + C_p \cdot C_{g_0} \cdot T_1 \cdot \left(1 - e^{-\frac{t_k}{T_1}}\right) + C_d \cdot T_2 \cdot \left(\frac{C_{gss} - C_{g_0}}{t_0}\right) \cdot \left(1 - e^{-\frac{t_k}{T_2}}\right)$$

for $t_0 < t < t_{\text{final}}$
$n := j \ldots 10$

$$r_2(n, T_1, T_2, C_p, C_d) := C_p \cdot C_{g_0} \cdot T_1 \cdot e^{-\frac{t_n}{T_1}} + C_p \cdot T_1^2 \cdot \left(\frac{C_{gss} - C_{g_0}}{t_0}\right) \cdot \left[e^{-\frac{t_n - t_0}{T_1}} \cdot \left(\frac{t_0}{T_1} - 1\right) + e^{-\frac{t_n}{T_1}}\right] \cdots + C_p \cdot C_{g_0} \cdot T_1 \cdot \left(e^{-\frac{t_n - t_0}{T_1}} - e^{-\frac{t_n}{T_1}}\right) + C_p \cdot C_{gss} \cdot T_1 \cdot \left(1 - e^{-\frac{t_n - t_0}{T_1}}\right) \cdots + C_d \cdot \left(\frac{C_{gss} - C_{g_0}}{t_0}\right) \cdot T_2 \cdot \left(e^{-\frac{t_n - t_0}{T_2}} - e^{-\frac{t_n}{T_2}}\right)$$

now define the sum of the error squared

$$\text{sse}\,(T_1,T_2,C_p,C_d) := \sum_{k=0}^{j}(r_{\text{exp}_k} - r_1(k,T_1,T_2,C_p,C_d))^2 \ldots$$

$$+ \sum_{n=j}^{10}\left(r_{\text{exp}_n} - r_2(n,T_1,T_2,C_p,C_d)\right)^2$$

given

sse $(T_1,T_2,C_p,C_d)=0$ 1=1   2=2   3=3

$$\begin{bmatrix} T_1 \\ T_2 \\ C_p \\ C_d \end{bmatrix} := \text{minerr}(T_1,T_2,C_p,C_d)$$

the parameters in the insulin release rate model

$T_1 = 10.11$ $T_2 = 1.881$

$C_p = 6.26336 \times 10^{-4}$ $C_d = 0.0106$

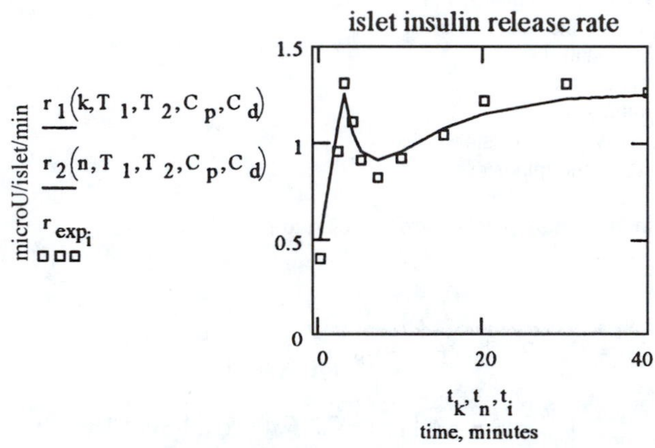

Rat islets have a very potent response to a step change in glucose concentration. Islets from dogs, pigs, and human pancreas typically show significantly lower insulin secretion rates on a per islet basis. The actual in vivo response of islets depends not only on glucose levels, but also depends in a significant way on the levels of other nutrients such as amino acids and gastrointestinal hormones[8]. These later substances, in the presence of rising glucose levels, can almost double the insulin secretion rate of an islet. However, no quantitative framework is available to describe the effect of these additional secretagogues on the insulin secretion rate.

Under physiological conditions, the variation in plasma glucose levels is less pronounced and the biphasic insulin release pattern is not as evident. In fact, the insulin secretion rate under these conditions displays a strong sigmoidal dependence on the

[8]Gastrointestinal hormones include gastrin, secretin, cholecystokinin, and gastric inhibitory peptide.

blood glucose levels (Guyton 1991; Sturis et al. 1991). The insulin secretion rate for the human pancreas saturates at a value of about 200,000 μU/min when the glucose concentration reaches 300 mg/dl. Since the human pancreas contains about 1 million islets, this saturation insulin release rate is 0.2 μU/islet/min.

Sturis et al. (1991) represented the insulin secretion rate under physiological conditions as a sigmoidal function of the plasma glucose concentration. The following equation proposed by Sturis et al. (1991) may then be used to describe the in vivo insulin response of an islet.

$$r_{\text{islet}} = \frac{0.209}{1 + \exp(-3.33G_B + 6.6)} \tag{8.7}$$

In this equation, the islet insulin release rate is in units of μU/islet/min and the glucose concentration is in mg/ml.

8.9 PHARMACOKINETIC MODELING OF GLUCOSE AND INSULIN INTERACTIONS

Although device testing in experimental diabetic animals provides conclusive proof of a device's efficacy and potential, mathematical models are also useful tools for investigating the myriad parameters that affect device performance. Mathematical models also provide the opportunity to perform scale-up studies. For example, they allow for critical examination of the results obtained in small laboratory animals and extrapolation of these results to understand how the device may perform in humans.

Several physiological pharmacokinetic models of glucose and insulin metabolism have been described (Guyton et al. 1978; Sorensen et al. 1982; Berger and Rodbard 1989; Sturis et al. 1991, 1995). These models can be combined with an islet glucose-insulin response model, and a device model, to assess quantitatively the level of glucose control that may be achieveable (Smith, Sarver, and Fournier 1991).

The pharmacokinetic model for glucose and insulin metabolism proposed by Sturis et al. (1991, 1995) is particularly attractive because of its relative simplicity and its ability to represent the experimentally observed temporal oscillations of insulin and glucose levels (Kraegen et al. 1972). Secretion of insulin in humans has been found to exhibit two distinct types of periodic oscillations. A rapid oscillation with a period of 10–15 minutes, and a longer or ultradian oscillation, with a period of 100–150 minutes. The rapid oscillations are of small amplitude (insulin $<$1–2 μU/ml and glucose $<$1mg/dl) and may be the result of an intrinsic pacemaker in the islets. The ultraradian oscillations exhibit a much larger amplitude and are self-sustained when the stimulus is continuously presented, for example by a constant infusion of glucose or by continuous enteral glucose feeding. On the other hand, the ultradian oscillations are damped when the glucose stimulus is presented as a discrete event, for example by either a meal or ingestion of a fixed quantity of glucose.

Figure 8.16 presents a diagram of the Sturis et al. (1991) model for describing the interactions between glucose and insulin. This model is based on four negative feedback loops involving interactions between glucose and insulin: (1) elevated glucose levels

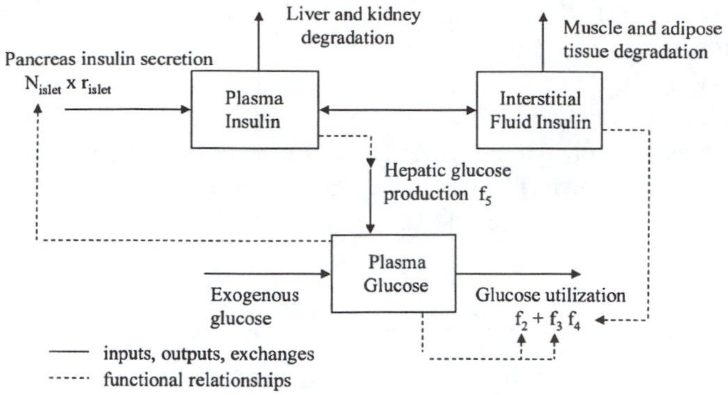

Figure 8.16 Model of glucose and insulin interactions in the body. (After Sturis et al. [1991].)

stimulates the secretion of insulin and the resulting increase in insulin levels decreases the endogenous production of glucose, which then has the effect of lowering glucose levels; (2) elevated glucose levels stimulates the secretion of insulin, the resulting increase in insulin levels enhances glucose utilization, which also has the effect of reducing glucose levels; (3) rising levels of glucose inhibits production of glucose; and (4) increasing levels of glucose stimulates its utilization. These four loops control the amount of glucose and insulin in the body. However, because of the nonlinear and dynamic interaction between glucose and insulin, the amounts of these substances in the body is never at a stable equilibrium.

The model also includes two time delays which are important for describing the observed oscillatory dynamics. The first delay affects the suppression of glucose production by insulin or the recovery of this process when insulin levels decrease. The second delay considers the fact that the biological action of insulin correlates better with the concentration of insulin in an intersititial compartment that equilibrates slowly with the plasma insulin concentration.

The three main variables in the model are the concentration of glucose in the plasma (G_B, mg/ml), the concentration of insulin in the plasma (I_B, μU/ml), and the concentration of insulin in the interstitial fluid (I_{IF}, μU/ml). There are also three additional variables that are introduced to account for the delay between the plasma insulin level and its effect on glucose production (x_1, x_2, x_3 with time lag τ_d).

A total of six differential equations are needed to describe the state of the system as a function of time (here in minutes). These equations are as follows:

$$V_{Bi}\frac{dI_B}{dt} = r_{\text{islet}}(G_B) \times 10^6 - E(I_B - I_{\text{IF}}) - \frac{I_B V_{Bi}}{\tau_p}$$

$$\frac{dI_{\text{IF}}}{dt} = E\left(\frac{I_B}{V_{\text{IF}}} - \frac{I_{\text{IF}}}{V_{\text{IF}}}\right) - \frac{I_{\text{IF}}}{\tau_i}$$

$$V_{Bg}\frac{dG_B}{dt} = r_{G\,\text{in}}(t) - f_2(G_B) - f_3(G_B)f_4(I_{\text{IF}}) + f_5(x_3) \tag{8.8}$$

$$\frac{dx_1}{dt} = 3\frac{I_B V_{Bi} - x_1}{\tau_d}, \qquad \frac{dx_2}{dt} = 3\frac{x_1 - x_2}{\tau_d}, \qquad \frac{dx_3}{dt} = 3\frac{x_2 - x_3}{\tau_d}$$

The insulin secretion rate of an islet, r_{islet}, was given earlier by Equation 8.6 and is multiplied by one million to account for all of the islets in the pancreas.

The insulin that is released distributes in the plasma space of volume given by V_{Bi}, or it enters the intersitial fluid space represented by V_{IF}. The insulin-transport rate into the interstitial fluid is proportional to the difference in the insulin concentration in the two compartments and is described by a rate constant E. Insulin is degraded within the plasma space by a first-order rate process with time constant given by τ_p. Interstitial insulin is also degraded with a time constant of τ_i and enhances glucose utilization as described by the function $f_4(I_{\text{IF}})$.

Glucose is assumed to distribute throughout a single compartment (V_{Bg}) and affects its own utilization through the functions f_2 and f_3. Glucose utilization represented by f_2 is insulin independent whereas f_3 is multiplied by an additional term f_4 that is dependent on the interstitial insulin concentration. The functions f_2 (mg/min) and $f_3 \times f_4$ (mg/min) are given by the following equations.

$$f_2 = 72[1 - \exp(-6.94 G_B)]$$

$$f_3 \times f_4 = 10 G_B \times \left[\frac{90}{1 + \exp\left(-1.772 \ln\left(I_{\text{IF}}\left(1 + \frac{V_{\text{IF}}}{E\tau_i}\right)\right) + 7.76\right)} + 4 \right] \tag{8.9}$$

Insulin also inhibits production of glucose by the liver (f_5) via a process that is dependent on a time delay (τ_d) between the appearance of insulin in the plasma and its inhibitory effect on glucose production. The function f_5 (mg/min) is given by the next equation and is dependent on a time delayed plasma insulin concentration represented by x_3.

$$f_5 = \frac{180}{1 + \exp\left(\frac{0.29 x_3}{V_{Bi}} - 7.5\right)} \tag{8.10}$$

Glucose can enter the body in an arbitrary time-varying fashion as given by the function $r_{G\,\text{in}}(t)$ (mg/min). For example, the glucose input function for a meal or an oral glucose tolerance test (OGTT) can be described by the following equation that describes the absorption of glucose from the gastrointestinal tract. In this equation, D (mg) represents the total amount of glucose ingested, k_a (1/min) is the absorption rate

Table 8.1 Parameter values for the Sturis et al. model of glucose and insulin interactions

Parameter	Value
E (rate constant for exchange of insulin between plasma and the interstitial fluid compartment)	200 ml/min
V_{Bi} (insulin plasma distribution volume)	3,000 ml
V_{IF} (insulin interstitial fluid distribution volume)	11,000 ml
V_{Bg} (glucose plasma distribution volume)	10,000 ml
τ_p (time constant for plasma insulin degradation)	6 min
τ_i (time constant for interstitial fluid insulin degradation)	100 min
τ_d (time delay between plasma insulin and glucose production)	36 min

From: Sturis et al. (1991, 1995).

constant, and k_e (1/min) is the elimination rate constant.

$$r_{G\,\text{in}}(t) = Dk_e\left(\frac{k_a}{k_a - k_e}\right)\left(e^{-k_e t} - e^{-k_a t}\right) \tag{8.11}$$

Table 8.1 summarizes the values of all the parameters used in the model of Sturis et al. (1991). For a given glucose input function, the above equations can be solved numerically to provide the time course of the plasma glucose and insulin levels.

8.10 USING THE PHARMACOKINETIC MODEL TO EVALUATE THE PERFORMANCE OF A BIOARTIFICIAL PANCREAS

We can now use the Sturis et al. (1991) pharmacokinetic model outlined above, along with an insulin-release model for an islet, to explore the level of glucose control that is possible with a bioartificial pancreas. There are two basic types of tests that can be used to measure the level of glucose control. One stringent measure of glucose control that is frequently used is called the *k-value*. The *k*-value represents the slope of the least squares fit of the logarithm of the plasma glucose concentration plotted as a function of time for the 50 minutes following the intravenous administration of a dose of glucose. This glucose dose is typically 0.5 g/kg of body weight and is referred to as an intravenous glucose tolerance test or IVGTT.

Following the glucose dose for the IVGTT, plasma glucose levels rapidly rise within a few minutes to levels exceeding 300 mg/dl from the initial fasting level of around 80 mg/dl. The rate of glucose decay following an IVGTT is indicative of the glucose and insulin responsiveness of the patient's pancreas. The *k*-value is expressed in units of %/min. If the *k*-value following an IVGTT is less than 1%/min, then the patient is assumed to have diabetes.

Normal *k*-values range from about 2–3%/min. A bioartificial pancreas would need to provide a *k*-value greater than about 1.5 %/min to be considered effective at controlling blood glucose levels in a patient with diabetes. This somewhat lower than normal k-value is acceptable because the IVGTT provides a major glucose challenge that is not normally observed following a meal.

The OGTT involves the oral administration of a beverage containing 50 grams of glucose. Glucose levels greater than 140 mg/dl two hours after administration of the glucose may be indicative of diabetes.

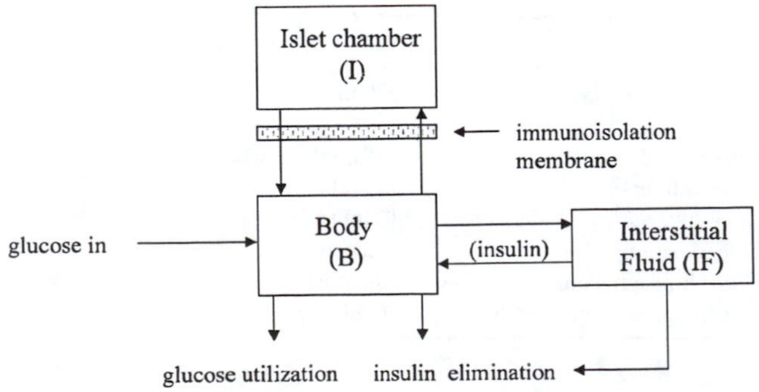

Figure 8.17 Compartmental model for evaluation of a bioartificial pancreas.

The simulation of blood glucose and insulin levels following an IVGTT or an OGTT using the pharmacokinetic model outlined above is straightforward. Assume that the immunoisolation membrane and the layer of islets is the controlling resistance for glucose and insulin transport. As shown in Figure 8.17, two-well-mixed compartments are used to define the distribution of glucose and insulin within the device and the rest of the body. The distribution of insulin also includes an additional interstitial fluid compartment to account for the delay in its action as discussed earlier. Unsteady mass balances for glucose and insulin can now be written for each of the compartments.

islet chamber:

$$V_I \frac{dI_1}{dt} = N_{\text{islets}} r_{\text{islet}} - P_{mI} S_m (I_I - I_B)$$
$$V_I \frac{dG_I}{dt} = P_{mG} S_m (G_B - G_I) \tag{8.12}$$

body distribution volume:

$$V_{Bi} \frac{dI_B}{dt} = P_{mI} S_m (I_I - I_B) - E(I_B - I_{\text{IF}}) - \frac{I_B V_{Bi}}{\tau_p}$$
$$V_{Bg} \frac{dG_B}{dt} = r_{G\,\text{in}}(t) - P_{mG} S_m (G_B - G_I) - f_2(G_B) - f_3(G_B) f_4(I_{\text{IF}}) + f_5(x_3) \tag{8.13}$$

interstitial fluid space for insulin:

$$V_{\text{IF}} \frac{dI_{\text{IF}}}{dt} = E(I_B - I_{\text{IF}}) - \frac{V_{\text{IF}} I_{\text{IF}}}{\tau_i} \tag{8.14}$$

equations for delayed insulin action on glucose production:

$$\frac{dx_1}{dt} = 3\frac{V_{Bi} I_B - x_1}{\tau_d}, \qquad \frac{dx_2}{dt} = 3\frac{x_1 - x_2}{\tau_d}, \qquad \frac{dx_3}{dt} = 3\frac{x_2 - x_3}{\tau_d} \tag{8.15}$$

For the special case of an IVGTT, the value of $r_{G\,\text{in}}(t)$ would be set equal to zero.

The initial conditions for the solution of the above equations for either an IVGTT or an OGTT are based on fasting levels. Fasting plasma glucose levels are typically about 80 mg/dl. Assuming negligible consumption of glucose by the tissue in the organoid, one may reasonably assume that the glucose concentrations in the organoid regions are the same as that found in the rest of the body. However, since the IVGTT glucose dose is rapidly distributed throughout the body distribution volume, in comparison to the length of time required for its subsequent removal, we assume that at $t = 0+$, the plasma glucose level would be given by the equation below.

$$G_B(t = 0+) = G_B(\text{fasting}) + \frac{500(\text{mg/kg bw}) \times (\text{body weight, kg})}{V_{Bg}} \tag{8.16}$$

Fasting plasma insulin levels are about 10 μU/ml. During the fasting period we assume that the organoid regions are at a steady-state equilibrium with the rest of the body as far as insulin is concerned. Accordingly, the islet insulin secretion rate based on the fasting glucose concentration must match the rate of insulin removal from the body. The steady state solution of the set of compartmental insulin equations provides the following relationships for the initial insulin concentration in each of the compartments.

$$I_B^0 = \frac{N_{\text{islets}} r_{\text{islet}}^0 \left(G_B^0\right) \tau_i}{V_{\text{IF}}} \times \frac{1}{\frac{V_{Bi}\tau_i}{V_{\text{IF}}\tau_p} - \frac{1}{\frac{V_{\text{IF}}}{\tau_i E}+1}}$$

$$I_{\text{IF}}^0 = \frac{I_B^0}{\left(\frac{1}{\tau_i E} + \frac{1}{V_{\text{IF}}}\right) V_{\text{IF}}} \qquad (8.17)$$

$$I_I^0 = I_B^0 + \frac{N_{\text{islet}} r_{\text{islet}}^0 \left(G_B^0\right)}{P_{mI} S_m}$$

Solution of the above set of equations provides the temporal change in the plasma glucose concentration following administration of either an OGTT or an IVGTT. The above equations can be easily modified to evaluate the glucose control for a variety of bioartificial pancreas approaches. The following example illustrates a prediction of the plasma glucose levels using a bioartificial pancreas following an OGTT.

Example 8.3 Estimate the glucose concentrations following an OGTT using islets that are macroencapsulated between two immunoisolation membrane disks. A total of 50 grams of glucose is taken orally. The glucose absorption rate constant is 0.042 1/min and the glucose elimination rate constant is 0.0083 1/min. Base the calculation on a 70-kg patient receiving a total of 750,000 EIN. Assume the islet insulin secretion rate is given by Equation 8.7. The half-thickness of the islet chamber is 75 microns. The void volume within the islet chamber must be at least 65% for sufficient oxygen transport. The immunoisolation membrane permeabilities for glucose and insulin are based on the membrane developed by Baker et al. (1997), i.e., 4×10^{-4} cm/sec for glucose and 8×10^{-5} cm/sec for insulin.

SOLUTION Mathcad® Chart 8.3 provides the solution for the glucose and insulin concentrations as a function of time. Note the damped ultradian oscillations of the glucose and insulin concentrations with time. The bioartificial pancreas provides a glucose and insulin response that is very similar to that observed in normal patients (Kraegen et al. 1972). We therefore would expect that the bioartificial pancreas has the potential to restore normoglycemia in patients with diabetes. The size of the device is 18 cm (7 inches) in diameter and only several millimeters in thickness. Rather than one large unit, several smaller units could be implanted. For example, four units each 9 cm (3.5 inches) in diameter would be equivalent.

Chart 8.3 for **Example 8.3:** Evaluation of the glucose control using a bioartificial pancreas.

$r_m := 9$ — cm, membrane radius

$\delta_I := .0075$ — cm, half-thickness of the islet chamber

$sf := 2$ — symmetry factor, 1 = single side device, 2 = dual side device

$V_I := \pi r_m^2 \delta_I sf$ — $V_I = 3.817$ cm^3, total volume of the islet chamber

$D := 50000$ — mg, ogtt glucose dose

$P_{mI} := 8 \times 10^{-5}$ — cm/sec, membrane insulin permeability

$P_{mI} := 60\, P_{mI}$

$P_{mG} := 4.0 \times 10^{-4}$ — cm/sec, membrane glucose permeability

$P_{mG} := 60\, P_{mG}$

$D_{\text{bulk}\,G} := .91 \times 10^{-5}$ — cm^2/sec, bulk glucose diffusivity

$D_{\text{bulk}\,I} := 2 \times 10^{-6}$ — cm^2/sec, bulk insulin diffusivity

$V_{Bg} := 10000$ — ml, glucose body distribution volume

$V_{Bi} := 3000$ — ml, insulin body distribution volume

$V_{\text{IF}} := 11000$ — ml, insulin interstitial fluid volume

$S_m := \pi r_m^2 sf$ — $S_m = 508.938$ cm^2, membrane surface area

$$P_{mI} := \frac{1}{\left(\frac{1}{P_{mI}} + \frac{\delta_I}{3 D_{\text{bulk}\,I} 60}\right)} \qquad P_{mG} := \frac{1}{\left(\frac{1}{P_{mG}} + \frac{\delta_I}{3 D_{\text{bulk}\,G} 60}\right)}$$

$P_{mI} = 0.004$ — $P_{mG} = 0.022$ cm/min, effective membrane permeabilities adjusting for diffusional resistance of the islet chamber

$\text{EIN} := 1.76 \times 10^{-6}$ — cm^3, volume of an equivalent islet

$N_{\text{islet}} := .75 \times 10^6$ — number of equivalent islets in the device

$$\varepsilon_I := \frac{V_I - N_{\text{islet}}\,\text{EIN}}{V_I}$$

$\varepsilon_I := 0.654$ — void fraction of the islet chamber

$\tau_p := 6$ — $\tau_i = 100 \quad \tau_d = 36$ minutes

$E := 200$ — ml/min, transfer coefficient

$k_a := .042$ — $k_e = .0083$ 1/min

$G_{B0} := 80$ — mg/dl, initial fasting glucose level in plasma just before the ivgtt

$G_{B0} := \frac{G_{B0}}{100}$ — $G_{I0} = G_{B0}$

Assume islet compartment equilibrated with the initial glucose in plasma prior to the OGTT.
Now we need to find the initial insulin concentrations for all the compartments.

$$r_{\text{islet}\,0} := \frac{.209}{1 + \exp((-.0333 \times 100 \times G_{B0} + 6.6))}$$ μU/islet/min, fasting islet insulin release rate

$$I_{B0} := \frac{N_{\text{islet}} r_{\text{islet}0} \tau_i}{V_{\text{IF}}} \frac{1}{\left(\frac{V_{Bi}\tau_i}{V_{\text{IF}}\tau_p}\right) - \frac{1}{\left(\frac{V_{\text{IF}}}{\tau_i E}\right) + 1}}$$ $I_{B0} = 6.997$ μU/ml, initial plasma insulin

$$I_{\text{IF}\,0} := \frac{I_{B0}}{\left(\frac{1}{\tau_i E} + \frac{1}{V_{\text{IF}}}\right) V_{\text{IF}}}$$

$I_{\text{IF}\,0} = 4.514$ — μU/ml, initial insulin level in the body interstitial fluid

$$I_{I0} := I_{B0} + \frac{N_{\text{islet}} r_{\text{islet}\,0}}{P_{mI} S_m} \qquad I_{I0} = 1.359 \times 10^3$$ μU/ml, initial insulin level in the device

$x_{10} := V_{Bi} I_{B0} \quad x_{20} := x_{10} \quad x_{30} := x_{20}$

the glucose absorption function for the ogtt

$$r_{G\text{in}}(t) := D k_e \left(\frac{k_a}{k_a - k_e}\right)(e^{-k_e t} - e^{-k_a t})$$

$$z := \begin{bmatrix} I_{I0} \\ G_{I0} \\ I_{B0} \\ G_{B0} \\ I_{IF0} \\ x_{10} \\ x_{20} \\ x_{30} \end{bmatrix}$$ the vector of initial conditions

the set of derivatives

$$D(t,z) := \begin{bmatrix} \frac{N_{islet}}{\varepsilon_I V_I} \frac{.209}{1+\exp((-3.33z_1+6.6))} - \frac{P_{mI}S_m}{\varepsilon_I V_I}\cdot(z_0-z_2) \\ \frac{P_{mG}S_m}{\varepsilon_I V_I}(z_3-z_1) \\ \frac{P_{mI}S_m}{V_{Bi}}\cdot(z_0-z_2) - \frac{E}{V_{Bi}}(z_2-z_4) - \frac{z_2}{\tau_p} \\ \frac{r_{Gin}(t)}{V_{Bg}} - \frac{P_{mG}S_m}{V_{Bg}}(z_3-z_1) - \left(\frac{72}{V_{Bg}}\right)(1-\exp(-6.94z_3))\cdots \\ + -\frac{10z_3}{V_{Bg}}\left[\frac{90}{1+\exp\left[-1.772\ln\left[z_4\left(1+\frac{V_{IF}}{E\tau_i}\right)\right]+7.76\right]} + -4\right]\cdots \\ + \frac{180}{V_{Bg}}\frac{1}{1+\exp\left(\frac{.29z_7}{V_{Bi}}-7.5\right)} \\ \left(\frac{E}{V_{IF}}\right)(z_2-z_4) - \frac{z_4}{\tau_i} \\ 3\frac{V_{Bi}z_2-z_5}{\tau_d} \\ 3\frac{z_5-z_6}{\tau_d} \\ 3\frac{z_6-z_7}{\tau_d} \end{bmatrix}$$

$Z :=$ rkfixed$(z,0,500,2000,D)$ the differential equation solver

$n := 0 \cdots 2000$

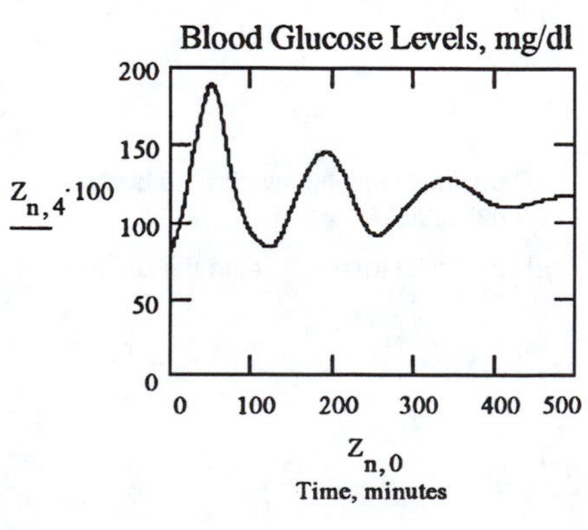

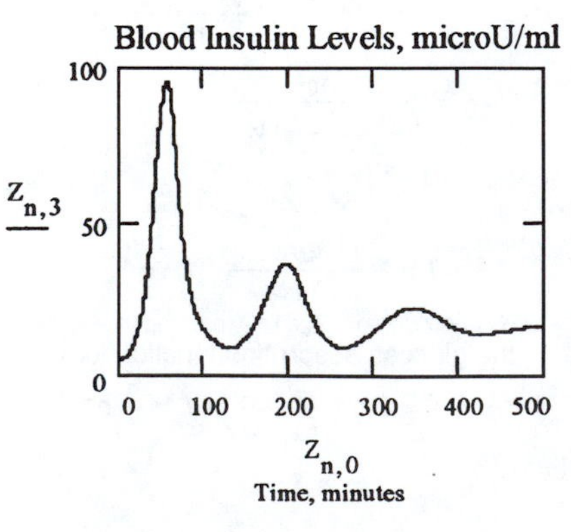

8.11 THE BIOARTIFICIAL LIVER

The liver is a very complex organ that performs a variety of life sustaining functions (Yarmush, Dunn, and Tompkins 1992; Galletti and Jauregui 1995). The human liver weighs about 1500 grams and is highly vascularized. From Table 4.2 we see that the liver receives over 25% of the cardiac output. The liver blood supply comes from two major sources. These are the hepatic artery and the portal vein. The portal vein collects the venous drainage from the spleen, the pancreas, and the intestines. The portal vein input from these regions passes through the liver before entering the rest of the systemic circulation. The liver therefore has the ability to detoxify potentially harmful substances that are absorbed from the gastrointestinal tract.

The basic functional cellular units of the liver are called the *hepatocytes*. Each hepatocyte is about 25 microns in diameter and there are close to 250 billion of them in the human liver, accounting for 75% of the liver volume. The hepatocytes are metabolically very active and provide an incredible variety of functions. Their role in carbohydrate metabolism includes the storage of excess glucose as *glycogen* and the release of this stored form of glucose (*glycogenolysis*) when blood glucose levels are low. In addition, the liver converts other sugars such as galactose and fructose to glucose. If blood glucose levels are low, and the glycogen stores are also depleted, then the liver performs a process known as *gluconeogenesis* wherein glucose is synthesized from amino acids.

The liver also plays a major role in fat metabolism. The liver is principally responsible for the body's ability to derive energy from fats. The liver can also convert excess carbohydrates and proteins into fat, it synthesizes cholesterol and phospholipids, and it forms the lipoprotein carrier molecules. The lipoproteins are responsible for transporting cholesterol, phospholipids, and fats to other tissues throughout the body. Cholesterol and phospholipids are important components of cellular membranes. The liver has an extremely important role in protein metabolism. These functions include the deamination of amino acids as a prelude to their use as an energy source, or their conversion to carbohydrates or fats. Deamination of amino acids produces large amounts of ammonia. The liver converts the ammonia into urea which is then excreted by the kidneys. Without this function, ammonia levels in the blood rapidly increase leading to hepatic coma and death.

With the exception of the gamma globulins (immunoglobulins), practically all of the plasma proteins are made in the liver. The most important being albumin. The liver also makes most of the substances found in the blood that are responsible for the clotting of blood. These include fibrinogen, prothrombin, and most of the other clotting factors. The liver also provides storage for a variety of nutrients such as vitamins and iron, which is stored as a protein complex called *ferritin*. The hepatocytes, because of their high enzyme content, also provide a major role in the detoxification and conjugation of a variety of materials such as drugs, environmental toxins, and hormones that are produced elsewhere in the body. These by-product materials are water-soluble and are excreted in the bile or by the kidneys.

The liver has the unique capability of being able to regenerate itself following tissue damage. This is extremely beneficial considering the role it plays in detoxifying materials from the gastrointestinal tract. However, if the tissue damage rate exceeds the liver's ability to regenerate, then liver failure results, leading to a life threatening situation. Liver failure accounts for over 30,000 deaths each year in the United States.

There are principally two types of liver failure: *cirrhosis* and *fulminant hepatic failure*. Cirrhosis of the liver accounts for over half of the deaths due to liver disease. In cirrhosis, fibrotic tissue forms in place of the damaged liver tissue, severely compromising the liver's ability to regenerate. Common causes of cirrhosis includes alcoholism and chronic hepatitis. Fulminant hepatic failure is a rapidly progressing failure of the liver that can lead to death within several weeks of onset. It can be caused by chemical and viral hepatitis.

Liver failure causes a variety of life-threatening abnormalities, including the accumulation in the plasma of ammonia, bilirubin, and decreased levels of albumin and clotting factors. There is also a build-up of toxins and overactivity of the hormonal systems that are believed to lead to a condition known as *hepatic encephalopathy* that can cause irreversible brain damage, coma, and death.

The only long-term and relatively successful treatment method for liver failure is transplantation. However, liver transplantation is severely limited by the shortage of donor organs. Many patients therefore die before a liver becomes available, or they succumb shortly after the transplant because of the complications and brain damage associated with liver failure.

8.11.1 Artificial Liver Systems

Treatment of liver failure by artificial means has been the focus of considerable research. Table 8.2 provides a summary of some of the artificial liver systems that have been evaluated (Yarmush et al. 1992). The primary goal of these artificial liver systems is to provide a means to maintain the patient in a stable state until a liver transplant is possible. In some cases, artificial liver support could provide the liver the chance to regenerate and the patient could return to a normal life without a transplant.

The difficulty in developing an artificial liver are the multitude of functions the liver performs. Initially it was thought that the best approach would be to minimize the build-up of toxins. Accordingly, efforts focused on such approaches as hemodialysis, hemoperfusion, and immobilized enzyme reactors for removal of these materials.

The *hemodialysis* systems were similar to those used in kidney dialysis, with the exception that they used membranes with a higher molecular weight cutoff to allow passage of the larger-sized toxic materials. This approach was still not effective at removing large protein bound toxins. *Hemoperfusion* employed beds of activated carbon that adsorbed the toxic molecules. A particular problem with this approach is its nonspecificity, removing beneficial substances as well. *Immobilized enzyme reactors* used

Table 8.2 Artificial liver systems

Hemodialysis	Cross circulation
Hemoperfusion	Extracorporeal perfusion
Immobilized enzymes	Cross hemodialysis
Plasma exchange	Hepatocyte hemoperfusion

liver enzymes to provide for more specific removal of the toxic molecules. The major difficulty is providing a complete set of liver enzymes. These three approaches also suffered from the disadvantage that they do not restore any of the synthetic functions of the liver.

Other approaches that were used attempted to both reduce the level of toxins and restore substances normally synthesized by the liver. These methods included plasma exchange, cross circulation, extracorporeal perfusion, and cross hemodialysis. *Plasma exchange* involves replacing the patients plasma with donor plasma. Provided the exchange rate is sufficiently high, this approach can replace the lost liver function. However, the major limitation is the large amount of donor plasma needed and the increased risk to the patient of viral infections. *Cross circulation* involved connecting the patient's circulation to that of another human. In this way the healthy liver is shared. However, the risk to the human donor limits the wide spread use of this approach. *Extracorporeal perfusion* of the patient's blood through a xenogeneic liver has also been tried. Some success with this approach has been achieved using pig and baboon livers. However, the liver function degrades quickly as a result of the host's immune rejection response. *Cross hemodialysis* attempts to minimize the host's immune response to the donor liver by perfusing the donor liver with a separate supply of blood. This blood is then sent through a hemodialyzer where it is contacted across a dialysis membrane with the patient's own blood. In this way toxins are removed and synthetic substances are added to the patient's blood. *Hemoperfusion of liver tissue* has also been tried. Liver tissue pieces are directly contacted with the patient's blood. Frozen or freeze-dried dead tissue was comparable in activity to fresh tissue indicating that only those few layers of cells near the surface of the fresh liver pieces are adequately oxygenated. The dead tissue still contained enzymes with some residual activity.

8.11.2 Bioartificial Livers

Hemoperfusion of liver tissue pieces is not capable of providing optimal conditions for the long-term function of the liver tissue. In the absence of an intact vasculature, the mass-transfer resistances within the liver pieces are just too large for simple diffusion to overcome. Recently, attention has focused on the development of *bioartificial livers* that are based on the use of isolated hepatocytes (Jauregui et al. 1997). Isolated hepatocytes would not be as severely affected by mass-transfer limitations and they can be immunoprotected. Isolated hepatocytes have been proposed for both implantable and extracorporeal systems. Impantable systems are based on either the tissue engineering concepts discussed in Chapter 7, or involve the encapsulation of the hepatocytes in a manner similar to the techniques already discussed for islets (Cai et al. 1988; Cima et al. 1991; Johnson et al. 1994; Yang et al. 1994).

The mass of hepatocytes that is needed is estimated to be on the order of 10% of the original liver mass (Yarmush et al. 1992; Rozga et al. 1994). This amounts to a cellular volume of nearly 100 ml. Attachment to a support structure, or encapsulation of this quantity of cells, while still maintaining a reasonable device volume, is a major obstacle with implantation approaches. For example, for hepatocytes attached to microcarriers[9], the estimated total volume would be 500 ml. For microencapsulation within a polymeric capsule, this volume is 2500 ml, whereas macroencapsulation in hollow fibers yields a volume that is estimated to be 1300 ml.

Because of these size constraints, extracorporeal bioartificial livers are an attractive approach for providing temporary liver function until an organ is available for transplantation or the patient's liver can regenerate. The patient's blood or plasma would flow through an external circuit that provides contact with the hepatocytes.

The hepatocytes can be presented to the patient's blood or plasma in a number of ways. The simplest would be as a suspension of cells. However, this provides no immunoprotection and like most mammalian cells, hepatocytes do better if they are attached to something. In some cases, the suspension of cells can be placed within a hemodialyzer and separated from the blood or plasma stream by a dialysis membrane. Although this approach provides immunoprotection, the low-molecular-weight cutoff of the dialysis membrane severely limits the molecular sizes that can be detoxified or synthesized. Other approaches involve perfusing the patient's blood or plasma through columns packed with microcarrier attached hepatocytes, or hepatocytes encapsulated within polymeric beads. Hepatocytes attached to microcarriers, however, provide no immunoprotection. Hollow-fiber units have also been proposed. In these systems the hepatocytes are within the shell space and attach themselves to the outer surface of the hollow fibers. Blood or plasma flows through the lumen of the hollow fibers.

One major limitation of the membrane-based approaches discussed so far involves the significant mass-transfer resistance created by the presence of the membrane. Solute transport through the membrane and into the space containing the hepatocytes is by diffusion only. This not only limits the supply of essential nutrients and oxygen to the hepatocytes, but also limits the detoxification and biosynthesis rates of the hepatocytes as well. Three particularly promising approaches attempt to overcome these mass-transfer limitations. All are based on the use of shell and tube hollow-fiber modules.

8.11.3 Three Extracorporeal Bioartificial Livers

Figure 8.18 ilustrates one approach that provides three distinct compartments (Nyberg et al. 1992, 1993a, 1993b). The first compartment consists of a gel-like collagen matrix that traps the hepatocytes within the hollow-fiber lumen. The second compartment consists of the space within the lumen of the hollow fiber that results after contraction of the original collagen suspension containing the hepatocytes. The third compartment is the shell space that surrounds the hollow fibers. The hollow fibers have a nominal molecular weight cutoff of 100,000 daltons and immunoprotect the hepatocytes from the patient's immune system.

In operation, the patient's blood flows through the shell space. Substances to be detoxified readily diffuse through the hollow-fiber membrane and into the collagen gel that contains the hepatocytes. Similarly, substances produced by the hepatocytes diffuse

[9] A microcarrier is a small particle about 200 microns in diameter, sometimes coated with ECM materials, that provides for cell attachments and formation of a confluent monolayre of cells.

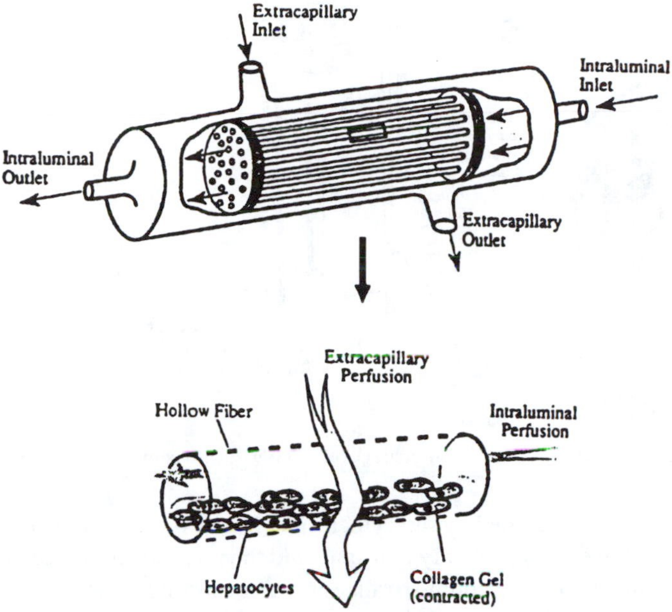

Figure 8.18 A three-compartment hollow-fiber bioartificial pancreas. (From Nyberg et al. [1993b], with permission.)

across the membrane and are carried away by the blood to the patient. Of particular note in this system is the presence of the space within the lumen that is formed after the gel contracts. This space provides an additional flow path that allows simultaneous perfusion of the hepatocytes with specific nutrients, growth factors, and hormones that are needed to maintain their differentiated function and viability. ECM-type materials can also be incorporated into the gel that contains the hepatocytes. In this way the hepatocytes can be presented with an optimal culture environment while performing their liver support functions.

This particular bioartificial liver was evaluated in vitro over a period of seven days. Approximately 5×10^7 cells were loaded into the device. Hepatocyte function was evaluated on the basis of albumin production, oxygen consumption, and lidocaine clearance. It is important to point out that unlike the bioartificial pancreas, there is no single measure of differentiated hepatocyte function. However, albumin was steadily produced during the 7-day period at rates comparable to that seen in static cultures. Furthermore, oxygen consumption remained relatively constant following an initial decline over the first two days of culture. This initial decrease in the oxygen consumption, although possibly being related to cell death, was also considered to be a result of the increased metabolic demands on the hepatocytes as a result of the trauma during their isolation and placement in a new environment.

The cytochrome P-450 system (Hantsen 1998) is a major pathway for the biotransformation in the liver of many substances and is thought to be an important function that must be provided by a bioartificial liver. Lidocaine clearance can be used as a

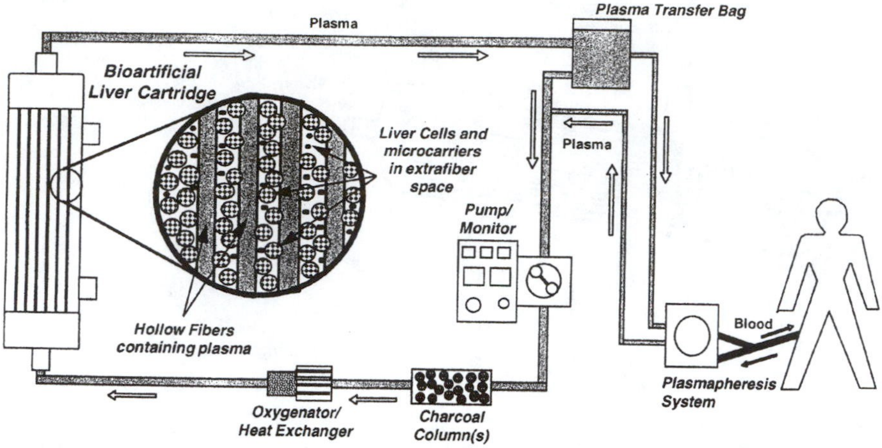

Figure 8.19 HepatAssist^{TM}2000 bioartificial liver. (From Jauregui et al. [1997], with permission.)

measure of oxidative metabolism provided by the cytochrome P-450 system. Lidocaine clearance in this bioartificial liver was relatively constant, and the presence of lidocaine metabolites demonstrated that lidocaine biotransformation was occurring. Electron micrographs after seven days of operation showed the presence of differentiated viable hepatocytes.

Figure 8.19 illustrates an extracorporeal bioartificial liver system that consists of several integrated components (Rozga et al. 1993, 1994; Giorgio et al. 1993; Jauregui et al. 1997). This system, the HepatAssist^{TM}2000, is now being developed by W.R. Grace & Co (Boca Raton, FL). A plasmapheresis unit is first used to form a plasma stream from arterial blood. This plasma stream feeds into a high-flow plasma recirculation loop that forms the core of the bioartificial liver support system. Within the recirculation loop, plasma first enters a column loaded with activated cellulose-coated charcoal. The activated charcoal column is used to enhance the detoxification capability of the overall system and to protect the hepatocytes from any toxic materials found in the patient's plasma. The detoxified plasma then enters a membrane oxygenator. This ensures an adequate supply of oxygen to maintain the viability and function of the hepatocytes. The oxygenated plasma then flows through the hollow-fiber module that contains the hepatocytes. The plasma flows through the hollow-fiber lumens. Plasma exiting the hollow-fiber cartridge is then recombined with the cellular components of the blood and returned to the patient's body.

The hepatocytes within the bioartificial liver cartridge are attached to collagen coated microcarrier beads that occupy the shell space. About 6 billion porcine hepatocytes are used. The hollow fibers have large pore sizes (0.2 microns) that allow for significant fluid convection through the hollow-fiber membrane as a result of the transfiber pressure drop. From the fiber entrance to its midpoint, there is a significant flow of plasma out of the fiber and into the shell space that contains the anchored hepatocytes. From the fiber midpoint to the fiber exit, this flow reverses itself and reenters the fiber lumen. This is much like the Starling flow phenomena discussed in Chapter 2 for capillaries. This convective flow into the shell space passes freely between the microcarrier beads that contain the hepatocytes on their surfaces. This convective flow through the shell space provides for very efficient contacting of the plasma solutes and the hepatocytes, and far surpasses what is achievable by diffusion alone.

One limitation of the approach shown in Figure 8.19 is the lack of complete immunoisolation of the xenogeneic hepatocytes. Plasma perfusion eliminates the cellular components of the immune system, but antibodies and complement will readily flow across the membrane and into the shell space that contains the hepatocytes. However, the treatment time proposed for this device is relatively short and the patient is immunosuppressed in anticipation of a liver transplant. So the trade-off between a high transmembrane flow resulting in good mass transport, and the resultant permeability to the humoral components of the immune system, seems appropriate. Certainly longer-term use of this device will require improved methods of immunoisolation that do not compromise the mass-transport properties of the hepatocyte bioreactor.

Figure 8.20 illustrates the Hepatix, Inc. (La Jolla, CA) Extracorporeal Liver Assist Device or ELAD (Sussman et al. 1992; Kelly and Sussman 1994). This is a hollow-fiber device that uses a cloned human cell line in place of hepatocytes. The cell line is derived from an hepatoblastoma and selected for liver specific functions. These cells exhibit the ability to synthesize protein, urea, and glucose, and are capable of detoxifying substances via the cytochrome P-450 system. The use of a human cell line provides an unlimited supply of cells and minimizes the immune problems that may result from the use of xenogeneic hepatocytes. The ELAD is seeded with 2 grams of cells and is placed in culture until 200 grams of cells are obtained. This quantity of cells is capable of producing therapeutic levels of plasma proteins. For example, 5 grams of albumin are produced each day, which is about one-half the normal daily production rate of a healthy adult human liver.

Blood flows from the patient's vein through a standard hemodialysis pump. The blood exiting the pump is mixed with heparin to prevent blood clots and then passes through the lumen of the hollow fibers of the ELAD device. A portion of the blood flowing through the hollow fibers forms a plasma ultrafiltrate that permeates the shell

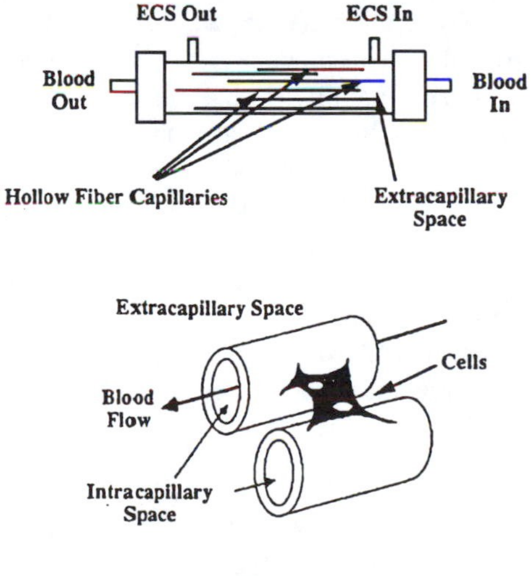

Figure 8.20 The extracorporeal liver assist device (ELAD). (From Sussman et al. [1992], with permission.)

space where it comes into direct contact with the cells. Ultrafiltration of the blood allows delivery of higher-molecular-weight solutes at rates significantly higher than that possible by diffusion alone. The ultrafiltrate then passes through a 0.45-micron filter before it is returned to the patient.

Evaluation of this device in dogs with fulminant hepatic failure provided rapid improvement in blood chemistry. There was also a rapid rise in plasma human albumin and α fetoprotein levels following connection of the device. Factor V–dependent clotting times showed a corresponding rapid decrease. After 48 hours of treatment, dogs with fulminant hepatic failure have sufficiently recovered so that they can be disconnected from the device. Compared to control animals that die, the treated animals regenerated their own livers.

These results are very promising and demonstrate that the bioartificial liver has the potential to provide a safe and effective means of stabilizing liver failure patients until a liver is available for transplantation.

8.12 THE BIOARTIFICIAL KIDNEY

Another potential area for the application of bioartificial organ technology is in the development of a bioartificial kidney (Ip and Aebischer 1989; Cieslinski and Humes 1994; Humes 1995, 1997). Patients with end-stage renal disease can be treated for many years either by hemodialysis or by the newer technique of *continuous ambulatory peritoneal dialysis* (CAPD). However, these are not permanent solutions and these patients ultimately require a kidney transplant in order to survive.

The kidney not only provides a filtration and waste removal function, but also provides several other important functions as well. For example, *erythropoietin* is released by specialized cells found in the kidney in response to hypoxia. Erythropoietin is a major stimulus for the production of red blood cells in the bone marrow. Uremic patients therefore suffer from *anemia.* Low blood pressure causes the release of *renin* from the juxtaglomerular cells that are found in the kidney. Renin initiates the formation of *angiotensin II*, a potent vasoconstrictor, that results in an increase in blood pressure. Angiotensin II also acts on the kidneys, decreasing the excretion of both salt and water. This expands the extracellular fluid volume with the result that the blood pressure is increased. The kidneys are also responsible for the conversion of vitamin D into a substance[10] that promotes the absorption of calcium from the intestine. Without this substance, the bones become severely weakened because of the loss of calcium. These other very important functions that are performed by the healthy kidney are therefore compromised as a result of kidney failure.

As shown earlier in Figure 5.3, the functional unit of the kidney is called the nephron. Recall that it consists of two major components, the glomerulus and the renal tubule. The glomerulus is responsible primarily for the selective ultrafiltration of waste products from the blood. It must perform this waste removal function and at the same time retain

[10] 1,25-dihydroxycholecalciferol

essential blood components such as albumin. The glomerular filtrate that is formed then passes through the various segments of the renal tubule. The specialized segments of the renal tubule regulate the amount of urine that is formed and its final solute composition. The renal tubule cells therefore have the ability to control both the fluid reabsorption rate and the transport rate of an individual solute. This is accomplished in such a manner so as to maintain homeostasis with regard to the body's fluid volume and overall composition. Because of the chemical sensing and selective transport ability of the renal tubule, it is unlikely that this sophisticated function could ever be reproduced artificially. Accordingly, artificial kidneys will primarily function at the level of simply removing waste products by dialysis.

A bioartificial kidney has the potential of reproducing many of the homeostatic, regulatory, and endocrine functions that are performed by the healthy kidney. This would far surpass the filtration function of existing dialysis systems and allow blood purification to occur in a more physiological fashion. A bioartificial kidney would consist of two main components, an artificial glomerulus and a bioartificial renal tubule (Ip and Aebischer 1989). The artificial glomerulus can be fabricated from polymeric membranes that have a high hydraulic conductance. The transmembrane pressure gradient can be controlled to provide the desired bulk flow rate of plasma across the membrane. This bulk flow, or ultrafiltration of the plasma, will also provide for significant convective transport of solutes across the membrane. The convective transport of all solutes will be essentially the same up to the molecular weight cutoff of the membrane. This is especially important for the larger size solutes where much more can be removed by convection than by simple diffusion. The artificial glomerular membranes could be configured as hollow fibers or perhaps spiral wound sheets. The major requirement being that the membrane arrangement is conducive to the production of a significant ultrafiltration flow.

Several problems still need to be addressed with regard to the fabrication of an artificial glomerulus. These include uncontrolled bleeding as a result of anticoagulation, a decreased ultrafiltration rate over time as a result of protein deposition in the membrane, and the challenge of developing a nonthrombogenic membrane. One exciting possibility for overcoming the clotting tendency of most polymeric materials is to cover the blood contacting surfaces with a monolayer of autologous endothelial cells. Endothelial cells form the blood contacting lining of the body's blood vessels and in the absence of injury are responsible for preventing the blood from clotting.

The bioartificial renal tubule could consist of viable renal epithelial cells supported within a tubular structure such as a hollow-fiber membrane. The hollow-fiber membrane, in addition to its role as a structural support, would also provide immunoprotection of these cells. The feasibility of creating a bioartificial tubule was demonstrated by Ip and Aebischer (1989). They grew renal epithelial cell lines to confluence on a permeable membrane. The cells inhibited the passive diffusive transport of solutes and exhibited active transport of specific solutes. Therefore it appears possible to construct a bioartificial renal tubule that can regulate the transport rates of a variety of solutes. This is an important first step towards the development of a bioartificial kidney.

Another possibility for creating a renal tubule is to use renal tubule stem cells (Cieslinski and Humes 1994; Humes and Cieslinsi 1992). *Stem cells* are a class of highly proliferative cells that give rise to cells that have a differentiated function. The resulting differentiated cells provide the ultimate physiological structure and function found in the body's tissues and organs. Perhaps the best known class of stem cells are the *pluripotential hemopoietic stem cells* (PHSC) (Koller and Palsson 1993) that are responsible for the formation of all the differentiated cells found in blood. The differentiated cells arising from the PHSC consist of the red blood cells, white blood cells, and platelets. In many cases the terminally differentiated state of the cell is incapable of reproducing. Therefore, the stem cell has the responsibility of producing and replacing those terminally differentiated cells that have become damaged or injured for a variety of reasons. For example, renal tubule stem cells would regenerate renal tubules following their injury. In the presence of the proper growth factors, renal tubule stem cells have the capability to form tubule-like structures in culture. The isolation and culture of human renal tubule stem cells could therefore be an important source of tissue for the formation of bioartificial renal tubules.

8.12.1 Two Configurations for a Bioartificial Kidney

Figure 8.21 illustrates two possible arrangements of the artificial glomerulus and the bioartificial tubule. In the first system, blood flows through the luminal spaces of the hollow fibers that comprise the hemofilter and the two tubule sections. A flow control valve placed after the hemofilter regulates the filtration rate of the hemofilter. The hyperosmotic blood then leaves the hemofilter and continues through the proximal and distal tubule modules. The ultrafiltrate generated in the hemofilter flows through the shell side of the two tubule modules. Renal tubule cells are grown on the outer surfaces of the hollow fibers in the tubule modules. Note the cells do not contact the blood and are immunoprotected by the hollow-fiber membrane. These cells selectively transport solutes from the shell side to the blood side or vice versa. The hyperosomotic blood that is formed in the hemofilter facilitates the reabsorption of water from the ultrafiltrate.

The second system has the advantage that blood only contacts an artificial membrane in the hemofiltration unit. The blood flows within the hollow fibers of the hemofilter and the ultrafiltrate that is formed collects within the hemofilter shell space. The ultrafiltrate then enters both the lumens and the shell spaces of the proximal and distal tubule hollow-fiber modules. In this case the renal tubule cells are grown on the luminal side of the hollow fiber membranes. Selective solute transport then occurs between the luminal and shell spaces of the tubule modules. The shell-side ultrafiltrate is then returned to the blood exiting the hemofilter. The ultrafiltrate that leaves the luminal side of the hollow fibers within the distal tubule module passes through a flow control valve that regulates the ultrafiltration rate of the hemofilter.

For an implantable device, the blood supply for both configurations could originate from an arteriovenous circuit that uses the common iliac artery and vein. The pseudo urine that is formed could be drained into a ureter for collection and final excretion via the bladder.

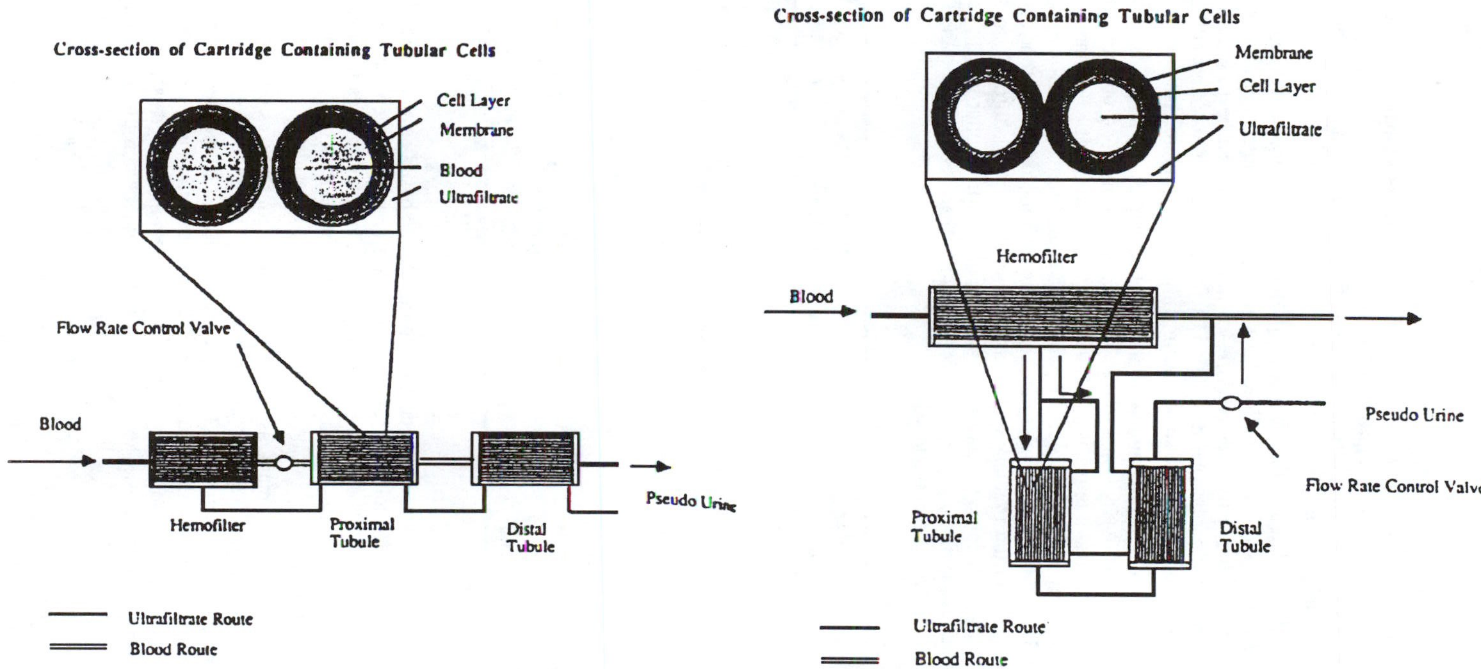

Fig. 8.21 Two proposed configurations for a bioartificial kidney. (From Ip and Aebischer [1989], with permission.)

PROBLEMS

1. Prepare a poster presentation that describes a specific application of a bioartificial organ. Include a description of the clinical need, the approach used, results that have been obtained (in vitro, in animals, and the status of any clinical trials), the potential clinical impact, the potential market impact, safety issues, and additional development needs.

2. In Example 8.1, the hindered diffusion model of Bungay and Brenner (1973) was used to estimate the effective pore size of the hollow fiber immunoisolation membrane based on the measured solute permeabilities. This hindered diffusion model provides at best a semi-quantitative description of the membrane's solute permeability. Discuss factors that may impact the ability of the hindered diffusion model to describe solute permeability data in immunoisolation membranes.

3. Baker et al. (1997) developed an immunoisolation membrane by incorporating a high water content polyvinyl alcohol (PVA) hydrogel into a thin disk-shaped microporous polyethersulfone (PES) filter. The PES filter had a thickness of 150 microns, a porosity of 81%, and a mean pore size of 0.2 microns. The hydrogel formed in the pores of the microporous membrane was found to contain 86% water by weight. The 0.2 micron pores of the microporous membrane should block the cellular components of the immune system and the PVA hydrogel, through control of the PVA cross-linking, can be tailored to control the passage of high-molecular-weight solutes. The table below summarizes the physical properties of the solutes that were used and their measured permeability in this composite membrane.

Solute	Molecular weight	Solute radius, nm	$D \cdot 10^6$, cm^2/sec, 23°C	$P_m \times 10^8$, cm/sec
Glucose	180	0.36	6.38	40000
Vitamin B12	1355	0.75	3.10	16000
Inulin	5500	1.41	1.65	8000
Lysozyme	14500	1.92	1.22	80
Myoglobin	16890	1.9	1.21	70
Chymo-Trypsinogen	25000	2.24	1.04	60
Ovalbumin	45000	2.73	0.851	25
Albumin	69000	3.55	0.655	4
IgG	160000	5.35	0.434	4.5
Blue Dextran	2000000	9.26	0.251	0.5

From Baker et al. (1997).

What is the effective pore size size of the composite membrane based on the hindered diffusion model? An alternative model for solute diffusion through this composite membrane can be based on the methods outlined in Chapter 2 for gels. In this case the solute moves through a random fibrous network consisting of the cross-linked polymer chains of the PVA hydrogel. Use Equations 2.15 and 2.16 or Equation 2.17 to describe the solute permeablity results.

4. It is estimated that perhaps as many as 500,000 patients with diabetes could benefit from a bioartificial pancreas. Assuming that each patient needs about 1 million islets, and that the islets need to be replaced each year, then about 1,000 pancreatic islet isolations need to be performed each day to meet this demand for islets. Prepare a report that describes a process for the massive isolation of islets of Langerhans from the pancreas of large mammals such as the dog or the pig. Be sure to describe any safety issues. The references should first be consulted for an assessment of the current state of islet isolation methods for a single pancreas.

5. Derive an expression for the steady-state islet insulin secretion rate based on the Nomura et al. (1984) model.

6. Derive Equations 8.5 and 8.6 for the islet insulin release rate following a step change in the glucose concentration.

7. The following perifusion data was obtained by Lakey et al. (1995) for human islets of Langerhans. The islets were challenged with glucose that was initially at 50 mg/100 ml, followed by a prompt increase of glucose to 500 mg/100 ml for 60 minutes, after which the glucose was returned to 50 mg/100 ml. Use the Nomura et al. (1984) islet insulin release model to describe this data.

Time, minutes	Islet insulin release rate, μU/islet/minute
0	0.10
5	0.83
8	0.75
28	0.61
50	0.68
60	0.50

8. Use the Sturis et al. (1991, 1995) pharmacokinetic model for glucose and insulin to describe an OGTT for a normal person. How do the normal person glucose and insulin profiles compare to those obtained in Example 8.3 using a bioartificial pancreas?

9. Repeat Example 8.3 assuming the 70-kg patient receives an IVGTT. What is the k-value?

10. Repeat Example 8.3, assuming human islets are used. Use the Nomura et al. (1984) islet model and the parameters obtained in Problem 7 for canine human islets.

11. Repeat Example 8.3 assuming the islets are macroencapsulated in the hollow fibers described in Example 8.2. These hollow-fiber membranes have a nominal external diameter of 950 microns. What is the entire length of fibers needed?

12. For the device described in Example 8.3, the oxygen permeability of the membrane is estimated to be 0.001 cm/sec. Based on the method outlined in Example 4.4, show that the maximum volume fraction of the islets is 0.35.

13. Repeat Example 8.3, assuming the half-thickness of the device is 350 microns. The membrane oxygen permeability is estimated to be 0.001 cm/sec. How are the glucose and insulin profiles affected by a change in device thickness? What volume fraction of islets is allowable based on oxygen transport limitations?

REFERENCES

Abbas, M., and V. P. Tyagi. 1987. Analysis of a hollow fiber artificial kidney performing simultaneous dialysis and ultrafiltration. *Chem. Eng. Sci.* 42:133–142.

Agrawal, C. M., G. G. Niederauer, and K. A. Athanasiou. 1995. Fabrication and characterization of PLA-PGA orthopedic implants. *Tissue Engineering* 1:241–252.

Alberts, B., D. Bray, J. Lewis, M. Raff, K. Roberts, and J. D. Watson. 1989. *The cell.* New York: Garland Publishing.

Alberts, B., D. Bray, J. Lewis, M. Raff, K. Roberts, and J. D. Watson. 1989. *Molecular biology of the cell,* 2nd ed. New York: Garland Publishing.

Al-Hendy, A., G. Hortelano, G. S. Tannenbaum, and P. L. Chang. 1995. Correction of the growth defect in dwarf mice with nonautologous microencapsulated myoblasts—an alternate approach to somatic gene therapy. *Hum. Gene Ther.* 6:165–175.

Altman, J. J., D. Houlbert, P. Callard, P. McMillan, B. A. Solomon, J. Rosen, and P. M. Galletti. 1986. Long-term plasma glucose normalization in experimental diabetic rats with microencapsulated implants of benign human insulinomas. *Diabetes* 35:625–633.

American Diabetes Association. 1993. Position statement. Implications of the diabetes control and complications trial. *Diabetes* 42:1555–1558.

Anderson, J. L., and J.A. Quinn. 1974. Restricted transport in small pores. *Biophys. J.* 14:130–149.

Anderson, J. M. 1994. The extracellular matrix and biomaterials. In *Implantation biology,* ed. by R. S. Greco. 113–130. Boca Raton, FL: CRC Press.

Armer, T. A., and T. R. Hanley. 1986. Characterization of mass transfer in the hollow fiber artificial kidney. *Chem. Eng. Commun.* 47:49–71.

Arnold, F., and D. C. West. 1991. Angiogenesis in wound healing. *Pharmacol. Therapy* 52:407–422.

Atkinson, M. A., and N. K. Maclaren. 1990. What causes diabetes? *Scientific American* 263:62–71.

Baker, A. R., R. L. Fournier, J. G. Sarver, J. L. Long, P. J. Goldblatt, J. M. Horner, and S. H. Selman. 1997. Evaluation of an immunoisolation membrane formed by incorporating a polyvinyl alcohol hydrogel within a microporous filter support. *Cell Transplant.* 6:585–595.

Barbee, J. H., and G. R. Cokelet. 1971a. Prediction of blood flow in tubes with diameters as small as 29 μ. *Microvasc. Res.* 3:17–21.

Barbee, J. H., and G. R. Cokelet. 1971b. The Fahraeus effect. *Microvasc. Res.* 3:6–16.

Beck, R. E., and J. S. Schultz. 1970. Hindered diffusion in microporous membranes with known pore geometry. *Science* 170:1302–1305.

Benjamin, E., and S. Leskowitz. 1991. *Immunology: a short course.* New York:Wiley-Liss.

Berger, M., and D. Rodbard. 1989. Computer simulation of plasma and glucose dynamics after subcutaneous insulin injection. *Diabetes Care* 12:725–736.

Bernstein, H., V. C. Yang, and R. Langer. 1987a. Distribution of heparinase covalently immobilized to agarose: experimental and theoretical studies. *Biotechnol. Bioeng.* 30:196–207.

Bernstein, H., V. C. Yang, and R. Langer. 1987b. Immobilized heparinase: in vitro reactor model. *Biotechnol. Bioeng.* 30:239–250.

Bird, R. B., W. E. Stewart, and E. N. Lightfoot. 1960. *Transport phenomena.* New York: John Wiley & Sons.

Brinkman, H. C. 1947. A calculation of the viscous force exerted by a flowing fluid on a dense swarm of particles. *Appl. Sci. Res.* A1:27–34.

Buladi, B. M., C. C. Chang, J. M. Belovich, and J. E. Gatica. 1996. Transport phenomena and kinetics in an extravascular bioartificial pancreas. *AIChE J.* 42:2668–2682.

Bungay, P. M., and H. Brenner. 1973. The motion of a closely fitting sphere in a fluid filled tube. *Int. J. Multiph. Flow.* 1:25.

Burton, A. C. 1972. *Physiology and biophysics of the circulation,* 2nd ed. Chicago: Year Book Medical Publishers.

Cai, Z., Z. Shi, G. M. O'Shea, and A. M. Sun. 1988. Microencapsulated hepatocytes for bioartificial liver support. *Artif. Orgs.* 12:388–393.

Cappello, A., F. Grandi, C. Lamberti, and A. Santoro. 1994. Comparative evaluation of different methods to estimate urea distribution volume and generation rate. *Int. J. Artif. Orgs.* 17:322–330.

Catapano, G., A. Wodetzki, and U. Baurmeister. 1992. Blood flow outside regularly spaced hollow fibers: the future concept of membrane devices? *Int. J. Artif. Orgs.* 15:327–330.

Cavallaro, J. F., P. D. Kemp, and K. H. Kraus. 1994. Collagen fabrics as biomaterials. *Biotechnol. Bioeng.* 43:781–791.

Chang, P. L. 1997. Nonautologous gene therapy with implantable devices. *IEEE Eng. Med. Biol.* September/October:145–150.

Chang, P. L., N. Shen, and A. J. Wescott. 1993. Delivery of recombinant gene products with microencapsulated cells in vivo. *Hum. Gene Ther.* 4:433–440.

Charm, S. E., and G. S. Kurland. 1974. *Blood flow and microcirculation.* New York: John Wiley & Sons.

Chick, W. L., J. J. Perna, V. Lauris, D. Low, P. M. Galletti, G. Panol, A. D. Whittemore, A. A. Like, C. K. Colton, and M. J. Lysaght. 1977. Artificial pancreas using living beta cells: effects on glucose homeostasis in diabetic rats. *Science* 197:780–782.

Cieslinski, D. A., and H. D. Humes. 1994. Tissue engineering of a bioartificial kidney. *Biotechnol. Bioeng.* 43:678–681.

Cima, L. G., D. E. Ingber, J. P. Vacanti, and R. Langer. 1991. Hepatocyte culture on biodegradable polymeric substrates. *Biotechnol. Bioeng.* 38:145–158.

Cima, L. G., J. P. Vacanti, C. Vacanti, D. Ingber, D. Mooney, R. Langer. 1991. Tissue engineering by cell transplantation using degradable polymer susbstrates. *J. Biomech. Eng.* 113:143–151.

Colton, C. K. 1995. Implantable biohybrid artificial organs. *Cell Transplant.* 4:415–436.

Colton, C. K., and E. G. Lowrie. 1981. Hemodialysis: physical principles and technical considerations. In *The kidney,* 2nd ed, Volume II, ed. by B. M. Brenner and F. C. Rector, Jr., Chapter 47. Philadelphia: W. B. Saunders.

Colton, C. K., and E. S. Avgoustiniatos. 1991. Bioengineering in development of the hybrid artificial pancreas. *J. Biomech. Eng.* 113:152–170.

Cooney, D. O. 1976. *Biomedical engineering principles.* New York: Marcel Dekker.

Curry, F. E., and C. C. Michel. 1980. A fiber matrix model of capillary permeability. *Microvas. Res.* 20:96–99.

Cussler, E. L. 1984. *Diffusion: Mass transfer in fluid systems.* Cambridge, UK: Cambridge University Press.

Deen, W. M. 1987. Hindered transport of large molecules in liquid-filled pores. *AIChE J.* 33:1409–1425.

Dionne, K. E., B. M. Cain, R. H. Li, W. J. Bell, E. J. Doherty, D. H. Rein, M. J. Lysaght, and F. T. Gentile. 1996. Transport characterization of membranes for immunoisolation. *Biomaterials.* 17:257–266.

Dionne, K. E., C. K. Colton, and M. L. Yarmush. 1989. Effect of oxygen on isolated pancreatic tissue. *ASAIO Trans.* 35:739–741.

Dionne, K. E., C. K. Colton, and M. L. Yarmush. 1991. A microperifusion system with environmental control for studying insulin secretion by pancreatic tissue. *Biotechnol. Prog.* 7:359–368.

Douglas, J. A., and M. V. Sefton. 1990. The permeability of EUDRAGIT RL and HEMA-MMA microcapsules to glucose and inulin. *Biotechnol. Bioeng.* 36:653–664.

Durbin, R. P. 1960. Osmotic flow of water across permeable cellulose membranes. *J. Gen. Physiol.* 44:315–326.

Fogler, H. S. 1992. *Elements of chemical reaction engineering,* 2nd ed. New York: Prentice-Hall.

Folkman, J. 1985. Tumor angiogenesis. *Can. Res.* 43:175–203.

Folkman, J., and M. Klagsbrun. 1987. Angiogenic factors. *Science* 235:442–447.

Foster et al. 1993. *Cancer Res.* 53:1249–1254.

Freed, L. E., and G. Vunjak-Novakovic. 1995. Tissue engineering of cartilage. In *The biomedical engineering handbook,* ed. J. D. Bronzino, 1788–1806. Boca Raton: CRC Press.

Friedman, D. W., P. J. Orland, and R. S. Greco. 1994. Biomaterials: an historical perspective. In *Implantation Biology,* ed. by R. S. Greco, 1–12. Boca Raton, FL: CRC Press.

Friedman, E. A. 1989. Toward a hybrid bioartificial pancreas. *Diabetes Care* 12:415–419.

Fung, Y. C. 1993. *Biomechanics: Mechanical properties of living tissues,* 2nd ed. New York: Springer-Verlag.

Gabriel, J. L., T. F. Miller, M. R. Wolfson, and T. H. Shaffer. 1996. Quantitative structure-activity relationships of perfluorinated hetero-hydrocarbons as a potential respiratory media. *ASAIO J.* 42: 968–973.

Gaehtgens, P. 1980. Flow of blood through narrow capillaries: Rheological mechanisms determining capillary hematocrit and apparent viscosity. *Biorheology* 17:183–189.

Galletti, P. M., and C. K. Colton. 1995. Artificial lungs and blood-gas exchange devices. In *The biomedical engineering handbook,* ed. J. D. Bronzino, 1879–1897. Boca Raton, FL: CRC Press.

Galletti, P. M., and H. O. Jauregui. 1995. Liver support systems. In *The biomedical engineering handbook,* ed. J. D. Bronzino, 1952–1966. Boca Raton, FL: CRC Press.

Galletti, P. M., C. K. Colton, and M. J. Lysaght. 1995. Artificial kidney. In *The biomedical engineering handbook,* ed. J. D. Bronzino, 1898-1922. Boca Raton, FL: CRC Press, Inc.

Garred, L. J., B. Canaud, and P. C. Farrell. 1983. A simple kinetic model for assessing peritoneal mass transfer in chronic ambulatory peritoneal dialysis. *ASAIO J.* 6:131–137.

Gharapetian, H., M. Maleki, G. M. O'Shea, R. C. Carpenter, and A. M. Sun. 1987. Polyacrylate microcapsules for cell encapsulation: effects of copolymer structure on membrane properties. *Biotechnol. Bioeng.* 29:775–779.

Gharapetian, H., N. A. Davies, and A. M. Sun. 1986. Encapsulation of viable cells within polyacrylate membranes. *Biotechnol. Bioeng.* 28:1595–1600.

Gibaldi, M., and D. Perrier. 1982. *Pharmacokinetics,* 2nd. ed. New York: Marcel Dekker.

Gibbon, J. H. 1954. Application of a mechanical heart and lung apparatus to cardiac surgery. *Minnesota Med.* 37:71.

Giorgio, T. D., A. D. Moscioni, J. Rozga, and A. A. Demetriou. 1993. Mass transfer in a hollow fiber device used as a bioartificial liver. *ASAIO J.* 39:886–892.

Gray, D. N. 1981. The status of olefin-SO_2 copolymers as biomaterials. In *Biomedical and dental applications of polymers,* ed. C. G. Gebelein and F. F. Koblitz, 21–27. New York: Plenum Publishers.

Gray, D. N. 1984. Polymeric Membranes for Artificial Lungs, ACS Symposium Series No. 256, *Polymer materials and artificial organs,* ed. C. G. Gebelein, 151–161. Washington, DC: American Chemical Society.

Grodsky, G. M. 1972. A threshold distribution hypothesis for packet storage of insulin and its mathematical modeling. *J. Clin. Investig.* 51:2047–2059.

Gu, Y. J., K. Inoue, S. Shinohara, R. Doi, M. Kogire, T. Aung, S. Sumi, M. Imamura, T. Fujisato, S. Maetani, and Y. Ikada. 1994. Xenotransplantation of bioartificial pancreas using a mesh reinforced polyvinyl alcohol bag. *Cell Transplant.* 3:S19–S21.

Guyton, A. C. 1991. *Textbook of medical physiology,* 8th. ed. Philadelphia: W. B. Saunders.

Guyton, J. R., R. O. Foster, J. S. Soeldner, M. H. Tan, C. B. Kahn, L. Koncz, and R. E. Gleason. 1978. A model of glucose insulin homeostasis in man that incorporates the heterogeneous fast pool theory of pancreatic insulin release. *Diabetes* 27:1027–1042.

Hachiya, H. L., P. A. Halban, and G. L. King. 1988. Intracellular pathways of insulin transport across vascular endothelial cells. *Am. J. Physiol.* (Cell Physiol. 24) 255:C459–C464.

Hammar, H. 1993. Wound Healing. *Int. J. Dermatol.* 32:6–15.

Hantsen, P. D. 1998. Understanding drug-drug interactions. *Sci. Med.* 5:16–25.

Hashimoto et al. 1997. *ASAIO J.* 43:84–88.

Hayashi, H., K. Inoue, T. Aung, T. Tun, G. Yuanjun, W. Wenjing, S. Shinohara, H. Kaji, R. Doi, H. Setoyama, M. Kato, M. Imamura, S. Maetani, N. Morikawa, H. Iwata, Y. Ikada, and J.-I. Miyazaki. 1996. Application of a novel B cell line MIN6 to a mesh reinforced polyvinyl alcohol hydrogel tube and three layer agarose microcapsules: an in vitro study. *Cell Transplant.* 5:S65–S69.

Haynes, R. F. 1960. Physical basis of the dependence of blood viscosity on tube radius. *Am. J. Physiol.* 198:1193.

Henderson and Doughherty, 1992. *Photochem. Photobiol.* 55:145–157.

Hill, R. S., L. A. Martinson, S. K. Young, and R. W. Dudek. 1994. Macroporous devices for diabetes correction. In *Immunoisolation of pancreatic islets,* ed. R. P. Lanza and W. L. Chick, Chapter 12. Boulder, CO: R. G. Landes.

Holder, W. D., H. E. Gruber, W. D. Roland, A. L. Moore, C. R. Culberson, A. B. Loebsack, K. J. L. Burg, D. J. Mooney. 1997. Increased vascularization and heterogeneity of vascular structures occurring in polyglycolide matrices containing aortic endothelial cells implanted in the rat. *Tissue Engineering* 3:149–160.

Horwitz, A. E. 1997. Integrins and health. *Scientific American* 276:68–75.

Hubbell, J. A. 1997. Matrix effects. In *Principles of tissue engineering,* ed. R. P. Lanza, R. Langer, and W. L. Chick, 247–262. Boulder, CO: R. G. Landes.

Hughes, M., A. Vassilakos, D. W. Andrews, G. Hortelano, J. Belmont, and P. L. Chang. 1994. Construction of a secretable adenosine deaminase for a novel approach to somatic gene therapy. *Hum. Gene Ther.* 5:1445–1454.

Humes, H. D. 1995. Tissue engineering of the kidney. In *The biomedical engineering handbook*, ed. J. D. Bronzino, 1807–1824. Boca Raton, FL: CRC Press.

Humes, H. D. 1997. Application of cell and gene therapies in the tissue engineering of renal replacement devices. In *Principles of tissue engineering,* ed. R. P. Lanza, R. Langer, and W. L. Chick, 577–589. Boulder, CO: R. G. Landes.

Humes, H. D., and D. A. Cieslinski. 1992. Interaction between growth factors and retinoic acid in the induction of kidney tubulogenesis. *Exp. Cell Res.* 201:8–15.

Inoue, K., T. Fujisato, Y. J. Gu, K. Burczak, S. Sumi, M. Kogire, T. Tobe, K. Uchida, I. Nakai, S. Maetani, and Y. Ikada. 1992. Experimental hybrid islet transplantation: application of polyvinyl alcohol membrane for entrapment of islets. *Pancreas2* 7:562–568.

Inoue, K., Y. Gu, S. Shinohara, M. Kogire, M. Mitsuo, I. Nakai, H. Hayashi, K. Uchida, S. Maetani, Y. Ikada, and T. Tobe. 1992. Isolation of adult pig islet. *Int. J. Pancreatol.* 12:173–180.

Intaglietta, M. 1997. Whitaker Lecture 1996. Microcirculation, biomedical engineering, and artificial blood. *Ann. Biomed. Eng.* 25:593–603.

Intaglietta, M., and R. M. Winslow. 1995. Artificial blood. In *The biomedical engineering handbook,* ed. J. D. Bronzino, 2011–2024. Boca Raton, FL: CRC Press.

Ip, T. K., and P. Aebischer. 1989. Renal epithelial-cell-controlled solute transport across permeable membranes as the foundation for a bioartificial kidney. *Artificial Organs* 13:58–65.

Iwata, H., N. Morikawa, and Y. Ikada. 1996. Permeability of filters used for immunoisolation. *Tissue Engineering* 2:289–298.

Jacobs et al. 1993. *Diabetes* 42:1649–1655.

James, K., and J. Kohn. 1996. New biomaterials for tissue engineering. *MRS Bulletin* 21:22–26.

Jarvik, R. K. 1981. The total artificial heart. *Scientific American* 244:74–80.

Jauregui, H. O., C. J. P. Mullon, and B. A. Solomon. 1997. Extracorporeal artificial liver support. In *Principles of tissue engineering,* ed. R. P. Lanza, R. Langer, and W. L. Chick, 463–479. Boulder, CO: R. G. Landes.

Johnson, L. B., J. Aiken, D. J. Mooney, B. L. Schloo, L. Griffith-Cima, R. Langer, and J. P. Vacanti. 1994. The mesentary as a laminated vascular bed for hepatocyte transplantation. *Cell Transplant.* 3:273–281.

Kaufmann, P. M., S. Heimrath, B. S. Kim, and D. J. Mooney. 1997. Highly porous polymer matrices as a three-dimensional culture system for hepatocytes. *Cell Transplant.* 6:463–468.

Kedem, O., and A. Katchalsky. 1958. Thermodynamic analysis of the permeability of biological membranes to nonelectrolytes. *Biochem. Biophys. Acta.* 27:229.

Kelly, J. H., and N. L. Sussman. 1994. The Hepatix extracorporeal liver assist device in the treatment of fulminant hepatic failure. *ASAIO J.* 40:83–85.

Kohn, M. C. 1997. The importance of anatomical realism for validation of physiological models of disposition of inhaled toxicants. *Toxicol. Appl. Pharmacol.* 147:448–458.

Kolff, W. J. 1947. *New ways of treating uremia: the artificial kidney, peritoneal lavage, intestinal lavage.* London: J. A. Churchill.

Koller, M. R., and B. O. Palsson. 1993. Tissue engineering: reconstitution of human hematopoiesis ex vivo. *Biotechnol. Bioeng.* 42:909–930.

Kordower, J., Y.-T. Liu, S. Winn, and D. F. Emerich. 1995. Encapsulated PC12 cell transplants into hemiparkinsonian monkeys: a behavioral, neuroanatomical, and neurochemical analysis. *Cell Transplant.* 4:155–171.

Koushanpour, E. 1976. *Renal physiology.* Philadelphia: W. B. Saunders.

Kraegen, E. W., J. D. Young, E. P. George, and L. Lazarus. 1972. Oscillations in blood glucose and insulin after oral glucose. *Horm. Met. Res.* 4:409–413.

Krewson, C. E., M. L. Klarman, and W. M. Saltzman. 1995. Distribution of nerve growth factor following direct delivery to brain interstitium. *Brain Research* 680:196–206.

Krogh, A. 1919. The number and distribution of capillaries in muscles with calculations of the oxygen pressure head necessary for supplying the tissue. *J. Physiol.* 52:409–415.

L'heureux, N., S. Paquet, R. Labbe, L. German, and F. A. Auger. 1998. A completely biological tissue-engineered human blood vessel. *Faseb J.* 12:47–56.

Lacy, P. E. 1995. Treating diabetes with transplanted cells. *Sci. Am.* 273:501–548.

Lacy, P. E., O. D. Hegre, A. Gerasimidi-Vazeou, F. T. Gentile, and K. E. Dionne. 1991. Maintenance of normoglycemia in diabetic mice by subcutaneous xenografts of encapsulated islets. *Science* 254:1782–1784.

Lagerlund, T. D., and P. A. Low. 1993. Mathematical modeling of time dependent oxygen transport in rat peripheral nerve. *Comput. Biol. Med.* 23:29–47.

Lakey et al. 1995. *Transplantation.* 59:689–694.

Lakey, J. R. T., G. L. Warnock, Z. Ao, and R. V. Rajotte. 1996. Bulk cryopreservation of isolated islets of Langerhans. *Cell Transplant.* 5:395–404.

Langer, R., and J. P. Vacanti. 1993. Tissue engineering. *Science* 260:920–926.

Langer, R., J. P., Vacanti, C. A. Vacanti, A. Atala, L. E. Freed, and G. Vunjak-Novakovic. 1995. Tissue engineering: biomedical applications. *Tissue Engineering* 1:151–161.

Lanza, R. P., and W. L. Chick. 1994b. *Immunomodulation of pancreatic islets,* vol. 2. Boulder, CO: R. G. Landes.

Lanza, R. P., A. M. Beyer, and W. L. Chick. 1994. Xenogeneic humoral responses to islets transplanted in biohybrid diffusion chambers. *Transplantation.* 57:1371–1375.

Lanza, R. P., A. M. Beyer, J. E. Staruk, and W. L. Chick. 1993. Biohybrid artificial pancreas. *Transplantation* 56:1067–1072.

Lanza, R. P., A. P. Monaco, S. J. Sullivan, and W. L. Chick. 1992d. Chapter 22. *Pancreatic islet cell transplantation,* ed. C. Ricordi, Chapter 22. Boulder, CO: R. G. Landes.

Lanza, R. P., and W. L. Chick. 1994a. (eds.) *Procurement of pancreatic islets,* vol. 1. Boulder, CO: R. G. Landes.

Lanza, R. P., and W. L. Chick. 1994b. (eds.) *Immunomodulation of pancreatic islets,* vol 2. Boulder, CO: R. G. Landes.

Lanza, R. P., and W. L. Chick. 1994c. (eds.) *Immunoisolation of pancreatic islets,* vol. 3. Boulder, CO: R. G. Landes.

Lanza, R. P., and W. L. Chick. 1997. Endocrinology: pancreas. In *Principles of tissue engineering,* ed. R. P. Lanza, R. Langer, and W. L. Chick, 405–425. Boulder, CO: R. G. Landes.

Lanza, R. P., D. H. Butler, K. M. Borland, J. E. Staruk, D. L. Faustman, B. A. Solomon, T. E. Muller, R. G. Rupp, T. Maki, A. P. Monaco, and W. L. Chick. 1991. Xenotransplantation of canine, bovine, and porcine islets in diabetic rats without immunosuppression. *Proc. Natl. Acad. Sci. USA* 88:11100–11104.

Lanza, R. P., D. H. Butler, K. M. Borland, J. M. Harvey, D. L. Faustman, B. A. Solomon, T. E. Muller, R. G. Rupp, T. Maki, A. P. Monaco, and W. L. Chick. 1992c. Successful xenotransplantation of a diffusion based biohybrid artificial pancreas: a study using canine, bovine, and porcine islets. *Transplant. Proc.* 24:669–671.

Lanza, R. P., K. M. Borland, J. E. Staruk, M. C. Appel, B. A. Solomon, and W. L. Chick. 1992b. Transplantation of encapsulated canine islets into spontaneously diabetic BB/Wor rats without immunosuppression. *Endocrinology* 131:637–642.

Lanza, R. P., K. M. Borland, P. Lodge, M. Carretta, S. J. Sullivan, T. E. Muller, B. A. Solomon, T. Maki, A. P. Monaco, and W. L. Chick. 1992a. Treatment of severely diabetic pancreatectomized dogs using a diffusion-based hybrid pancreas. *Diabetes* 41:886–889.

Lanza, R. P., S. J. Sullivan, and W. L. Chick. 1992e. Islet transplantation with immunoisolation. *Diabetes* 41:1503–1510.

Lanza, R. P., W. M. Kuhtreiber, and W. L. Chick. 1995. Encapsulation technologies. *Tissue engineering.* 1:181–196.

Lauffenburger, D. A., and J. J. Linderman. 1993. *Receptors.* New York: Oxford University Press.

Lavin, A., C. Sung, A. L. Klibanov, and R. Langer. 1985. A potential treatment for neonatal jaundice. *Science* 230:543–545.

Lee, C.-J., and Y.-L. Chang. 1988. Theoretical evaluation of the performance characteristics of a combined hemodialysis/hemoperfusion system. I. Single pass flow. *Can. J. Chem. Eng.* 66:263–270.

Lee, C.-J., S.-T. Hsu, and S.-C. Hu. 1989. A one compartment single pool model for extracorporeal hemoperfusion. *Comput. Biol. Med.* 19:83–94.

Lepeintre, J., H. Briandet, F. Moussy, D. Chicheportiche, S. Darquy, J. Rouchette, P. Imbaud, J. J. Duron, and G. Reach. 1990. Ex vivo evaluation in normal dogs of insulin released by a bioartificial pancreas containing isolated rat islets of Langerhans. *Artificial Organs.* 14:20–27.

Levick, J. R. 1991. *An introduction to cardiovascular physiology.* Boston, MA: Butterworths & Co.

Lewis, R. 1997. New directions in research on blood substitutes. *Genetic Eng. News* 17:1.

Li, Z., T. Yipintsoi, and J. B. Bassingthwaighte. 1997. Nonlinear model for capillary-tissue oxygen transport and metabolism. *Ann. Biomed. Eng.* 25:604–619.

Lifson, N., K. G. Kramlinger, R. R. Mayrand, and E. J. Lender. 1980. Blood flow to the rabbit pancreas with special reference to the islets of Langerhans. *Gastroenterology* 79:466–473.

Lightfoot, E. N. 1995. The roles of mass transfer in tissue function. In *The Biomedical engineering handbook,* ed. J. D. Bronzino, 1656–1670. Boca Raton, FL: CRC Press, Inc.

Long, M. W. 1995. Tissue microenvironments. In *The biomedical engineering handbook,* ed. J. D. Bronzino, 1692–1709. Boca Raton, FL: CRC Press.

Lu, L., and A. G. Mikos. 1996. The importance of new processing techniques in tissue engineering. *MRS Bulletin.* 21:28–32.

Lum, Z.-P., I. T. Tai, M. Krestow, J. Norton, I. Vacek, and A. M. Sun. 1991. Prolonged reversal of diabetic state in NOD mice by xenografts of microencapsulated rat islets. *Diabetes* 40:1511–1516.

Lysaght, M. J., and J. Moran. 1995. Peritoneal dialysis equipment. In *The biomedical engineering handbook,* ed. J. D. Bronzino, 1923–1935. Boca Raton, FL: CRC Press.

Lysaght, M. J., and P. C. Farrell. 1989. Membrane phenomena and mass transfer kinetics in peritoneal dialysis. *J. Mem. Sci.* 44:5–33.

Maki, T., A. P. Monaco, C. J. P. Mullon, and B. A. Solomon. 1996. Early treatment of diabetes with porcine islets in a bioartificial pancreas. *Tissue Engineering* 2:299–306.

Maki, T., C. S. Ubhi, H. Sanchez-Farpon, S. J. Sullivan, K. Borland, T. E. Muller, B. A. Solomon, W. L. Chick, and A. P. Monaco. 1991. Successful treatment of diabetes with the biohybrid artificial pancreas in dogs. *Transplantation* 51:43–51.

Maki, T., J. J. O'Neil, J. Porter, C. J. P. Mullon, B. A. Solomon, and A. P. Monaco. 1996. Porcine islets for xenotransplantation. *Transplantation* 62:136–138.

Maki, T., J. P. A. Lodge, M. Carretta, H. Ohzato, K. M. Borland, S. J. Sullivan, J. Staruk, T. E. Muller, B. A. Solomon, W. L. Chick, and A. P. Monaco. 1993. Treatment of severe diabetes mellitus for more than one year using a vascularized hybrid artificial pancreas. *Transplantation* 55:713–718.

Mann, G. E., L. H. Smaje, and D. L. Yudilevich. 1979. Permeability of the fenestrated capillaries in the cat submandibular gland to lipid-insoluble molecules. *J. Physiol.* 297:335–354.

Marchant, R. E., and I. Wang. 1994. Physical and chemical aspects of biomaterials used in humans. In *Implantation biology,* ed. by R. S. Greco, 13–38. Boca Raton, FL: CRC Press.

Margiotta, M. S., L. D. Benton, R. S. Greco. 1994. Integrins, adhesion molecules, and biomaterials. In *Implantation biology,* ed. by R. S. Greco. 149-163. Boca Raton, FL: CRC Press.

Matthew, H. W., S. O. Salley, W. D. Peterson, and M. D. Klein. 1993. Complex coacervate microcapsules for mammalian cell culture and artificial organ development. *Biotechnol. Prog.* 9:510–519.

Maxwell, J. C. 1873. *A treatise on electricity and magnetism,* volume 1. Oxford, UK: Clarendon Press.

McCabe, W. L., J. C. Smith, and P. Harriott. 1985. *Unit operations of chemical engineering*, 4th ed. New York: McGraw-Hill.

McNabb, M. E., R. V. Ebert, and K. McCusker. 1982. Plasma nicotine levels produced by chewing nicotine gum. *JAMA*. 248:865–868.

Merrill, E. W., A. M. Benis, E. R. Gilliland, T. K. Sherwood, and E. W. Salzman. 1965. Pressure flow relations of human blood in hollow fibers at low flow rates. *J. Appl. Physiol.* 20:954–967.

Mikos, A. G., G. Sarakinos, M. D. Lyman, D. E. Ingber, J. P. Vacanti, and R. Langer. 1993. Pevascularization of porous biodegradable polymers. *Biotechnol. Bioeng.* 42:716–723.

Mikos, A. G., M. G. Papadaki, S. Kouvroukoglou, S. L. Ishaug, and R. C. Thomson. 1994. Mini-review: islet transplantation to create a bioartificial pancreas. *Biotechnol. Bioeng.* 43:673–677.

Mirhashemi, S., K. Messmer, and M. Intaglietta. 1987. Tissue perfusion during normovolemic hemodilution investigated by a hydraulic model of the cardiovascular system. *Int. J. Microcirc. Clin. Exp.* 6:123–136.

Mochizuki, S., and A. L. Zydney. 1992. Effect of protein adsorption on the transport characteristics of asymmetric ultrafiltration membranes. *Biotechnol. Prog.* 8:553–561.

Mooney, D. J., C. Breuer, K. McNamara, J. P. Vacanti, and R. Langer. 1995. Fabricating tubular devices from polymers of lactic and glycolic acid for tissue engineering. *Tissue engineering* 1:107–117.

Mooney, D. J., G. Organ, J. P. Vacanti, and R. Langer. 1994. Design and fabrication of biodegradable polymer devices to engineer tubular tissues. *Cell Transplant.* 3:203–210.

Mooney, D. J., P. M. Kaufmann, K. Sano, S. P. Schwendeman, K. Majahod, B. Schloo, J. P. Vacanti, and R. Langer. 1996. Localized delivery of epidermal growth factor improves the survival of transplanted hepatocytes. *Biotech. Bioeng.* 50:422–429.

Mooney, D. J., S. Park, P. M. Kaufmann, K. Sano, K. McNamara, J. P. Vacanti, and R. Langer. 1995. Biodegradable sponges for hepatocyte transplantation. *J. Biomed. Mater. Res.* 29:959–965.

Morgan, J. R., and M. L. Yarmush. 1997. Bioengineered skin substitutes. *Sci. Med.* 4:6–15.

Morgan, J. R., R. G. Tompkins, and M. L. Yarmush. 1994. Genetic engineering and therapeutics. In *Implantation biology*, ed. R. S Greco, 387–400. Boca Raton, FL: CRC Press.

Moussy, F., J. Rouchette, G. Reach, R. Cannon, and M. Y. Jaffrin. 1989. In vitro evaluation of a bioartificial pancreas under various hemodynamic conditions. *Artificial Organs* 13:109–115.

Naughton, B. A. 1995. The importance of stromal cells. In *The biomedical engineering handbook*, ed. J. D. Bronzino, 1710–1727. Boca Raton, FL: CRC Press.

Naughton, G. K., W. R. Tolbert, and T. M. Grillot. 1995. Emerging developments in tissue engineering and cell technology. *Tissue Engineering* 1:211–219.

Nomura, N., M. Schihiri, R. Kawamori, Y. Yamasaki, N. Iwama, and H. Abe. 1984. A mathematical insulin secretion model and its validation in isolated rat pancreatic islets perifusion. *Comput. Biomed. Res.* 17:570–579.

Notari, R. E. 1987. *Biopharmaceutics and clinical pharmacokinetics*, 4th ed. New York: Marcel Dekker.

Notkins, A. L. 1979. The causes of diabetes. *Scientific American* 241:62–73.

Nucci, M. L., and A. Abuchowski. 1998. The search for blood substitutes. *Sci. Am.* 278:72–77.

Nugent, L. J., and R. K. Jain. 1984a. Extravascular diffusion in normal and neoplastic tissues. *Cancer Res.* 44:238–244.

Nugent, L. J., and R. K. Jain. 1984b. Pore and fiber-matrix models for diffusive transport in normal and neoplastic tissues. *Microvas. Res.* 28:270–274.

Nyberg, S. L., H. J. Mann, R. P. Remmel, W.-S. Hu, and F. B. Cerra. 1993a. Pharmacokinetic analysis verifies P450 function during in vitro and in vivo application of a bioartificial liver. *ASAIO J.* 39:M252–M256.

Nyberg, S. L., J. L. Platt, K. Shirabe, W. D. Payne, W.-S. Hu, and F. B. Cerra. 1992. Immunoprotection of xenocytes in a hollow fiber bioartificial liver. *ASAIO J.* 38:M463–2M467.

Nyberg, S. L., R. A. Shatford, M. V. Peshwa, J. G. White, F. B. Cerra, and W.-S. Hu. 1993b. Evaluation of a hepatocyte entrapment hollow fiber bioreactor: a potential bioartificial liver. *Biotechnol. Bioeng.* 41:194–203.

Ogston, A. G. 1958. The spaces in a uniform random suspension of fibers. *Trans. Faraday Soc.* 54:1754–1757.

Opong, W. S., and A. L. Zydney. 1991. Diffusive and convective protein transport through asymmetric membranes. *AIChE J.* 37:1497–1510.

Pachence, J. M., and J. Kohn. 1997. Biodegradable polymers for tissue engineering. In *Principles of tissue engineering*, ed. R. P. Lanza, R. Langer, and W. L. Chick, 273–293. Boulder, CO: R. G. Landes.

Parsons-Wingerter, P., and E. H. Sage. 1997. Regulation of cell behavior by extracellular proteins. In *Principles of tissue engineering,* ed. R. P. Lanza, R. Langer, and W. L. Chick, 111–131. Boulder, CO: R. G. Landes.

Pathak, C. P., A. S. Sawhney, and J. A. Hubbell. 1992. Rapid photopolymerization of immunoprotective gels in contact with cells and tissues. *J. Am. Chem. Soc.* 114:8311–8312.

Perez, E. P., E. W. Merrill, D. Miller, L. G. Cima. 1995. Corneal epithelial wound healing on bilayer composite hydrogels. *Tissue Engineering* 1:263–277.

Petley et al. 1995. *Diabetes* 44:152–155.

Petruzzo, P., L. Pibiri, M. A. DeGiudici, G. Basta, R. Calafiore, A. Falorni, P. Brunetti, and G. Brotzu. 1991. Xenotransplantation of microencapsulated pancreatic islets contained in a vascular prosthesis: preliminary results, *Transplant Int.* 4:200–204.

Ramachandran, P. A., and R. A. Mashelkar. 1980. Lumped parameter model for hemodialyzer with application to simulation of patient-artificial kidney system. *Med. Biol. Eng. Comput.* 18:179–188.

Renkin, E. M. 1954. Filtration, diffusion, and molecular sieving through porous cellulose membranes. *J. Gen. Physiol.* 38:225–243.

Renkin, E. M. 1977. Multiple pathways of capillary permeability. *Circ. Res.* 41:735–743.

Renkin, E. M., and F. E. Curry. 1979. Transport of water and solutes across capillary endothelium. In *Membrane transport in biology,* volume 4, ed. G. Giebisch and D. C. Tosteson. New York: Springer-Verlag.

Replogle, R. L., H. J. Meiselman, and E. W. Merrill. 1967. Clinical implications of blood rheology studies. *Circulation* 36:148 –160.

Richardson, P. D. 1987. Artificial lungs and oxygenation devices. In *Handbook of bioengineering,* ed. R. Skalak and S. Chien, Chapter 27. New York: McGraw-Hill.

Ricordi, C. 1992. (ed.) *Pancreatic islet cell transplantation.* Boulder, CO: R. G. Landes.

Ricordi, C., C. Socci, A. Davalli, A. Vertova, P. Baro, I. Sassi, S. Braghi, N. Giuzzi, G. Pozza, and V. DiCarlo. 1990b. Application of the automated method to islet isolation in swine. *Transplant. Proc.* 22:784–785.

Ricordi, C., C. Socci, A. Davalli, C. Staudacher, A. Vertova, P. Baro, M. Freschi, F. Gavazzi, F. Bertuzzi, G. Pozza, V. DiCarlo. 1990a. Swine islet isolation and transplantation. *Horm. Metabol. Res.* 25:26–30.

Ricordi, C., P. E. Lacy, E. H. Finke, B. J. Olack, and D. W. Scharp. 1988. Automated method for isolation of human pancreatic islets. *Diabetes* 37:413–420.

Riley, M. R., F. J. Muzzio, and H. M. Buettner. 1995b. The effect of structure on diffusion and reaction in immobilized cell systems. *Chem. Eng. Sci.* 50:3357–3367.

Riley, M. R., F. J. Muzzio, H. M. Buettner, and S. C. Reyes. 1994. Monte Carlo calculation of effective diffusivities in two- and three-dimensional heterogeneous materials of variable structure. *Physical Rev. E.* 49:3500–3503.

Riley, M. R., F. J. Muzzio, H. M. Buettner, and S. C. Reyes. 1995a. Diffusion in heterogeneous media: application to immobilized cell systems. *AIChE J.* 41:691–700.

Riley, M. R., F. J. Muzzio, H. M. Buettner, and S. C. Reyes. 1996. A simple correlation for predicting effective diffusivities in immobilized cell systems. *Biotech. Bioeng.* 49:223–227.

Riley, M. R., H. M. Buettner, F. J. Muzzio, and S. C. Reyes. 1995. Monte Carlo simulation of diffusion and reaction in two-dimensional cell structures. *Biophysical J.* 68:1716–1726.

Robertson, B. C., and A. L. Zydney. 1990a. Hindered protein diffusion in asymmetric ultrafiltration membranes with highly constricted pores. *J. Membr. Sci.* 49:287–303.

Robertson, B. C., and Zydney, A. L. 1990b. Protein adsorption in asymmetric ultrafiltration membranes with highly constricted pores. *J. Col. Int. Sci.* 134:563–575.

Rosen, S. L. 1993. *Fundamental principles of polymeric materials,* 2nd ed. New York: John Wiley & Sons.

Rosenberg, G. 1995. Artificial heart and circulatory assist devices. In *The biomedical engineering handbook,* ed. J. D. Bronzino, 1839–1846. Boca Raton, FL: CRC Press.

Rozga, J., E. Morsiani, E. LePage, A. D. Moscioni, T. Giorgio, and A. A. Demetriou. 1994. Isolated hepatocytes in a bioartificial liver: a single group view and experience. *Biotechnol. Bioeng.* 43:645–653.

Rozga, J., F. Williams, M.-S. Ro, D. F. Neuzil, T. D. Giorgio, G. Backfisch, A. D. Moscioni, R. Hakim, and A. A. Demetriou. 1993. Development of a bioartificial liver: properties and function of a hollow fiber module inoculated with liver cells. *Hepatology* 17:258–265.

Rubin, E., and J. L. Farber. 1994. *Pathology,* 2nd ed. Phildaelphia: J. B. Lippincott.

Saltzman, W. M. 1997. Cell interactions with polymers. In *Principles of tissue engineering,* ed. R. P. Lanza, R. Langer, and W. L. Chick, 225–246. Boulder, CO: R. G. Landes.

Sandler, S. I. 1989. *Chemical and engineering thermodynamics.* New York: John Wiley & Sons.

Sarver, J. G., 1994. Development of a tracer technique that utilizes a physiological pharmacokinetic model to quantitatively assess the change in mass transfer rates associated with vascular growth and the application of this technique to the evaluation of a novel bioartificial organoid. PhD dissertation. Toledo, OH: The University of Toledo.

Sarver, J. G., and R. L. Fournier. 1990. Numerical investigation of a novel spiral wound membrane sandwich design for an implantable bioartificial pancreas. *Comput. Biol. Med.* 20:105–119.

Sarver, J. G., R. L. Fournier, P. J. Goldblatt, T. L. Phares, S. E. Mertz, A. R. Baker, R. J. Mellon, J. M. Horner, and S. H. Selman. 1995. Tracer technique to measure in vivo chemical transport rates within an implantable cell transplantation device. *Cell Transplant.* 4:201–217.

Scharp, D. W., C. J. Swanson, B. J. Olack, P. P. Latta, O. D. Hegre, E. J. Doherty, F. T. Gentile, K. S. Flavin, M. F. Ansara, and P. E. Lacy. 1994. Protection of encapsulated human islets implanted without immunosuppression in patients with type I or type II diabetes and in nondiabetic control subjects. *Diabetes* 43:1167–1170.

Schrezenmeir, J., J. Kirchgessner, L. Gero, L. A. Kunz, J. Beyer, and W. Mueller-Klieser. 1994. Effect of microencapsulation on oxygen distribution in islets organs. *Transplantation* 57:1308–1314.

Scott, M. D., K. L. Murad, F. Koumpouras, M. Talbot, and J. W. Eaton. 1997. Chemical camouflage of antigenic determinants: Stealth erythrocytes. *Proc. Natl. Acad. Sci. USA* 94:7566–7571.

Secomb, T. W., R. Hsu, M. W. Dewhirst, B. Klitzman, and J. F. Gross. 1993. Analysis of oxygen transport to tumor tissue by microvascular networks. *Int. J. Radiation Oncology Biol. Phys.* 25:481–489.

Sefton, M. V., R. M. Dawson, R. L. Broughton, J. Blysniuk, and M. E. Sugamori. 1987. Microencpsulation of mammalian cells in a water insoluble polyacrylate by coextrusion and interfacial precipitation. *Biotechnol. Bioeng.* 29:135–1143.

Selman, S. L., M. Kreimer-Birnbaum, J. E. Klaunig, P. J. Goldblatt, R. W. Keck, and S. L. Britton. 1984. Blood flow in transplantable bladder tumors treated with hematoporphyrin derivative and light. *Cancer Res.* 44:1924–1927.

Shah, N., and A. Mehra. 1996. Modeling of oxygen uptake in perfluorocarbon emulsions. *ASAIO J.* 42:181–189.

Shaoting, W., Z. Fengbao, M. Tengxiang, and G. Hanqing. 1990. Investigation on patient-artificial kidney system using compartment models. *Chem. Eng. Sci.* 45:2943–2948.

Shiroki, R., T. Mohanakumar, and D. W. Scharp. 1995. Analysis of the serological and cellular sensitization induced by encapsulated human islets transplantation in type I and type II diabetes patients. *Cell Transplant.* 4:535–538.

Smith, B., J. G. Sarver, and R. L. Fournier. 1991. A comparison of islet transplantation and subcutaneous insulin injections for the treatment of diabetes mellitus. *Comput. Biol. Med.* 21:417–427.

Soon-Shiong, P. 1994. Encapsulated islet transplantation: pathway to human clinical trials. In *Immunoisolation of pancreatic islets,* vol. 3, ed. R. P. Lanza and W. L. Chick, Chapter 6. Boulder, CO: R. G. Landes.

Soon-Shiong, P., E. Feldman, R. Nelson, J. Komtebedde, O. Smidsrod, G. Skjak-Braek, T. Espevik, R. Heintz, and M. Lee. 1992b. Successful reversal of spontaneous diabetes in dogs by intraperitoneal microencapsulated islets. *Transplantation* 54:769–774.

Soon-Shiong, P., E. Feldman, R. Nelson, R. Heintz, N. Merideth, P. Sandford, T. Zheng, and J. Komtebedde. 1992a. *Trans. Proc.* 24:2946–2947.

Soon-Shiong, P., E. Feldman, R. Nelson, R. Heintz, Q. Yao, O. Smidsrod, and P. Sandford. 1993. Longterm reversal of diabetes by the injection of immunoprotected islets. *Proc. Nat. Acad. Sci. USA* 90:5843–5847.

Sorenson, J. T., C. K. Colton, R. S. Hillman, and J. S. Soeldner. 1982. Use of a pharmacokinetic model of glucose homeostasis for assessment of performance requirements for improved insulin therapies. *Diabetes Care* 5:148–157.

Spotnitz, H. M. 1987. Circulatory assist devices. In *Handbook of bioengineering, ed.* R. Skalak and S. Chien, 38.1–38.18. New York: McGraw-Hill.

Staverman, A. J. 1948. *Trans. Faraday Soc.* 48:176.

Stryer, L. 1988. *Biochemistry.* New York: W. H. Freeman.

Sturis, J., C. Knudsen, N. M. O'Meara, J. S. Thomsen, E. Mosekilde, E. Van Cauter, and K. S. Polonsky. 1995. Phase-locking regions in a forced model of slow insulin and glucose oscillations. In *Dynamical disease: mathematical analysis of human illness.* ed. J. Belair, L. Glass, U. An der Heiden, and J. Milton. Woodbury, NY: AIP Press.

Sturis, J., K. S. Polonsky, E. Mosekilde, and E. Van Cauter. 1991. Computer model for mechanisms underlying ultradian oscillations of insulin and glucose. *Am. J. Physiol.* 260:E801–E809.

Sugamori, M. E., and M. V. Sefton. 1989. Microencapsulation of pancreatic islets in a water insoluble polyacrylate. *Trans. Am. Soc. Artif. Organs* 35:791–799.

Sullivan, S. J., T. Maki, K. M. Borland, M. D. Mahoney, B. A. Solomon, T. E. Muller, A. P. Monaco, and W. L. Chick. 1991. Biohybrid artificial pancreas: longterm implantation studies in diabetic, pancreatectomized dogs. *Science* 252:718–721.

Sun, A. M., W. Parisius, G. M. Healy, I. Vacek, and H. G. Macmorine. 1977. The use, in diabetic rats and monkeys, of artificial capillary units containing cultured islets of Langerhans. *Diabetes* 26:1136–1139.

Sun, Y., X. Ma, D. Zhou, I. Vacek, and A. M. Sun. 1996. Normalization of diabetes in spontaneously diabetic cynomologus monkeys by xenografts of microencapsulated porcine islets without immunosuppression. *J. Clin. Invest.* 98:1417–1422.

Sung, C., A. Lavin, A. M. Klibanov, and R. Langer. 1986. An immobilized enzyme reactor for the detoxification of bilirubin. *Biotechnol. Bioeng.* 28:1531–1539.

Sussman, N. L., M. G. Chong, T. Koussayer, D. He, T. A. Shang, H. H. Whisennand, and J. H. Kelly. 1992. Reversal of fulminant hepatic failure using an extracorporeal liver assist device. *Hepatology* 16:60–65.

Suzuki, K., S. Bonner-Weir, J. Hollister-Lock, C. K. Colton, and G. C. Weir. 1998. Number and volume of islets transplanted in immunobarrier devices. *Cell Transplant.* 7:47–52.

Takeda, T., S. Murphy, S. Uyama, G. M. Organ, B. L. Schloo, and J. P. Vacanti. 1995. Hepatocyte transplantation in swine using prevascularized polyvinyl alcohol sponges. *Tissue Engineering* 1:253–262.

Thomas, L. C. 1992. *Heat transfer.* New York: Prentice-Hall.

Thompson, J. A., C. C. Haudenschild, K. D. Anderson, J. M. DiPietro, W. F. Anderson, and T. Maciag. 1989. Heparin-binding growth factor 1 induces the formation of organoid neovascular structures in vivo. *Proc. Nat. Acad. Sci. USA* 86:7928–7932.

Thompson, J. A., K. D. Anderson, J. M. DiPietro, J. A. Zwiebel, M. Zametta, W. F. Anderson, and T. Maciag. 1988. Site-directed neovessel formation in vivo. *Science* 241:1349–1352.

Thomson, R. C., M. J. Yaszemski, and A. G. Mikos. 1997. Polymer scaffold processing. In *Principles of tissue engineering,* ed. R. P. Lanza, R. Langer, and W. L. Chick, 263–272. Boulder, CO: R. G. Landes.

Tong, J., and J. L. Anderson. 1996. Partitioning and diffusion of proteins and linear polymers in polyacrylamide gels. *Biophysical J.* 70:1505–1513.

Tsai, A. G., K.-E. Arfors, and M. Intaglietta. 1990. Analysis of oxygen transport to tissue during extreme hemodilution. *Adv. Exp. Med. Biol.* 277:881–887.

Tun, T., H. Hayashi, T. Aung, Y.-J. Gu, R. Doi, H. Kaji, Y. Echigo, W.-J. Wang, H. Setoyama, M. Imamura, S. Maetani, N. Morikawa, H. Iwata, and Y. Ikada. 1996. A newly developed three layer agarose microcapsule for a promising biohybrid artificial pancreas: rat to mouse xenotransplantation. *Cell Transplant.* 5:S59–S63.

Tyn, M. T., and T. W. Gusek. 1990. Prediction of diffusion coefficients of proteins. *Biotech. Bioeng.* 35:327–338.

Vaslef, S. N., L. F. Mockros, R. W. Anderson, and R. J. Leonard. 1994. Use of a mathematical model to predict oxygen transfer rates in hollow fiber membrane oxygenators. *ASAIO J.* 40:990–996.

Wake, M. C., C. W. Patrick, Jr., and A. G. Mikos. 1994. Pore morphology effects on the fibrovascular tissue growth in porous polymer substrates. *Cell Transplant.* 3:339–343.

Warnock, W. L., and R. V. Rajotte. 1988. Critical mass of purified islets that induce normoglycemia after implantation into dogs. *Diabetes* 37:467–470.

Warnock, W. L., D. Ellis, R. V. Rajotte, I. Dawidson, S. Baekkeskov, and J. Egebjerg. 1988. Studies of the isolation and viability of human islets of Langerhans. *Transplantation* 45:957–963.

Welling, P. G. 1986. *Pharmacokinetics: Processes and mathematics.* ACS Monograph 185. Washington, DC: American Chemical Society.

Winslow, R. M. 1997. Blood substitutes. *Sci. Med.* 4:54–63.

Yang, M. B., J. P. Vacanti, and D. E. Ingber. 1994. Hollow fibers for hepatocyte encapsulation and transplantation: studies of survival and function in rats. *Cell Transplant.* 3:373–385.

Yang, M.-C., and E. L. Cussler. 1986. Designing hollow-fiber contactors. *AIChE J.* 32:1910–1916.

Yarmush, M. L., J. C. Y. Dunn, and R. G. Tompkins. 1992. Assessment of artificial liver support technology. *Cell Transplant.* 1:323–341.

Zekorn, T., U. Siebers, R. G. Bretzel, M. Renardy, H. Planck, P. Zschocke, and K. Federlin. 1990. Protection of islets of Langerhans from interleukin-1 toxicity by artificial membranes. *Transplantation* 50:391–394.

Zelman, A. The artificial kidney. In *Handbook of bioengineering,* ed. R. Skalak and S. Chien, Chapter 39. New York: McGraw-Hill Co.

Zielinski, B. A., M. B. Goddard, and M. J. Lysaght. 1997. Immunoisolation. In *Principles of tissue engineering,* ed. R. P. Lanza, R. Langer, and W. L. Chick, 323–332. Boulder, CO: R. G. Landes.

Zydney, A. L. 1995. Therapeutic apheresis and blood fractionation. In *The biomedical engineering handbook,* ed. J. D. Bronzino, 1936–1951. Boca Raton, FL: CRC Press, Inc.

Zydney, A. L., and C. K. Colton. 1986. A concentration polarization model for the filtrate flux in cross-flow microfiltration of particulate suspensions. *Chem. Eng. Comm.* 47:1–21.

SUGGESTED READINGS

Aebischer, P., M. Goddard, A. P. Signore, and R. L. Timpson. 1994. Functional recovery in hemiparkinsonian primates transplanted with polymer-encapsulated PC12 cells. *Exp. Neurology* 126:151–158.

Ahrendt, G., D. E. Chickering, and J. P. Ranieri. 1998. Angiogenic growth factors: A review for tissue engineering. *Tissue Engineering* 4:117–130.

Aung, T., K. Inoue, M. Kogire, S. Sumi, T. Fujisato, Y. J. Gu, S. Shinohata, H. Hayashi, R. Doi, M. Imamura, M. Mitsuo, I. Nakai, S. Maetani, and Y. Ikada. 1994. Improved insulin release from a bioartificial pancreas using mesh reinforced polyvinyl alcohol hydrogel tube: immobilization of islets in agarose gel. *2Transplant. Proc.* 26:790–791.

Aung, T., M. Kogire, K. Inoue, T. Fujisato, Y. Gu, K. Burczak, S. Shinohara, M. Mitsuo, S. Maetani, Y. Ikada, and T. Tobe. 1993. Insulin release from a bioartificial pancreas using a mesh reinforced polyvinyl alcohol hydrogel tube. *ASAIO J.* 39:93–96.

Avgoustiniatos, E. S. and C. K. Colton. 1997. Design considerations in immunoisolation. In *Principles of tissue engineering*, ed. R. P. Lanza, R. Langer, and W. L. Chick, 333–346. Boulder, CO: R.G. Landes.

Bell, E. 1995. Strategy for the selection of scaffolds for tissue engineering. *Tissue Engineering* 1:163–179.

Bellamkonda, R. and P. Aebischer. 1995. Tissue engineering in the nervous system. In *The biomedical engineering handbook*, ed. J. D. Bronzino, 1754–1773. Boca Raton, FL: CRC Press.

Berthiaume, F., M. Toner, R. G. Tompkins, and M. L. Yarmush. 1994. Tissue engineering. In *Implantation biology*, ed. by R. S. Greco, Chapter 21. Boca Raton, FL: CRC Press.

Brooks, S. V., N. M. Cole, and J. A. Faulkner. 1995. Tissue engineering of skeletal muscle. In *The biomedical engineering handbook*, ed. J. D. Bronzino, 1774–1787. Boca Baton, FL: CRC Press.

Caplan, A. I. and S. P. Bruder. 1997. Cell and molecular engineering of bone regeneration. In *Principles of tissue engineering*, ed. R. P. Lanza, R. Langer, and W. L. Chick, 603–618. Boulder, CO: R.G. Landes.

Chen, U. 1997. Lymphocyte engineering, its status of art and its future. In *Principles of tissue engineering*, ed. R. P. Lanza, R. Langer, and W. L. Chick, 527–561. Boulder, CO: R.G. Landes.

Desai, T. A., W. H. Chu, J. K. Tu, G. M. Beattie, A. Hayak, and M. Ferrari. 1998. Microfabricated immunoisolating biocapsules. *Biotechnol. Bioeng.* 57:118–120.

Friedman, E. A. 1989. Toward a hybrid bioartificial pancreas. *Diabetes Care* 12:415–419.

Friedman, M. H. 1986. *Principles and Models of Biological Transport*. Berlin: Springer-Verlag.

Galletti, P. M., C. K. Colton, M. Jaffrin, and G. Reach. 1995. Artificial pancreas. In *The biomedical engineering handbook*, ed. J. D. Bronzino, 1967–1984. Boca Raton, FL: CRC Press.

Galpern, W. R., D. M. Frim, S. B. Tatter, C. A. Altar, M. F. Beal, and O. Isacson. 1996. Cell mediated delivery of brain-derived neurotrophic factor enhances dopamine levels in an MPP+ rat model of substantia nigra degeneration. *Cell Transplant.* 5:225–232.

Goulet, F., L. Germaine, D. Rancourt, C. Caron, A. Normand, and F. A. Auger. 1997. Tendons and ligaments. In *Principles of tissue engineering*, ed. R. P. Lanza, R. Langer, and W. L. Chick, 633–644. Boulder, CO: R.G. Landes.

Hammang, J. P., D. F. Emerich, S. R. Winn, A. Lee, M. D. Lindner, F. T. Gentile, E. J. Doherty, J. H. Kordower, and E. E. Baetge. 1995. Delivery of neurotrophic factors to the CNS using encapsulated cells: developing treatments for neurodegenerative diseases. *Cell Transplant.* 4:S27–S28.

Hsu, B. R. S., Y. S. Ho, S. H. Fu, Y. Y. Huang, S. C. Chiou, and H. S. Huang. 1995. Membrane compactness affects the integrity and immunoprotection of alginate-poly-L- lysine-alginate microcapsules. *Transplant. Proc.* 27:3227–3231.

Humes, H. D. 1995. Tissue engineering of the kidney. In *The biomedical engineering handbook*, ed. J. D. Bronzino, 1807–1824. Boca Raton, FL: CRC Press.

Ijima, H., T. Matsushita, K. Nakazawa, Y. Fujii, and K. Funatsu. 1998. Hepatocyte spheroids in polyurethane foams: Functional analysis and application for a hybrid artificial liver. *Tissue Engineering* 4:213–226.

Iwata, H., N. Morikawa, T. Fujii, T. Takagi, T. Samejima, and Y. Ikada. 1995. Does immunoisolation need to prevent the passage of antibodies and complement. *Transplant. Proc.* 27:3224–3226.

Joseph, J. M., M. B. Goddard, J. Mills, V. Padrun, A. Zurn, B. Zielinski, J. Favre, J. P. Gardaz, F. Mosimann, J. Sagen, L. Christenson, and P. Aebischer. 1994. Transplantation of encapsulated bovine chromaffin cells in the sheep subarachnoid space: a preclinical study for the treatment of cancer pain. *Cell Transplant.* 3:355–364.

Joseph, J. M., M. B. Goddard, J. Mills, V. Padrun, A. Zurn, B. Zielinski, J. Favre, J. P. Gardaz, F. Mosimann, J. Sagen, L. Christenson, and P. Aebischer. 1994. Transplantation of encapsulated bovine chromaffin cells in the sheep subarachnoid space: a preclinical study for the treatment of cancer pain. *Cell Transplant.* 3:355–364.

Kessinger, A. and G. Sharp. 1997. Tissue engineering of the hematopoietic stem cell. In *Principles of tissue engineering*, ed. R. P. Lanza, R. Langer, and W. L. Chick, 563–573. Boulder, CO: R.G. Landes.

Kim, B.-S., A. J. Putnam, T. J. Kulik, and D. J. Mooney. 1998. Optimizing seeding and culture methods to engineer smooth muscle tissue on biodegradable polymer matrices. *Biotechnol. Bioeng.* 57:46–54.

Kim, T. H. and J. P. Vacanti. 1995. Tissue engineering of the liver. In *The biomedical engineering handbook*, ed. J. D. Bronzino, 1745–173. Boca Raton, FL: CRC Press, Inc.

Koller, M. R. and B. O. Palsson. 1995. Tissue engineering of bone marrow. In *The biomedical engineering handbook*, ed. J. D. Bronzino, 1728–1744. Boca Raton, FL: CRC Press.

Lamberton, P., M. Lipsky, and P. McMillan. 1988. Use of semipermeable polyurethane hollow fibers for pituitary organ culture. *In Vitro Cell. Dev. Biol.* 24:500–504.

Lanza, R. P. and W. L. Chick. 1994a. *Procurement of pancreatic islets*, volume 1. Boulder, CO: R.G. Landes.

Lavoie, M. P., M. Palmatier, F. T. Gentile, F. A. Kaplan, D. M. Fiore, T. F. Hazlett, W. J. Bell, and T. R. Flanagan. 1993. Two PC12 pheochromocytoma lines sealed in hollow fiber based capsules tonically release L-dopa in vitro. *Cell Transplant.* 2:163–173.

Liu, H., F. A. Ofosu, and P. L. Chang. 1993. Expression of human factor IX by microencapsulated recombinant fibroblasts. *Hum. Gene Ther.* 4:291–301.

Maki, T., I. Otsu, J. J. O'Neil, K. Dunleavy, C. J. P. Mullon, B. A. Solomon, and A. P. Monaco. 1996. Treatment of diabetes by xenogeneic islets without immunosuppression. *Diabetes* 45:342–347.

Miller, N. A., M. C. Bene, P. Ambrosini, J. Penaud, and G. C. Faure. 1997. Tissue engineering: periodontal applications. In *Principles of tissue engineering*, ed. R. P. Lanza, R. Langer, and W. L. Chick, 709–725. Boulder, CO: R.G. Landes.

Mitsuo, M., K. Inoue, I. Nakai, T. Oda, Y. Gu, S. Shinohara, M. Kogire, T. Fujisato, S. Maetani, Y. Ikada, T. Tobe, and T. Oka. 1992. Efficacy of mesh reinforced polyvinylalcohol tube as a novel device for bioartificial pancreas: a functional study of rat islets in vivo. *Transplant. Proc.* 24:2939–2940.

Miyoshi, H., K. Yanagi, H. Fukuda, and N. Ohshima. 1994. Long-term continuous culture of hepatocytes in a packed bed reactor using porous resins. *Biotechnol. Bioeng.* 43:635–2644.

Monaco, A. P., T. Maki, H. Ozato, M. Carretta, S. J. Sullivan, K. M. Borland, M. D. Mahoney, W. L. Chick, T. E. Muller, J. Wolfrum, and B. Solomon. 1991. Transplantation of islet allografts and xenografts in totally pancreatectomized diabetic pigs using the hybrid artificial pancreas. *Ann. Sur.* 214: 339–362.

Mooney, D. J., B.-S. Kim, J. P. Vacnati, R. Langer, and A. Atala. 1997. Tissue engineering: genitourinary system. In *Principles of tissue engineering*, ed. R. P. Lanza, R. Langer, and W. L. Chick, 591–600. Boulder, CO: R.G. Landes.

Natsume, T., O. Ike, T. Okada, N. Takimoto, Y. Shimizu, and Y. Ikada. 1993. Porous collagen sponge for esophageal replacement. *J. Biomed. Mater. Res.* 27:867–875.

Naughton, G. K. 1997. Skin and epithelia. In *Principles of tissue engineering*, ed. R. P. Lanza, R. Langer, and W. L. Chick, 769–782. Boulder, CO: R.G. Landes.

Nehrer, S., H. H. Breinan, S. Ashkar, S. Shortkoff, T. Minas, C. B. Sledge, I. V. Yannas, and M. Spector. 1998. Characteristics of articular chondrocytes seeded in collagen matrices in vitro. *Tissue Engineering* 4:175–184.

Ohgawara, H., J.-I. Miyazaki, Y. Nakagawa, S. Sato, S. Karibe, and T. Akaike. 1996. Xenotransplantation using a diffusion chamber with a B cell line (MIN6) as a bioartificial pancreas. *Cell Transplant.* 5:S71–S73.

Ojcius, D. M., C.-C. Liu, and J. D. Young. 1998. Pore-forming proteins. *Sci. Med.* 5:44–53.

Organ, G. M. and J. P. Vacanti. 1997. Tissue engineering neointestine. In *Principles of tissue engineering*, ed. R. P. Lanza, R. Langer, and W. L. Chick, 441–462. Boulder, CO: R.G. Landes.

Organ, G. M., D. J. Mooney, L. K. Hansen, B. Schloo, and J. P. Vacanti. 1992. Transplantation of enterocytes utilizing polymer-cell constructs to produce a neointestine. *Transplant. Proc.* 24:3009–3011.

Renkin, E. M. 1977. Multiple pathways of capillary permeability. *Circulation Res.* 41:735–743.

Saitoh, Y., T. Taki, N. Arita, T. Ohnishi, and T. Hayakawa. 1995. Cell therapy with encapsulated xenogeneic tumor cells secreting β-endorphin for treatment of peripheral pain. *Cell Transplant.* 4:S13–S17.

Sarver, J. G. 1994. Development of a tracer technique that utilizes a physiological pharmacokinetic model to quantitatively assess the change in mass transfer rates associated with vascular growth and the application of this technique to the evaluation of a novel bioartificial organoid. PhD dissertation. Toledo, OH: The University of Toledo.

Shatford, R. A., S. L. Nyberg, S. J. Meier, J. G. White, W. D. Payne, W.-S. Hu, and F. B. Cerra. 1992. Hepatocyte function in a hollow fiber bioreactor: a potential bioartificial liver. *J. Surg. Res.* 53:549–557.

Sun, A. M. 1997. Parathryroid. In *Principles of tissue engineering*, ed. R. P. Lanza, R. Langer, and W. L. Chick, 427–437. Boulder, CO: R.G. Landes.

Sutherland, D. E. R. 1989. Current status of pancreas transplantation. *ASAIO J.* 35:14–16.

Tresco, P. A., S. R. Winn, S. Tan, C. B. Jaeger, L. A. Greene, and P. Aebischer. 1992. Polymer encapsulated PC12 cells: long-term survival and associated reduction in lesion induced rotational behavior. *Cell Transplant.* 1:255–264.

Trinkaus-Randall, V. 1997. Cornea. In *Principles of tissue engineering*, eds. R. P. Lanza, R. Langer, and W. L. Chick, 383–402. Boulder, Co.: R.G. Landes.

Tziampazis, E. and A. Sambanis. 1995. Tissue engineering of a bioartificial pancreas: modeling the cell environment and device function. *Biotechnol. Prog.* 11:115–126.

Uludag, H. and M. V. Sefton. 1993. Metabolic activity and proliferation of CHO cells in hydroxyethyl methacrylate-methyl methacrylate (hema-mma) microcapsules. *Cell Transplant.* 2:175–182.

Uyama, S., P.-M. Kaufmann, T. Takeda, and J. P. Vacanti. 1993. Delivery of whole liver equivalent hepatocyte mass using polymer devices and hepatotropic stimulation. *Transplantation* 55:932–935.

Vacanti, C. A. and J. P. Vacanti. 1997. Bone and cartilage reconstruction. In *Principles of tissue engineering*, ed. R. P. Lanza, R. Langer, and W. L. Chick, 619–631. Boulder, CO: R.G. Landes.

Vacanti, J. P., M. A. Morse, W. M. Saltzman, A. J. Domb, A. Perez-Atayde, and R. Langer. 1988. Selective cell transplantation using bioabsorbable artificial polymers as matrices. *J. Ped. Sur.* 23:3–9.

Valentini, R. F. and P. Aebischer. 1997. Strategies for the engineering of peripheral nervous tissue regeneration. In *Principles of tissue engineering*, ed. R. P. Lanza, R. Langer, and W. L. Chick, 671–684. Boulder, CO: R.G. Landes.

Villadsen, J. and M. L. Michelsen. 1978. *Solution of differential equation models by polynomial approximation.* New York: Prentice-Hall.

Voytik-Harbin, S. L., A. O. Brightman, B. Z. Waisner, J. P. Robinson, and C. H. Lamar. 1998. Small intestinal submucosa: A tissue-derived extra cellular matrix that promotes tissue-specific growth and differentiation of cells in vitro. *Tissue Engineering* 4:157–174.

Warnock, W. L., D. K. Ellis, M. Cattral, D. Untch, N. M. Kneteman, and R.V. Rajotte. 1989. Viable purified islets of Langerhans from collagenase perfused human pancreas. *Diabetes* 38:136–139.

Yannas, I. V. 1995. Artificial skin and dermal equivalents. In *The biomedical engineering handbook*, ed. J. D. Bronzino, 2025–2038. Boca Raton, FL: CRC Press.

INDEX